GLOBAL DIFFERENTIAL GEOMETRY

MAA STUDIES IN MATHEMATICS

Published by
THE MATHEMATICAL ASSOCIATION OF AMERICA

———

Committee on Publications
Donald J. Albers, Chairman

Subcommittee on MAA Studies in Mathematics
Meyer Jerison

Studies in Mathematics

Volume 1: STUDIES IN MODERN ANALYSIS
       2: STUDIES IN MODERN ALGEBRA
       3: STUDIES IN REAL AND COMPLEX ANALYSIS
       4: STUDIES IN GLOBAL GEOMETRY AND ANALYSIS
       5: STUDIES IN MODERN TOPOLOGY
       6: STUDIES IN NUMBER THEORY
       7: STUDIES IN APPLIED MATHEMATICS
       8: STUDIES IN MODEL THEORY
       9: STUDIES IN ALGEBRAIC LOGIC
     10: STUDIES IN OPTIMIZATION
     11: STUDIES IN GRAPH THEORY, PART I
     12: STUDIES IN GRAPH THEORY, PART II
     13: STUDIES IN HARMONIC ANALYSIS
     14: STUDIES IN ORDINARY DIFFERENTIAL EQUATIONS
     15: STUDIES IN MATHEMATICAL BIOLOGY, PART I
     16: STUDIES IN MATHEMATICAL BIOLOGY, PART II
     17: STUDIES IN COMBINATORICS
     18: STUDIES IN PROBABILITY THEORY
     19: STUDIES IN STATISTICS
     20: STUDIES IN ALGEBRAIC GEOMETRY
     21: STUDIES IN FUNCTIONAL ANALYSIS
     22: STUDIES IN COMPUTER SCIENCE
     23: STUDIES IN PARTIAL DIFFERENTIAL EQUATIONS
     24: STUDIES IN NUMERICAL ANALYSIS
     25: STUDIES IN MATHEMATICAL ECONOMICS
     26: STUDIES IN THE HISTORY OF MATHEMATICS
     27: GLOBAL DIFFERENTIAL GEOMETRY

Lamberto Cesari
University of Michigan

S. S. Chern
Mathematical Sciences Research Institute and
University of California, Berkeley

Patrick Eberlein
University of North Carolina

Harley Flanders
University of Michigan

Hermann Karcher
University of Bonn

Shoshichi Kobayashi
University of California, Berkeley

Marston Morse

Robert Osserman
Stanford University

L. A. Santalo
University of Buenos Aires

Studies in Mathematics

Volume 27

GLOBAL DIFFERENTIAL GEOMETRY

S. S. Chern, editor
Mathematical Sciences Research Institute

Contributors:
Lamberto Cesari
Patrick Eberlein
Harley Flanders
Hermann Karcher
Shoshichi Kobayashi
Marston Morse
Robert Osserman
L. A. Santalo

Published and distributed by
The Mathematical Association of America

1989 by
The Mathematical Association of America (Incorporated)
Library of Congress Catalog Card Number 88-081000

Complete Set ISBN 0-88385-100-8
Vol. 27 ISBN 0-88385-129-6

Printed in the United States of America

Current Printing (last digit):
10 9 8 7 6 5 4 3 2 1

CONTENTS

INTRODUCTION   i
S. S. Chern

VECTOR BUNDLES WITH A CONNECTION   1
S. S. Chern

DIFFERENTIAL FORMS   27
Harley Flanders

MINIMAL SURFACES IN R^3 73
Robert Osserman

CURVES AND SURFACES IN
   EUCLIDEAN SPACE   99
S. S. Chern

ON CONJUGATE AND CUT LOCI   140
Shoshichi Kobayashi

RIEMANNIAN COMPARISON
   CONSTRUCTIONS   170
Hermann Karcher

MANIFOLDS OF NONPOSITIVE
   CURVATURE   223
Patrick Eberlein

WHAT IS ANALYSIS IN THE LARGE?   259
Marston Morse

SURFACE AREA   270
Lamberto Cesari

INTEGRAL GEOMETRY   303
L. A. Santalo

INDEX   351

INTRODUCTION

This is essentially a second edition of the MAA Studies Volume 4 entitled *Global Geometry and Analysis*, which appeared in 1967 and has been out of print for a long time. In those twenty years the subject has undergone great developments and remains one of the major areas of mathematics.

Most of the original chapters have been revised and updated. In particular, Professor Harley Flanders has made some interesting additions to his chapter.

Four new chapters are included. Two of them are concerned with Riemannian geometry, which is fundamental for differential geometry. As in non-Euclidean geometry the properties of the space depend very much on the sign of the (sectional) curvature. Professor Karcher's chapter, "Riemannian Comparison Constructions," deals with Riemannian manifolds with curvature ≥ 0, whose study has led to many amazing results. Professor Karcher showed his authority on the subject by being able to pick the important results which can be proved completely in his exposition.

On the other side are the Riemannian manifolds of negative or nonpositive curvature, which are perhaps more important for applications. There is a wide gulf between Riemannian manifolds of positive and nonpositive curvature, and the problems and methods are quite different. Professor Eberlein's chapter, "Manifolds of Nonpositive Curvature," gives an introduction to its fundamental concepts and contains many valuable examples.

Harmonic mappings have been an active topic in recent years and will undoubtedly remain so for years to come. In the ordinary Euclidean space R^3 this leads to the classical subject of minimal surfaces. Two of the fundamental results are the Plateau problem (existence) and the Bernstein theorem (uniqueness). Osserman interpreted the latter result as an equidistribution assertion, viz., a statement on the size of the image of the Gauss map of a complete minimal surface, which is not a plane. After the striking result of F. Xavier in 1981, it was proved by H. Fujimoto in 1986 that the Gauss map of a complete minimal surface in R^3, not a plane, can omit at most four points. Combined with a classical example of Osserman, this number is the best possible. Osserman's chapter gives an exposition of this latest result.

The Editor is convinced that the notion of a connection in a vector bundle will soon find its way into a class on advanced calculus, as it is a fundamental notion and its applications are wide-spread. His chapter, "Vector Bundles with a Connection," hopefully will show that it is basically an elementary concept.

S. S. Chern

VECTOR BUNDLES WITH A CONNECTION

Shiing-shen Chern[1]

1. INTRODUCTION

Riemannian geometry, which was the high-dimensional general-ization of Gauss' intrinsic surface theory, gives a geometrical structure which is entirely local. It was later realized that many of its geometrical properties derive from the Levi-Civita parallelism, i.e., the connection in the tangent bundle. Recent developments in mathematics and physics have shown the importance of the notion of a "vector bundle with a connection" over a manifold. An introductory account of this notion and some of its applications will be given in this chapter.

2. VECTOR BUNDLES

We will be dealing with C^∞-manifolds and their C^∞-mappings. A vector bundle over a manifold M is a mapping

$$\pi: E \to M \qquad (1)$$

such that the following conditions are satisfied.

[1] Work done under partial support of NSF Grant No. DMS84-03201.

B1) $\pi^{-1}(x)$, $x \in M$, is a (real or complex) vector space Y of dimension q. Either $\pi^{-1}(x)$ or Y will be called a *fiber*.

B2) E is locally a product, i.e., every point $x \in M$ has a neighborhood U such that there is a diffeomorphism

$$\varphi_U: U \times Y \to \pi^{-1}(U) \tag{2}$$

satisfying the condition

$$\pi \circ \varphi_U(x, y_U) = x, \qquad y_U \in Y; \tag{3}$$

(x, y_U) are the local coordinates of E relative to U.

B3) For two neighborhoods U, V, with $U \cap V \neq \varnothing$, the relation

$$\varphi_U(x, y_U) = \varphi_V(x, y_V), \qquad x \in U \cap V; \quad y_U, y_V \in Y, \tag{4}$$

holds if and only if

$$y_U = g_{UV}(x) y_V, \tag{5}$$

where $g_{UV}(x)$ is a nondegenerate endomorphism of Y. If y_U, y_V are expressed as one-columned matrices, then $g_{UV}(x)$ is a nonsingular $(q \times q)$-matrix. The condition B3) means that the linear structure on the fiber has a meaning.

E is called a *vector bundle* over the *base manifold M*. A vector bundle can be viewed as a family of vector spaces parametrized by a manifold such that it is locally trivial and the linear structure on the fiber is defined.

A *section* is a mapping $s: M \to E$ such that $\pi \circ s = $ identity. Two sections can be added, and a section can be multiplied by a (real or complex-valued) function, remaining a section. All the sections form a vector space which we will denote by $\Gamma(E)$.

The functions $g_{UV}(x)$, $x \in U \cap V \neq \phi$, have values in $GL(q; \mathbf{R})$ or $GL(q; \mathbf{C})$, and are called *transition functions*. If $\{U, V, W, \dots\}$ form a covering of M, the transition functions satisfying the conditions

$$g_{UU}(x) = \text{identity}$$
$$g_{UV}(x)g_{VU}(x) = \text{identity}, \qquad \text{in } U \cap V \neq \phi \qquad (6)$$
$$g_{UV}(x)g_{VW}(x)g_{WU}(x) = \text{identity}, \qquad \text{in } U \cap V \cap W \neq \phi.$$

It can be proved that a family of transition functions relative to the covering, which satisfy the conditions (6) define a vector bundle.

We give some examples of vector bundles:

EXAMPLE 1. $E = M \times Y$, and π is the projection to the first factor. A section is a graph in $M \times Y$ and is exactly a vector-valued function on M (see the following figure).

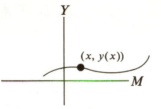

EXAMPLE 2. $E = TM$, the tangent bundle of M. Relative to the local coordinates u^i of M, $1 \leqslant i \leqslant n$ ($= \dim M$), a tangent vector can be written as $\sum y_U^i(\partial/\partial u^i)$. In an open set where the coordinates u^i and v^j are both valid, we have

$$\sum y_U^i \frac{\partial}{\partial u^i} = \sum y_V^j \frac{\partial}{\partial v^j}, \qquad 1 \leqslant i, j \leqslant n, \qquad (7a)$$

if and only if

$$y_U^i = \sum_j \frac{\partial u^i}{\partial v^j} y_V^j. \qquad (7b)$$

Thus the transition functions are given by the Jacobian matrices. A section of $TM \to M$ is a vector field on M.

EXAMPLE 3. If M is immersed in the Euclidean space E^{n+q} of dimension $n + q$, the normal vectors to M form its *normal bundle*.

3. CONNECTIONS

Consider the trivial bundle $E = M \times Y \to M$ in Example 1. A section $s \in \Gamma(E)$ is a Y-valued function on M. If X is a vector field on M, the directional derivative of s along X, which we denote by ∇_{X^s}, is again a section. We wish to generalize this "differentiation" to any vector bundle.

A *connection* is a map

$$D: \Gamma(E) \to \Gamma(E \otimes T^*M), \qquad (8)$$

where T^*M is the cotangent bundle of M, such that the following conditions are satisfied:

D1) $D(s_1 + s_2) = Ds_1 + Ds_2$,
D2) $D(fs) = fDs + s \otimes df, s, s_1, s_2 \in \Gamma(E)$,

where f is a (real or complex-valued) function on M. From Ds we get the directional derivative by the pairing

$$\nabla_{X^s} = \langle X, Ds \rangle. \qquad (9)$$

To describe a connection analytically we restrict to a neighborhood and take in it a *frame field*, i.e., q sections e_i, $1 \leqslant i \leqslant q$, which are everywhere linearly independent. Since the e_i's form a basis on each fiber, we can write

$$De_i = \sum \omega_i^j \otimes e_j, \qquad (10)$$

when ω_i^j, $1 \leqslant i$, $j \leqslant q$, are linear differential forms. The matrix of linear differential forms:

$$\omega = \left(\omega_i^j \right) \qquad (11)$$

is called the *connection matrix* relative to the frame field e_i.

The connection matrix determines the connection completely. For any section can be written

$$s = \sum s^i e_i,$$ (12)

and we have, by the Properties D1) and D2),

$$Ds = \sum \left(ds^i + s^j \omega_j^i \right) \otimes e_i.$$ (13)

The section s is said to be *parallel*, if $Ds = 0$.

Our fundamental formula relates the connection matrices under a change of the frame field. In fact, let

$$e_i' = \sum a_i^j e_j$$ (14)

be a new frame field, where $\det(a_i^j) \neq 0$. Let

$$De_i' = \sum \omega_i'^j \otimes e_j', \qquad \omega' = \left(\omega_i'^j \right).$$ (15)

Here it will be most easy to compute with matrix equations. We write

$$e = \begin{pmatrix} e_1 \\ \vdots \\ e_q \end{pmatrix}, \qquad e' = \begin{pmatrix} e_1' \\ \vdots \\ e_q' \end{pmatrix}, \qquad A = \left(a_i^j \right).$$ (16)

Then we have the matrix equations

$$De = \omega e, \qquad De' = \omega' e', \qquad e' = Ae.$$ (17)

Differentiating the last equation and using the properties D1), D2), we get immediately the fundamental formula

$$\omega' A = dA + A\omega.$$ (18)

This relation satisfies the condition of consistency, i.e., if e'' is a third frame field and ω'' is the connection matrix relative to e'', the relation between ω and ω'' follows from the relations between ω, ω', and ω', ω''.

A connection is also called a *covariant* or *absolute differentiation*. In physics it is called a *gauge potential*, for a reason to be given below. It is given relative to a frame field by a matrix of linear differential forms, with the transformation law (18).

REMARK. From (18) we see that connections exist in a vector bundle. The main idea of the proof is as follows: Take an open covering $\{U_\alpha\}$ of M, by coordinate neighborhoods, in each of which choose a frame field. Take any connection matrix in U_1. It suffices to define the connection in $U_{1 \leqslant \alpha \leqslant m}U_\alpha$ by induction on m. Suppose the connection be defined in $V = U_{1 \leqslant \alpha \leqslant m-1}U_\alpha$. Relative to the frame field in U_m the connection matrix is given in $V \cap U_m$. It remains to extend it over U_m, using standard extension theorems in Euclidean space. The complete argument needs a little care, because the extension theorem holds only for functions defined on closed sets. We leave the details to the reader.

Taking the exterior derivative of (18) and using the equation itself, we get

$$\Omega' A = A\Omega, \tag{19}$$

where Ω is a $(q \times q)$-matrix of exterior two-forms, given by

$$\Omega = d\omega - \omega \wedge \omega. \tag{20}$$

In the last term we use the multiplication of matrices, while the multiplication of differential forms is exterior multiplication. The matrix Ω is called the *curvature matrix*, relative to the frame field e_i.

Exterior differentiation of (20) gives

$$d\Omega = \omega \wedge \Omega - \Omega \wedge \omega. \tag{21}$$

This is called the *Bianchi identity*.

EXERCISE. Let X, Y be two vector fields on M, and let

$$R(X, Y) = \langle X \wedge Y, \Omega \rangle, \qquad R'(X, Y) = \langle X \wedge Y, \Omega' \rangle. \quad (22)$$

Then $R(X, Y)$ and $R'(X, Y)$ are matrices satisfying the equation

$$R'(X, Y) = A R(X, Y) A^{-1}. \quad (23)$$

For the section s in (12) let

$$\sigma = (s^1, \ldots, s^q). \quad (24)$$

Then $\sigma R e = \sigma' R' e'$ is independent of the choice of the frame field. The map

$$K(X, Y): s = \sigma e \rightarrow \sigma R(X, Y) e \quad (25)$$

is called the *curvature transformation*.

Curvature measures the noncommutativity of covariant differentiation. Prove that, as an operator on $\Gamma(E)$,

$$K(X, Y) = \nabla_X \nabla_Y - \nabla_Y \nabla_X - \nabla_{[X, Y]}. \quad (26)$$

4. CHERN FORMS AND CHERN CLASSES

We consider complex vector bundles. Let

$$\det\left(I + \frac{\sqrt{-1}}{2\pi} \Omega \right) = 1 + c_1(\Omega) + \cdots + c_q(\Omega), \quad (27)$$

where I is the unit matrix. Then $c_i(\Omega)$, $1 \leqslant i \leqslant q$, is an exterior differential form of degree $2i$. By (19) they are independent of the choice of the frame field, i.e.,

$$c_i(\Omega) = c_i(\Omega'). \quad (28)$$

It follows that $c_i(\Omega)$ is globally defined on M. For we can take an open covering $\{U_\alpha\}$ of M and choose in each U_α a frame field.

The resulting c_i's must agree in the intersection of any two members of the covering. We write

$$c_i(E; D) = c_i(\Omega), \qquad (29)$$

indicating that the forms now only depend on the connection D (and of course on the bundle E). These c_i are called the *Chern forms*.

We introduce another set of forms in M as follows:

$$b_i(\Omega) = \text{Tr}\left(\frac{\sqrt{-1}}{2\pi}\Omega^i\right), \qquad 1 \leqslant i \leqslant q. \qquad (30)$$

When Ω is diagonal, both $c_i(\Omega)$ and $b_i(\Omega)$ are symmetrical functions of its diagonal elements. Between them Newton's identities are valid:

$$b_i - c_1 b_{i-1} + c_2 b_{i-2} + \cdots + (-1)^{i-1} c_{i-1} b_1 + (-1)^i i c_i = 0,$$

$$1 \leqslant i \leqslant q. \quad (31)$$

Since both $c_i(\Omega)$ and $b_i(\Omega)$ are invariant under the change (19), we can use the latter to put Ω in a normal form, and it is not hard to see that (31) is true in general. It follows that b_i (resp. c_i) is a polynomial in $c_1, \ldots, c_i$ (resp. $b_1, \ldots, b_i$) with integral (resp. rational) coefficients.

THEOREM 1. *The forms b_i and c_i, $1 \leqslant i \leqslant q$, are closed.*

It suffices to show that b_i is closed. In fact, we have, by (21),

$$dr \text{Tr}\, \Omega^i = i \text{Tr}(d\Omega \wedge \Omega^{i-1})$$

$$= i \text{Tr}(\omega \wedge \Omega^i - \Omega \wedge \omega \wedge \Omega^{i-1}) = 0.$$

This proves the theorem.

The following theorem gives the effect on these forms of changing the connection.

THEOREM 2.   *Let D and D_1 be two connections on the bundle E. Then the forms*

$$b_i(E; D) - b_i(E; D_1), \qquad c_i(E; D) - c_i(E; D_1), \qquad 1 \leqslant i \leqslant q,$$

are exact.

Again it is sufficient to prove the theorem for b_i. We observe that

$$D_t = (1 - t)D_0 + tD_1, \qquad 0 \leqslant t \leqslant 1, \qquad D_0 = D, \qquad (32)$$

is also a connection. Relative to a frame field e_i let ω_t be the connection matrix of D_t. Its curvature matrix is given by

$$\Omega_t = d\omega_t - \omega_t \wedge \omega_t, \qquad (33)$$

and, by exterior differentiation, we have the Bianchi identity

$$d\Omega_t = -\Omega_t \wedge \omega_t + \omega_t \wedge \Omega_t. \qquad (34)$$

Let

$$\eta = \omega_1 - \omega_0. \qquad (35)$$

By (18) we have, under a change of the frame field,

$$\eta' A = A\eta. \qquad (36)$$

Hence the form

$$\alpha = \text{Tr}(\eta \wedge \Omega_t^{i-1}) \qquad (37)$$

is globally defined on M. By using (33), we find

$$d\alpha = \text{Tr}\{(d\eta - \eta \wedge \omega_t - \omega_t \wedge \eta) \wedge \Omega_t^{i-1}\}.$$

(Differentiate each of the factors and telescope!) The expression between the parentheses is equal to

$$\beta = d\eta - \eta \wedge \omega_0 - \omega_0 \wedge \eta - 2t\eta \wedge \eta.$$

On the other hand, expansion of (33) in t gives

$$\Omega_t = d\omega_0 - \omega_0 \wedge \omega_0 + t(d\eta - \omega_0 \wedge \eta - \eta \wedge \omega_0) - t^2 \eta \wedge \eta,$$

so that

$$\frac{d}{dt}\Omega_t = \beta.$$

It follows that

$$\frac{1}{i}\frac{d}{dt}\text{Tr}(\Omega_t^i) = d\alpha.$$

Integrating with respect to t, we get

$$\text{Tr}(\Omega_1^i) - \text{Tr}(\Omega_0^i) = id\int_0^1 \alpha\,dt. \tag{38}$$

This proves the statement in the theorem for b_i, and the theorem is proved.

The de Rham cohomology on a differentiable manifold can be summarized as follows. Let A^r be the space of all C^∞ differential forms of degree r on M, and C^r its subspace of forms which are closed. The quotient space

$$H^r(M; C) = \frac{C^r}{dA^{r-1}} \tag{39}$$

is called the r-dimensional *de Rham cohomology group* of M. An element of $H^r(M; C)$ is called an r-dimensional de Rham cohomology class. If $\alpha \in C^r$, we shall denote its de Rham class by $\{\alpha\}$.

Theorem 2 says that the de Rham cohomology class $\{c_i(E; D))$ is independent of the connection D. It is called the *ith Chern class* of the bundle E and will be denoted by $c_i(E)$, $1 \leqslant i \leqslant q$.

If M is a compact oriented manifold of even dimension $2m$ and $i_1 + \cdots + i_p = m$, then

$$\int_M c_{i_1}(E) \cdots c_{i_p}(E) = c_{i_1 \cdots i_p}(E) \tag{40}$$

is called a *Chern number* of E.

This is a remarkable story. From its very definition it is not clear whether there are nontrivial vector bundles, i.e. whether every vector bundle must be globally a product. The necessity of introducing a covariant differentiation and the possibility that it is noncommutative lead to the curvature. The latter should be considered as a two-form, and some elementary combinations of it, based essentially on the eigenvalues of a matrix and their elementary symmetric functions, lead to the first and most important invariants of a vector bundle.

The vector bundle E is called *hermitian*, if there is a C^∞-field of positive definite hermitian scalar products $(,)$, such that

$$\overline{(s,t)} = (t,s), \qquad s,t \in \pi^{-1}(x), \qquad x \in M. \tag{41}$$

Let e_i be a frame field and let

$$(e_i, e_j) = g_{ij}, \qquad 1 \leqslant i, j \leqslant q. \tag{42}$$

Then the matrix

$$(g_{ij}) \tag{43}$$

is positive-definite hermitian. If

$$s = \sum s^i e_i, \qquad t = \sum t^i e_i, \tag{44}$$

then

$$(s,t) = \sum g_{ij} s^i t^j, \qquad t^j = \overline{t^j}. \tag{45}$$

The connection D is called *admissible*, if (s, t) remains constant, when s and t are parallelly displaced, i.e., when $Ds = Dt = 0$. In fact, under these conditions we find

$$d(s, t) = \sum (dg_{ij} - \omega_{ij} - \omega_{ji}) s^i t^j,$$

where

$$\omega_{ij} = \sum \omega_i^k g_{kj}, \qquad \omega_{ji} = \bar{\omega}_{ji}. \tag{46}$$

Hence the conditions for an admissible connection are

$$dg_{ij} = \omega_{ij} + \omega_{ji}. \tag{47}$$

It follows, by an extension argument (see the Remark in Section 3), that on a given hermitian vector bundle admissible connections exist.

The frame e_i is called *unitary*, if

$$(e_i, e_j) = \delta_{ij} \qquad (=1, \text{ if } i = j, \text{ and } = 0 \text{ otherwise}). \tag{48}$$

On an hermitian vector bundle with an admissible connection we can restrict ourselves to unitary frames. Then we have, by (47),

$$\omega_{ij} + \omega_{ji} = 0, \tag{49}$$

i.e., the connection matrix is skew-hermitian. By (20) it follows that the same is true of the curvature matrix. This implies that $\det(I + (i/2\pi)\Omega_{ij})$ in (27) is real, and so are the forms $c_i(\Omega)$, $1 \leqslant i \leqslant q$. This differential-geometric argument shows that the Chern classes $c_i(E)$ are real cohomology classes. It can be proved that they are integral cohomology classes, i.e., elements of $H^{2i}(M; \mathbf{Z})$.

Similar notions can be introduced for real vector bundles, so that we have Riemannian vector bundles and their admissible connections. By restricting to orthonormal frames which satisfy the conditions

$$(e_i, e_j) = \delta_{ij}, \tag{50}$$

we find that both the connection matrix and the curvature matrix are skew-symmetric. It follows that $c_i(\Omega) = 0$ if i is odd. The form

$$p_i(\Omega) = (-1)^i c_{2i}(\Omega) \tag{51}$$

of degree $4i$, is called a *Pontrjagin form* of a real vector bundle E, and the class $p_i(E) = \{ p_i(\Omega)\}$ is a cohomology class of dimension $4i$, to be called a *Pontrjagin class*.

5. SUBMANIFOLDS IN EUCLIDEAN SPACE

We consider an immersion

$$x: M^n \to E^{n+q} \tag{52}$$

of a manifold $M^n = M$ of dimension n into the Euclidean space of dimension $n + q$. The notation x will denote both a point $x \in M$ and its position vector from a fixed point of E^{n+q}. Over M we consider orthonormal frames xe_A, such that $x \in M$, e_α are tangent vectors to M at x, and hence e_i are normal vectors to M. Throughout this section we will use the following ranges of indices:

$$1 \leqslant A, B, C \leqslant n + q,$$

$$1 \leqslant \alpha, \beta, \gamma \leqslant n,$$

$$n + 1 \leqslant i, j, k \leqslant n + q. \tag{53}$$

We can write

$$dx = \sum \omega_\alpha e_\alpha,$$
$$de_A = \sum \omega_{AB} e_B, \tag{54}$$

where

$$\omega_{AB} + \omega_{BA} = 0. \tag{55}$$

Taking the exterior derivative of the first equation of (54) and equating to zero the coefficient of e_i, we get

$$\sum_\alpha \omega_\alpha \wedge \omega_{\alpha i} = 0. \tag{56}$$

Since x is an immersion, the ω_α are linearly independent and it follows by Cartan's lemma that

$$\omega_{\alpha i} = \sum h_{i\alpha\beta}\omega_\beta, \tag{57}$$

where

$$h_{i\alpha\beta} = h_{i\beta\alpha}. \tag{58}$$

The (ordinary) quadratic differential forms

$$\prod_i = \sum \omega_\alpha \omega_{\alpha i} = \sum h_{i\alpha\beta}\omega_\alpha\omega_\beta \tag{59}$$

are the *second fundamental forms* of M. The second fundamental form, also to be denoted by $\sum\prod_i e_i$, is a quadratic differential form with value in the normal bundle.

The exterior differentiation of the second equation of (54) gives

$$d\omega_{AB} = \sum \omega_{AC} \wedge \omega_{CB}. \tag{60}$$

From (54) we get, by projection from the normal bundle,

$$De_\alpha = \sum \omega_{\alpha\beta}e_\beta. \tag{61}$$

This defines a connection in the tangent bundle, as the conditions D1) and D2) in Section 3 are satisfied. It is a fundamental theorem of local Riemannian geometry that the connection D depends only on the induced metric on M. This is the Levi-Civita connection, as originally defined by him.

Similarly, there is a connection

$$D^\perp e_i = \sum \omega_{ij}e_j, \tag{62}$$

the normal connection, in the normal bundle.

We consider the case $n = 2$, $q = 1$, i.e., surfaces in the ordinary Euclidean space E^3. Suppose M be oriented, so that the unit normal vector e_3 is well-defined, and the orthonormal frame e_1, e_2 in the tangent plane is defined up to the transformation

$$\begin{pmatrix} e_1' \\ e_2' \end{pmatrix} = \begin{pmatrix} \cos\theta & \sin\theta \\ -\sin\theta & \cos\theta \end{pmatrix} \begin{pmatrix} e_1 \\ e_2 \end{pmatrix}. \tag{63}$$

The connection matrix relative to e_1, e_2 is

$$\omega = \begin{pmatrix} 0 & \omega_{12} \\ -\omega_{12} & 0 \end{pmatrix}. \tag{64}$$

By (18) we find that under the change of frame (63),

$$\omega_{12}' = \omega_{12} + d\theta. \tag{65}$$

The curvature matrix is

$$\Omega = \begin{pmatrix} 0 & \Omega_{12} \\ -\Omega_{12} & 0 \end{pmatrix} \tag{66}$$

and is invariant under a change of frame.

The curvature form Ω_{12} has a simple geometrical interpretation. In fact, by (20) we have

$$\Omega_{12} = d\omega_{12}. \tag{67}$$

The latter is, by (60), equal to

$$d\omega_{12} = -\omega_{13} \wedge \omega_{23} = -\left(h_{311}h_{322} - h_{312}^2\right)\omega_1 \wedge \omega_2, \tag{68}$$

where the expression between the parentheses is the determinant of the second fundamental form Π_3 and is equal to the Gaussian curvature K of M. Thus we have

$$d\omega_{12} = \Omega_{12} = -K\,dA, \tag{69}$$

where $dA = \omega_1 \wedge \omega_2$ is the element of area. This leads immediately to the following:

GAUSS-BONNET THEOREM.   *Let M be a compact oriented surface in E^3. Then*

$$\frac{1}{2\pi} \int_M K\,dA = \chi(M), \qquad (70)$$

where $\chi(M)$ is the Euler characteristic of M.

To prove this theorem let E_0 be the space of unit tangent vectors xe_1' of M. E_0 can be identified with the space of orthonormal frames of the tangent bundle of M, because e_1' determines e_2' as the unit vector orthogonal to e_1' such that $e_1'e_2'$ agrees with the orientation of M. E_0 is a three-dimensional manifold. As its local coordinates we can take those of M and the angle θ in (63). In E_0 we have the global one-form

$$\omega_{12}' = (de_1', e_2') = \omega_{12} + d\theta, \qquad (71)$$

whose exterior derivative is

$$d\omega_{12}' = -K\,dA. \qquad (72)$$

We define a vector field on M with singularities at a finite number of points x_k, $1 \leqslant k \leqslant r$; such a vector field always exists. Let Δ_k be a small disk with x_k as center. The vector field lifts $M - \bigcup_{1 \leqslant k \leqslant r} \Delta_k$ into a surface in E_0 with the boundaries $\partial \Delta_k$. Applying Stokes' Theorem to (72) we have

$$\frac{1}{2\pi} \int_{M - \cup \Delta_k} K\,dA = \frac{1}{2\pi} \sum \int_{-\partial \Delta_k} d\theta + \omega_{12}.$$

As the Δ_k's tend to the points x_k, we have at the limit,

$$\frac{1}{2\pi} \int_M K\,dA = \sum_{1 \leqslant k \leqslant r} I_k, \qquad (73)$$

where I_k is the *index* of the vector field at x_k. Thus we have proved that the integral at the left-hand side is equal to the sum of indices, which is therefore independent of the choice of the vector field.

For another discussion of the Gauss-Bonnet theorem we refer the reader to Chapter 4, "Curves and Surfaces in Euclidean Space," of this book.

By choosing a special vector field, we show that it is equal to $\chi(M)$. Our proof demonstrates at the same time the theorem that the sum of indices of a continuous vector field on M with a finite number of singularities is equal to $\chi(M)$.

Even the case $n = 1$, $q = 2$, i.e., curves in E^3, leads to interesting conclusions. In this case the normal planes xe_2e_3 form the normal bundle. Its normal connection $D^\perp$ is given by the matrix

$$\omega = \begin{pmatrix} 0 & \omega_{23} \\ -\omega_{23} & 0 \end{pmatrix}. \tag{74}$$

As above, when the normal frames undergo the change

$$\begin{pmatrix} e_2' \\ e_3' \end{pmatrix} = \begin{pmatrix} \cos\theta & \sin\theta \\ -\sin\theta & \cos\theta \end{pmatrix} \begin{pmatrix} e_2 \\ e_3 \end{pmatrix}, \tag{75}$$

the connection form is modified according to:

$$\omega_{23}' = \omega_{23} + d\theta. \tag{76}$$

There is no curvature form, because the base manifold M is one-dimensional.

However, an interesting invariant can be introduced as follows: Let M be a closed curve, and let e_2 be a continuous and smooth normal vector field. Then

$$T(e_2) = \frac{1}{2\pi} \int_M \omega_{23} \tag{77}$$

is a real number. From (76) we see that if e_2' is another smooth normal vector field, $T(e_2')$ differs from $T(e_2)$ by an integer. Thus $T(e_2) \bmod 1$ is an invariant of the closed curve M, to be called the *total twist* of M. If the curvature of M never vanishes and M is C^3, we can choose e_2 to be the principal normal vector and $T(e_2)$ becomes the total torsion. But the total twist is defined for C^1-curves.

The total twist plays an important role in molecular biology. T. Banchoff and J. White proved that it is invariant under conformal transformations [2].

6.   COMPLEX LINE BUNDLES

A complex vector bundle with $q = 1$ is called a *complex line bundle*. It plays an important role in various parts of mathematics.

For a complex line bundle the connection and curvature matrices ω, Ω are one- and two-forms respectively, and (20) becomes

$$\Omega = d\omega. \tag{78}$$

By (27) we have

$$c_1(\Omega) = \frac{i}{2\pi}\Omega. \tag{79}$$

If the bundle is hermitian and the connection is admissible, we will restrict ourselves to unitary frames. Then both ω and Ω are skew-hermitian:

$$\omega + \bar{\omega} = 0, \qquad \Omega + \bar{\Omega} = 0, \tag{80}$$

i.e., both $\sqrt{-1}\,\omega$ and $\sqrt{-1}\,\Omega$ are real. We emphasize that ω and Ω in this section are forms, not matrices.

Perhaps the most important complex line bundle is the Hopf bundle, defined as follows. The map

$$\pi\colon C_{n+1} - \{0\} \to P_n(\mathbf{C}) \tag{81}$$

defines the complex projective space $P_n(\mathbf{C})$ of dimension n as the space of all lines through the origin of C_{n+1}. If $(z_0, z_1, \ldots, z_n) = Z \in \mathbf{C}_{n+1} - \{0\}$, $\pi Z = [Z]$ is the point in P_n with Z as the homogeneous coordinate vector.

To study the geometry in $P_n(\mathbf{C})$ we introduce in $\mathbf{C}_{n+1}$ the hermitian scalar product

$$(Z, W) = \sum z_A \bar{w}_A, \tag{82}$$

where

$$Z = (z_0, \ldots, z_n), \qquad W = (w_0, \ldots, w_n), \tag{83}$$

and we use in this section the range of indices:

$$0 \leqslant A, B, C \leqslant n. \tag{84}$$

Z_A is called a unitary frame, if we have

$$(Z_A, Z_B) = \delta_{A\bar{B}}. \tag{85}$$

The space of all unitary frames can be identified with the unitary group $U(n + 1)$. We have the diagram

$$
\begin{array}{ccc}
\{Z_0, Z_1, \ldots, Z_n\} & \in & U(n+1) \\
\downarrow & & \pi_0 \downarrow \\
Z_0 & \in & S^{2n+1} = \{Z \in C_{n+1} | (Z, Z) = 1\} \\
\downarrow & & \pi \downarrow \\
[Z_0] & \in & P_n(\mathbf{C})
\end{array}
\tag{86}
$$

The map π defines a circle bundle over $P_n(\mathbf{C})$, called the *Hopf bundle*. For $n = 1$, $P_1(\mathbf{C}) = S^2$ and the resulting map

$$\pi: S^3 \to S^2, \tag{87}$$

the *Hopf map*, was the first map discovered from a manifold to one of lower dimension, which is not homotopic to a constant map (which collapses the manifold to a point of the image space).

In the space of unitary frames Z_A we can write

$$dZ_A = \sum \omega_{A\bar{B}} Z_B, \tag{88}$$

where

$$\omega_{A\bar{B}} + \omega_{\bar{B}A} = 0, \qquad \omega_{\bar{B}A} = \bar{\omega}_{B\bar{A}}. \tag{89}$$

The $\omega_{A\bar{B}}$ are the Maurer-Cartan forms of $U(n+1)$. They satisfy the following Maurer-Cartan equations obtained by exterior differentiation of (88):

$$d\omega_{A\bar{B}} = \sum_C \omega_{A\bar{C}} \wedge \omega_{C\bar{B}}. \tag{90}$$

The method of moving frames uses the equations in $U(n+1)$ to study the geometry of $P_n(\mathbf{C})$ or S^{2n+1}.

S^{2n+1} consists of the unit vectors of the complex line bundle (81). By (88),

$$DZ_0 = \omega_{0\bar{0}} Z_0 \tag{91}$$

defines an admissible connection in this bundle. It is a connection because the conditions D1) and D2) in Section 3 are satisfied, and is admissible because (89) implies that $\omega_{0\bar{0}}$ is skew-hermitian. The connection form is

$$\omega = \omega_{0\bar{0}}. \tag{92}$$

By (90) the curvature form of this connection is

$$\Omega = d\omega = \sum \omega_{0\bar{\alpha}} \wedge \omega_{\alpha\bar{0}} = -\sum \omega_{0\bar{\alpha}} \wedge \omega_{\bar{0}\alpha}, \qquad 1 \leqslant \alpha \leqslant n. \tag{93}$$

To this corresponds the hermitian differential form

$$\sum \omega_{0\bar{\alpha}} \omega_{\bar{0}\alpha} = (dZ_0, dZ_0) - (dZ_0, Z_0)(Z_0, dZ_0). \tag{94}$$

The last expression remains invariant, when Z_0 undergoes the change $Z_0 \to \lambda Z_0$, $|\lambda| = 1$. Hence it is a positive-definite hermitian differential form in $P_n(\mathbf{C})$, and defines an hermitian metric in $P_n(\mathbf{C})$. Its Kähler form is

$$K = \frac{i}{2} \sum \omega_{0\bar{a}} \wedge \omega_{\bar{0}a} = -\frac{i}{2}\Omega, \qquad (95)$$

which is clearly closed. Thus this metric is Kählerian. It is called the *Study-Fubini metric* on $P_n(\mathbf{C})$.

A fundamental fact is that, because of the complex structures involved, the connection form ω has a "potential", that is, it can be "integrated." In fact, let $Z \in C_{n+1} - \{0\}$, and let

$$Z_0 = Z/|Z|, \qquad |Z|^2 = (Z, Z). \qquad (96)$$

Then we find

$$\pi_0^* \omega_{0\bar{0}} = (DZ_0, Z_0) = (dZ_0, Z_0)$$

$$= \frac{1}{2|Z|^2} \{(dZ, Z) - (Z, dZ)\} = (\partial - \bar{\partial})\log|Z|, \qquad (97)$$

where ∂ and $\bar{\partial}$ are differentiations in C_{n+1} with respect to the holomorphic coordinates z_A and the anti-holomorphic coordinates $\bar{z}_A$ respectively. But the last expression is an expression in $C_{n+1} - \{0\}$. To get one in $P_n(\mathbf{C})$ we take a fixed vector $A \in C_{n+1} - \{0\}$. Then the equation

$$(Z, A) = 0 \qquad (98)$$

defines a hyperplane L in $P_n(\mathbf{C})$. The function

$$\frac{|Z, A|^2}{|Z|^2}, \qquad |Z, A|^2 = |(Z, A)|^2 \qquad (99)$$

is well-defined in $P_n(\mathbf{C}) - L$, because it remains unchanged when Z is multiplied by a factor. Clearly

$$\partial\bar{\partial} \log|Z, A|^2 = 0. \qquad (100)$$

(This can be described by saying that the logarithm of the absolute value of a holomorphic function is harmonic.) We have therefore the relation

$$\Omega = \partial\bar{\partial} \log\frac{|Z, A|^2}{|Z|^2} \tag{101}$$

in $P_n(\mathbf{C}) - L$. This contains the analytic content of the following theorem.

THEOREM (WIRTINGER).   *Let M be a compact Riemann surface and f: $M \to P_n(\mathbf{C})$ be a holomorphic mapping. The image f(M) meets all hyperplanes of $P_n(\mathbf{C})$ the same number of times, which is equal to the area of f(M), properly normalized.*

We normalize the element of area of $P_n(\mathbf{C})$ to be K/π, K being the Kähler form defined in (95). Then

$$\frac{K}{\pi} = -\frac{i}{2\pi}\Omega = -\frac{i}{\pi}\partial\bar{\partial}\log\frac{|Z, A|}{|Z|} = \frac{i}{2\pi}d(\partial - \bar{\partial})\log\frac{|Z, A|}{|Z|} \tag{102}$$

We leave it to the reader to show that the total area of $P_1(\mathbf{C})$ is 1.

By hypothesis $f(M)$ is a compact holomorphic curve. We suppose that it does not lie in a hyperplane L. Then it meets L in a finite number of points, say $x_k \in M$, $1 \leqslant k \leqslant r$. About each x_k we take a disc Δ_k. Formula (102) allows us to apply Stokes' Theorem to $M - \cup\Delta_k$. It is easy to see that the boundary term will tend, as Δ_k shrinks to x_k, to the number of zeros of the holomorphic function (Z, A), and the theorem is proved.

Wirtinger's Theorem could also be proved by a topological argument. The great advantage of our approach is that it relates the local and global aspects of the problem in the clearest way and extends to the case when M is noncompact. The ideas introduced in our differential-geometric treatment is at the basis of the value distribution theory of noncompact holomorphic curves. Needless to say, the noncompact case is much more delicate. For its development see [6], [9].

The tangent bundle of a surface in E^3 and the normal bundle of a curve in E^3, as discussed in Section 5, can also be considered complex line bundles: the fibers are oriented planes with an inner product and a complex structure is defined in which multiplication by i is rotation by 90°. The results at the end of Section 5 can be put in the notation of this section. We will leave it as an exercise.

7. MAXWELL'S AND YANG-MILLS' EQUATIONS

Maxwell's equations in the space-time $(x^1, x^2, x^3, t = x^0)$, as commonly given, can be written as

$$\operatorname{div} \vec{E} = 4\pi\rho,$$

$$\operatorname{curl} \vec{B} - \frac{\partial}{\partial t} \vec{E} = 4\pi \vec{j}, \tag{103a}$$

$$\operatorname{div} \vec{B} = 0,$$

$$\operatorname{curl} \vec{E} + \frac{\partial}{\partial t} \vec{B} = 0, \tag{103b}$$

where

$$\vec{E} = (E_1, E_2, E_3) = \text{electric field},$$

$$\vec{B} = (B_1, B_2, B_3) = \text{magnetic field},$$

$$\rho = \text{charge density}, \tag{104}$$

$$\vec{j} = (j_1, j_2, j_3) = \text{current vector}.$$

These can be written in a different form, which will be susceptible to natural and important generalizations.

In fact, introduce the antisymmetric matrix

$$(F_{\alpha\beta}) = \begin{pmatrix} 0 & E_1 & E_2 & E_3 \\ -E_1 & 0 & -B_3 & B_2 \\ -E_2 & B_3 & 0 & -B_1 \\ -E_3 & -B_2 & B_1 & 0 \end{pmatrix}, \tag{105}$$

and let

$$J = \rho\, dx^0 + j_1 dx^1 + j_2 dx^2 + j_3 dx^3$$
$$F = \sum F_{\alpha\beta}\, dx^\beta \wedge dx^\alpha. \tag{106}$$

In the space-time we make use of the Lorentz-metric

$$ds^2 = -(dx^0)^2 + (dx^1)^2 + (dx^2)^2 + (dx^3)^2 \tag{107}$$

and the corresponding $*$-operator in the sense of Hodge. With these notations it is immediately verified that (103a), (103b) can be written respectively as

$$d*F = 4\pi J \tag{108a}$$

$$dF = 0, \tag{108b}$$

where $d*$ is the codifferential defined by

$$d* = *d*. \tag{109}$$

This is exactly the situation discussed in the last section, when M is a four-dimensional Lorentzian manifold and E is an hermitian line bundle over M. There is a discrepancy between our notation and terminology and those of the physicists, and we make the following table:

mathematics	physics
connection form ω	gauge potential A
curvature form Ω	Faraday F or strength

Equation (103b) or (108b) is the Bianchi identity. It is a consequence of the equation

$$dA = F, \tag{110}$$

which rewrites (78). It is for this fact that A is called a gauge potential.

It has recently been realized that in order to describe all phenomena in electricity and magnetism one should use (110) instead of the classical equation (108b), which is a consequence of (110) and has been adequate for most applications. This was the result of an important experiment proposed by Y. Aharanov and D. Böhm in 1959 and carried out by R. G. Chambers in 1960; see [10]. In other words, the unknown in Maxwell's equations should be the connection form or the gauge potential A and not the curvature form or the strength F, and the equations should be

$$d*F = 4\pi J, \qquad dA = F. \tag{111}$$

This becomes manifest when one generalizes to the Yang-Mills equations; see [1]. The object here is an SU(2)-bundle over M, again a four-dimensional Lorentzian manifold. As the gauge group is nonabelian, one has to use covariant differentiation D_A. The Yang-Mills equations are

$$D_A^* F = 4\pi J, \qquad D_A A = F, \tag{112}$$

where

$$D_A A = dA - A \wedge A. \tag{113}$$

Thus we see that the Yang-Mills equations are a straightforward generalization of Maxwell's equations, being the cases when the fiber dimensions are $q = 1$ and $q = 2$ respectively. The Yang-Mills equations are playing a far-reaching role in the study of four-dimensional manifolds (work of S. Donaldson), the main reason being that their solutions give manifolds which have an important geometrical meaning.

BIBLIOGRAPHY

(Standard books on differential geometry are not included.)

1. M. F. Atiyah, *Geometry of Yang-Mills Fields*, Pisa, 1979.
2. T. Banchoff and J. White, The behaviour of the total twist and the self-linking number of a closed space curve under inversions. *Math. Scandinavica* 36, (1975) 254–262.
3. S. Chern, *Complex Manifolds Without Potential Theory*, 2nd edition, Springer, 1979.
4. ____, *Circle Bundles, Geometry and Topology*, III Latin Amer. School of Math., Lecture Notes in Math., No. 597, 114–131, Springer, 1977.
5. Johan L. Dupont, *Curvature and Characteristic Classes*, Lecture Notes in Math. 640, Springer, 1978.
6. P. Griffiths, *Entire Holomorphic Mappings in One and Several Complex Variables*, Annals of Math. Studies, No. 85, Princeton Univ. Press, 1976.
7. J. W. Milnor and J. Stasheff, *Characteristic Classes*, Annals of Math. Studies 76, Princeton, 1974.
8. H. V. Pittie, *Characteristic Classes of Foliations*, Pitman, San Francisco, 1976.
9. H. Wu, *The Equidistribution Theory of Holomorphic Curves*, Annals of Math. Studies, No. 64, Princeton Univ. Press, 1970.
10. T. T. Wu and C. N. Yang, Concept of non-integrable phase-factors and global formulation of gauge fields, *Physical Review D*, 12, (1975) 3845–57.

DIFFERENTIAL FORMS

Harley Flanders

1. INTRODUCTION

It is the purpose of this exposition to discuss the calculus of differential forms and several aspects of differential geometry in which differential forms are a natural tool. In this chapter, I shall emphasize a number of explicit examples, hoping to inspire students to pursue differential geometry in one of the many modern texts now available, some of which are listed in the references ([**Au1**], [**Bi1**], [**Bl2**], [**Ch1**], [**Go1**], [**Gr1**], [**Gu1**], [**He1**], [**Kl1**], [**Ko1**], [**Ko2**], [**La1**], [**Li1**], [**Li2**], [**Lo1**], [**M1**], [**Rh1**], [**Sp1**], [**St1**], [**Su1**], [**Wi1**], [**Y1**]). I also list books that particularly emphasize differential forms ([**Ca6**], [**Fl1**], [**Gr2**], [**Sc1**], [**Sl1**], [**Va1**]).

Much of modern differential geometry has its origin in classical mathematics. It is always important for the student to know the sources of the material he studies today. For this reason, we list two older sources of classical differential geometry ([**Bl1**], [**Da1**]).

Without any question, the most important geometer of the first half of this century was Elie Cartan. He developed differential forms into a powerful tool in differential geometry, and his books and papers will inspire mathematical research for years to come. His pertinent work may be found in [**Ca1–Ca5**].

Finally, we mention some sources for applications of differential forms outside of mathematics. There are many, and the ones listed here include many other references ([**Ch2**], [**Ed1**], [**We1**]).

2. DIFFERENTIAL FORMS IN $\mathbf{E}^n$

In this section, we shall not worry much about technical details, but shall try to develop a feel for the object, a differential form, in $\mathbf{E}^n$. Let $(x^1, \ldots, x^n)$ denote Cartesian coordinates of a point in $\mathbf{E}^n$. Our first approximation is that a *differential form ω of degree p* is the expression

$$\omega = \sum A_{i_1 \cdots i_p}(x^1, \ldots, x^n)\, dx^{i_1} \wedge dx^{i_2} \wedge \cdots \wedge dx^{i_p}$$

that occurs under the integral sign

$$\int_{\mathbf{c}^p} \omega,$$

where $\mathbf{c}^p$ is a piece of p-dimensional surface in $\mathbf{E}^n$.

The familiar line integral

$$\int_{\mathbf{c}^1} \left(A_1\, dx^1 + A_2\, dx^2 + A_3\, dx^3 \right)$$

and surface integral

$$\int_{\mathbf{c}^2} \left(B_{23}\, dx^2 \wedge dx^3 + B_{31}\, dx^3 \wedge dx^1 + B_{12}\, dx^1 \wedge dx^2 \right)$$

are examples in $\mathbf{E}^3$.

Why is the wedge $\wedge$ placed between the differentials? First, our integrals are always *oriented* integrals. We remember that in a surface integral there is always an outward normal to the surface; reverse the normal and the integral changes sign. Likewise, the curve in a line integral has a sense of direction, the reversal of which changes the sign of the integral.

Second, we want the rule for change of variables to be valid for these oriented integrals, which says, for example,

$$dx^1 \wedge dx^2 = \frac{\partial(x^1, x^2)}{\partial(u^1, u^2)} du^1 \wedge du^2. \tag{2.1}$$

Since the Jacobian changes sign when its rows are interchanged, we are forced to write,

$$dx^2 \wedge dx^1 = -dx^1 \wedge dx^2. \tag{2.2}$$

Thus, the wedge, or *exterior product*, of linear differential forms is a skew-symmetric (or alternating) product. We are also forced to have

$$dx \wedge dx = 0, \tag{2.3}$$

because the Jacobian vanishes when its rows are equal. For more motivation, see [**Fl1**, pp. 1–2] and [**St1**, pp. 97–99].

In $\mathbf{E}^4$ we have 0-forms, or functions:

$$f = f(x^1, \ldots, x^4) \qquad \text{(also called } scalars);$$

1-forms, or linear forms:

$$A_1 dx^1 + A_2 dx^2 + A_3 dx^3 + A_4 dx^4 \qquad \text{(also called } Pfaffians);$$

2-forms:

$$B_{12} dx^1 \wedge dx^2 + B_{13} dx^1 \wedge dx^3 + \cdots + B_{34} dx^3 \wedge dx^4;$$

3-forms:

$$C_1 dx^2 \wedge dx^3 \wedge dx^4 + \cdots + C_4 dx^1 \wedge dx^2 \wedge dx^3;$$

and 4-forms:

$$D dx^1 \wedge dx^2 \wedge dx^3 \wedge dx^4 \qquad \text{(also called } densities).$$

Note that a 0-form involves one function, a 1-form involves 4 functions, a 2-form involves $_4C_2 = 6$ functions, a 3-form involves 4 functions, and a 4-form involves 1 function.

Since we must soon discuss differentiation of differential forms, we simplify matters by assuming that the functions involved have derivatives of arbitrarily high order—that they are functions of class C^∞ (infinitely differentiable).

In $\mathbf{E}^n$, we use the relations (2.2) and (2.3) to express the general p-form as

$$\omega = \sum_{1 \leqslant i_1 < i_2 < \cdots < i_p \leqslant n} A_{i_1 \cdots i_p} dx^{i_1} \wedge \cdots \wedge dx^{i_p}, \qquad (2.4)$$

so that there are

$$_nC_p = \frac{n!}{p!(n-p)!}$$

functions $A_{(i)}$ involved. An alternative expression is

$$\omega = \frac{1}{p!} \sum A_{i_1 \cdots i_p} dx^{i_1} \wedge \cdots \wedge dx^{i_p}, \qquad (2.5)$$

where $A_{i_1 \cdots i_p}$ is skew-symmetric—that is, it vanishes whenever two indices are equal and changes sign whenever two indices are transposed.

The *exterior product* of any two differential forms is defined as:

$$\left(\sum A_{i_1 \cdots i_p} dx^{i_1} \wedge \cdots \wedge dx^{i_p} \right) \wedge \left(\sum B_{j_1 \cdots j_q} dx^{j_1} \wedge \cdots \wedge dx^{j_q} \right)$$

$$= \sum \left(A_{i_1 \cdots} B_{j_1 \cdots} \right) \left(dx^{i_1} \wedge \cdots \wedge dx^{i_p} \wedge dx^{j_1} \wedge \cdots \wedge dx^{j_q} \right). \qquad (2.6)$$

The rules (2.2) and (2.3) are used to wipe out any terms that have a repetition of dx's, to arrange the dx's in the remaining terms in a suitable order, and then to collect coefficients. (Note that $\mathbf{E}^n$ does not have forms of degree greater than n, so that if α is a p-form and β is a q-form in $\mathbf{E}^n$, and $p + q > n$, then automatically, $\alpha \wedge \beta = 0$.)

In $\mathbf{E}^3$ we have

$$\left(A_1\,dx_1 + A_2\,dx^2 + A_3\,dx^3\right) \wedge \left(B_1\,dx^1 + B_2\,dx^2 + B_3\,dx^3\right)$$

$$= (A_2 B_3 - A_3 B_2)\,dx^2 \wedge dx^3 + (A_3 B_1 - A_1 B_3)\,dx^3 \wedge dx^1 \quad (2.7)$$

$$+ (A_1 B_2 - A_2 B_1)\,dx^1 \wedge dx^2,$$

and

$$\left(A_1\,dx^1 + A_2\,dx^2 + A_3\,dx^3\right)$$

$$\wedge \left(C_1\,dx^2 \wedge dx^3 + C_2\,dx^3 \wedge dx^1 + C_3\,dx^1 \wedge dx^2\right) \quad (2.8)$$

$$= (A_1 C_1 + A_2 C_2 + A_3 C_3)\,dx^1 \wedge dx^2 \wedge dx^3.$$

In $\mathbf{E}^4$, if for instance

$$\omega = dx^1 \wedge dx^2 + dx^3 \wedge dx^4,$$

then

$$\omega \wedge \omega = 2\,dx^1 \wedge dx^2 \wedge dx^3 \wedge dx^4.$$

In general, exterior multiplication is associative and is distributive with respect to addition. It is anticommutative in precisely this sense: If α is a p-form and β is a q-form then

$$\beta \wedge \alpha = (-1)^{pq} \alpha \wedge \beta. \quad (2.9)$$

In $\mathbf{E}^3$, Stokes' theorem says that if $\mathbf{c}^2$ is a piece of oriented surface whose boundary curve $\partial \mathbf{c}^2$ is directed accordingly, then

$$\int_{\partial \mathbf{c}^2}\left(A_1\,dx^1 + A_2\,dx^2 + A_3\,dx^3\right)$$

$$= \int_{\mathbf{c}^2}\left[\left(\frac{\partial A_3}{\partial x^2} - \frac{\partial A_2}{\partial x^3}\right)dx^2 \wedge dx^3 \right. \qquad (2.10)$$

$$\left. + \left(\frac{\partial A_1}{\partial x^3} - \frac{\partial A_3}{\partial x^1}\right)dx^3 \wedge dx^1 + \left(\frac{\partial A_2}{\partial x^1} - \frac{\partial A_1}{\partial x^2}\right)dx^1 \wedge dx^2\right].$$

Thus, with each 1-form

$$\omega = A_1 \, dx^1 + A_2 \, dx^2 + A_3 \, dx^3,$$

there is associated a 2-form

$$d\omega = \left(\frac{\partial A_3}{\partial x^2} - \frac{\partial A_2}{\partial x^3} \right) dx^2 \wedge dx^3 + \cdots, \tag{2.11}$$

so that the formula

$$\int_{\partial \mathbf{c}^2} \omega = \int_{\mathbf{c}^2} d\omega \tag{2.12}$$

is valid for each piece of oriented surface $\mathbf{c}^2$.

This carries over to all dimensions, and is one of the beauties of the calculus of exterior forms. It is possible to define for each p-form ω a $(p+1)$-form $d\omega$, called the *exterior derivative* of ω, in such a way that Stokes' theorem holds: If $\mathbf{c}^{p+1}$ is an oriented $(p+1)$-dimensional domain of integration in $\mathbf{E}^n$, and if $\mathbf{c}^p = \partial \mathbf{c}^{p+1}$ is its p-dimensional boundary (also a domain of integration) which is oriented in a way coherent with the orientation of $\mathbf{c}^{p+1}$, then

$$\int_{\partial \mathbf{c}^{p+1}} \omega = \int_{\mathbf{c}^{p+1}} d\omega. \tag{2.13}$$

Exterior differentiation d has a simple definition. It is additive, so that we need only write it down for the individual summands in (2.4):

$$d(A \, dx^{i_1} \wedge \cdots \wedge dx^{i_p})$$

$$= \left(\sum \frac{\partial A}{\partial x^i} dx^i \right) \wedge (dx^{i_1} \wedge \cdots \wedge dx^{i_p}). \tag{2.14}$$

Example in $\mathbf{E}^4$:

$$d\left(A\,dx^1 \wedge dx^2 + B\,dx^1 \wedge dx^4 \right)$$

$$= (dA) \wedge dx^1 \wedge dx^2 + (dB) \wedge dx^1 \wedge dx^4$$

$$= \left(\frac{\partial A}{\partial x^1} dx^1 + \frac{\partial A}{\partial x^2} dx^2 + \frac{\partial A}{\partial x^3} dx^3 + \frac{\partial A}{\partial x^4} dx^4 \right) \wedge dx^1 \wedge dx^2$$

$$+ \left(\frac{\partial B}{\partial x^1} dx^1 + \frac{\partial B}{\partial x^2} dx^2 + \frac{\partial B}{\partial x^3} dx^3 + \frac{\partial B}{\partial x^4} dx^4 \right) \wedge dx^1 \wedge dx^4$$

$$= \frac{\partial A}{\partial x^3} dx^1 \wedge dx^2 \wedge dx^3 + \left(\frac{\partial A}{\partial x^4} - \frac{\partial B}{\partial x^2} \right) dx^1 \wedge dx^2 \wedge dx^4$$

$$- \frac{\partial B}{\partial x^3} dx^1 \wedge dx^3 \wedge dx^4.$$

The exterior derivative has the following characteristic properties:

$$\begin{cases} dA = \sum \dfrac{\partial A}{\partial x^i} dx^i & \text{for a scalar } A; \\[2mm] d(\alpha \wedge \beta) = (d\alpha) \wedge \beta + (-1)^{\deg \alpha}\alpha \wedge (d\beta) \\[2mm] d(d\alpha) = 0 & \text{for any } \alpha. \end{cases} \qquad (2.15)$$

Let us note in passing another familiar case of the general Stokes' theorem (2.13). We let $\mathbf{c}^3$ be a three-dimensional domain of integration in $\mathbf{E}^3$. Then,

$$\int_{\partial \mathbf{c}^3} \left(A_1\,dx^2 \wedge dx^3 + A_2\,dx^3 \wedge dx^1 + A_3\,dx^1 \wedge dx^2 \right)$$

$$= \int_{\mathbf{c}^3} \left(\frac{\partial A_1}{\partial x^1} + \frac{\partial A_2}{\partial x^2} + \frac{\partial A_3}{\partial x^3} \right) dx^1 \wedge dx^2 \wedge dx^3, \qquad (2.16)$$

the theorem of Gauss (or Ostrogradsky).

The third formula in (2.15), $d(d\alpha) = 0$, is known as the *Poincaré lemma*. It is very important, first because it allows us to think of d as a coboundary operation with topological implications, and second, because it is the source of what are called integrability conditions in differential geometry. This relation, $d(d\alpha) = 0$, is nothing more than the equality of mixed second partials:

$$\frac{\partial^2 A}{\partial x \, \partial y} = \frac{\partial^2 A}{\partial y \, \partial x}.$$

From Equations (2.7), (2.8), (2.10), and (2.16), we see that the calculus of exterior differential forms includes the standard operations of vector analysis: vector product, scalar product, curl (rotation), and divergence.

We come now to a most striking property of differential forms—their behavior under mappings. We take two Euclidean spaces—$\mathbf{E}^m$ with coordinates $x^1, \ldots, x^m$, and $\mathbf{E}^n$ with coordinates $y^1, \ldots, y^n$. We assume an infinitely differentiable mapping,

$$\phi: \mathbf{E}^m \to \mathbf{E}^n,$$

which we may express in coordinates by

$$y^i = y^i(x^1, \ldots, x^m) \qquad (i = 1, \ldots, n). \tag{2.17}$$

If

$$\omega = \sum A_{i_1 \cdots i_p}(y^1, \ldots, y^n) \, dy^{i_1} \wedge \cdots \wedge dy^{i_p}$$

is any *p*-form on $\mathbf{E}^n$, we may associate with it an *induced p-form* $\phi^*\omega$ on $\mathbf{E}^m$ by the following substitution rules. In the expression for ω, replace each coefficient $A(y^1, \ldots, y^n)$ by the composite function

$$A\big(y^1(x^1, \ldots, x^m), \ldots, y^n(x^1, \ldots, x^m)\big),$$

and replace each dy^i by the linear differential

$$\frac{\partial y^i}{\partial x^1} dx^1 + \cdots + \frac{\partial y^i}{\partial x^m} dx^m.$$

Then use exterior multiplication and consolidate terms.

Thus, from the mapping ϕ: $\mathbf{E}^m \to \mathbf{E}^n$, we have constructed a mapping ϕ^* which takes each p-form ω on $\mathbf{E}^n$ to a p-form $\phi^*\omega$ on $\mathbf{E}^m$. The operations on differential forms are preserved in precisely this sense:

$$\begin{cases} \phi^*(\omega_1 + \omega_2) = \phi^*\omega_1 + {}^*\phi\omega_2, \\ \phi^*(\omega \wedge \eta) = (\phi^*\omega) \wedge (\phi^*\eta), \\ \phi^*(d\omega) = d(\phi^*\omega). \end{cases} \qquad (2.18)$$

Proving these relations is an excellent test of your grasp of this subject so far and is recommended as an exercise.

Suppose we are given a composition of infinitely differentiable mappings:

if ω is a p-form on $\mathbf{E}^r$, then we have

$$(\psi \circ \phi)^*\omega = \phi^*(\psi^*\omega), \qquad (2.19)$$

which we may write simply as

$$(\psi \circ \phi)^* = \phi^* \circ \psi^*.$$

On checking this, we recognize that it is essentially the chain rule.

We have now come far enough so that we may save writing by dropping the wedge, $\wedge$. Henceforth, we understand $dx\,dy$ means $dx \wedge dy$, etc.

3. THE POINCARÉ LEMMA

The Poincaré lemma, $d(d\alpha) = 0$, may be stated as follows.

LEMMA. *If ω is a p-form on $\mathbf{E}^n$ for which there exists a $(p-1)$-form α such that $d\alpha = \omega$, then $d\omega = 0$.*

It is remarkable, and not at all obvious, that the converse is true.

CONVERSE OF THE POINCARÉ LEMMA.   *If ω is a p-form on $\mathbf{E}^n$ such that $d\omega = 0$, then there is a $(p-1)$-form α such that $\omega = d\alpha$. (Exception: $p = 0$. Then $\omega = f$ is a scalar, and $df = 0$ simply implies f is constant.)*

A standard proof, motivated by the cylinder construction of topology, is given in [**Fl1**, pp. 27–29]. Here we shall give a more direct proof, by induction. As is often the case with induction, it pays to strength the conclusion so that the induction hypothesis is stronger. We do so by allowing parameters. We shall prove:

Let ω be a p-form in $dx^1, \ldots, dx^n$ whose coefficients are smooth functions of $x = (x^1, \ldots, x^n)$ and also of $t = (t^1, t^2, \ldots)$. Let $d = d_x$ denote the exterior derivative with respect to the space variables x alone. Assume $d\omega = 0$. Then $\omega = d\eta$, where η is a $(p-1)$-form in $dx^1, \ldots, dx^n$ whose coefficients are smooth functions of x and t. (Exception: $p = 0$. Then $\omega = f(x, t)$ is a scalar, and $df = 0$ simply implies $f = f(t)$ is independent of x.)

Proof.   We go by induction on n. If $n = 1$, then only $p = 1$ is possible, and any indefinite integral η of $A(x, t)$ with respect to x satisfies $d\eta = A\,dx$.

We digress slightly: If a q-form α happens to omit dx^n, then we can add x^n into the parameter space, so that in addition to the exterior differentiation operator d we also have d', exterior differentiation with respect to $x^1, \ldots, x^{n-1}$. Their relation is readily seen to be

$$d\alpha = d'\alpha + \beta\,dx^n, \qquad (3.1)$$

where the $(q-1)$-form β, like α, omits dx^n.

To return to the proof, consider the special case

$$\omega = \alpha\,dx^n.$$

Then $d\omega = 0$ implies

$$(d'\alpha + \beta\,dx^n)\,dx^n = 0,$$
$$d'\alpha\,dx^n = 0.$$

But $d'\alpha$ omits dx^n, so we have $d'\alpha = 0$. By the induction hypothesis, $\alpha = d'\gamma$. However, by (3.1),

$$d\gamma = d'\gamma + \delta\,dx^n = \alpha + \delta\,dx^n,$$

so

$$\omega = \alpha\,dx^n = (d\gamma - \delta\,dx^n)\,dx^n$$

$$= d\gamma\,dx^n = d(\gamma\,dx^n) = d\eta,$$

precisely what we wanted to prove.

In the general case, we write

$$\omega = \alpha + \sigma\,dx^n \tag{3.2}$$

where α omits dx^n. Then $d\omega = 0$ implies

$$d\alpha + d\sigma\,dx^n = 0,$$

$$d'\alpha + \beta\,dx^n + d\sigma\,dx^n = 0,$$

$$d'\alpha + (\beta + d\sigma)\,dx^n = 0.$$

It follows that $d'\alpha = 0$, so by induction

$$\alpha = d'\gamma = d\gamma - \delta\,dx^n. \tag{3.3}$$

Therefore by (3.2) and (3.3),

$$\omega = d\gamma - \delta\,dx^n + \sigma\,dx^n = d\gamma + \lambda\,dx^n.$$

By the Poincaré lemma, the form

$$\omega_1 = \omega - d\gamma = \lambda\,dx^n$$

satisfies $d\omega_1 = 0$. By the special case, $\omega_1 = d\eta_1$, so finally

$$\omega = \omega_1 + d\gamma = d\eta_1 + d\gamma$$

$$= d(\eta_1 + \gamma) = d\eta.$$

4. MANIFOLDS

If we look critically at which properties of Euclidean space E^n we have used to set up the calculus of differential forms, we see that the coordinate system $x^1, \ldots, x^n$ is the only one we have used. However, according to the behavior of differential forms under mappings, (2.18) and (2.19), any other coordinate system could have been used as long as the two systems were related by C^∞ functions. In particular, we automatically have a theory of differential forms on any open subset of E^n simply by restricting the Euclidean coordinate functions to the open subset.

But this method is far from adequate, since we want to study geometric spaces that cannot be considered open subsets of E^n as such. For example, the 2-sphere S^2 is a two-dimensional surface in E^3 which cannot be imbedded in the plane. We need a theory of differential forms on such surfaces.

It might seem adequate to extend the definition of a (nonsingular) surface in E^3 to a definition of a (nonsingular) variety of dimension p in E^n. Actually, it is adequate, but only as a result of a deep imbedding theorem of H. Whitney. However, there are many rather simple spaces which appear to have sufficient structure for study by analytic methods, but which we do not identify in any apparent fashion with subspaces of E^n. Two examples will suffice.

1. *The space of all directed lines in* E^3. A minute's reflection will convince us that this is a four-dimensional space. It appears to be smooth—no corners, edges, etc. but it is not obvious how to consider it as a subspace of E^n. (The Whitney theorem just mentioned shows that this space may be considered a subspace of E^8, but does not do so explicitly. See [St1, p. 63], [Wh1, p. 236], and [Wh2, pp. 111, ff].)

2. *Projective space* P^n. This space may be considered the set of all undirected lines through the origin in E^{n+1}. Alternatively, it is obtained by identifying antipodal points on the sphere S^n. (The projective plane, P^2, can be imbedded in E^4 but not in E^3.)

Now we set down the formal definition of a (differentiable) manifold, probably the most general kind of structure on which the

differential form calculus is possible. The idea is to have a topological space, each point of which has a neighborhood with coordinate
functions making said neighborhood indistinguishable from a
neighborhood in $\mathbf{E}^n$.

We fix our object, $\mathbf{M}$, a connected topological space, and we fix
what will be its dimension, n.

DEFINITION (4.1).   A *chart* on $\mathbf{M}$ is a pair $(\mathbf{U}, f)$ where $\mathbf{U}$ is an
open subset of $\mathbf{M}$ and f is a homeomorphism on $\mathbf{U}$ onto an open
subset $f(\mathbf{U})$ of $\mathbf{E}^n$.

DEFINITION (4.2).   An *atlas* on $\mathbf{M}$ is a set $\mathscr{A}$ of charts such
that: (1) the sets $\mathbf{U}$ coming from the charts $(\mathbf{U}, f)$ of $\mathscr{A}$ cover $\mathbf{M}$;
(2) if $(\mathbf{U}, f)$ and $(\mathbf{V}, g)$ are in the atlas, and if $\mathbf{U} \cap \mathbf{V} \neq \varnothing$, then the
mapping $g \circ f^{-1}$: $f(\mathbf{U} \cap \mathbf{V}) \to g(\mathbf{U} \cap \mathbf{V})$ is of class C^∞; and (3) $\mathscr{A}$
is maximal with respect to properties (1) and (2).

DEFINITION (4.3).   A *differentiable manifold* of dimension n is a
triple $(\mathbf{M}, n, \mathscr{A})$ where $\mathbf{M}$ is a connected topological space and $\mathscr{A}$
is an atlas of charts $(\mathbf{U}, f)$, $f : \mathbf{U} \to \mathbf{E}^n$.

We commonly refer to the space $\mathbf{M}$ itself as the *differentiable
manifold*, keeping in mind the structure imposed on this space.
(This is no worse than saying "the group $\mathbf{G}$" or "the topological
space $\mathbf{X}$," etc.)

Let $p \in \mathbf{M}$ and let $(\mathbf{U}, f)$ be a chart such that $p \in \mathbf{U}$. Now, $f(\mathbf{U})$
is a subset of $\mathbf{E}^n$ with its Cartesian coordinates $x^1, \ldots, x^n$. Since
$f(\mathbf{U})$ is homeomorphic to $\mathbf{U}$, we may think of $x^1, \ldots, x^n$, as functions on $\mathbf{U}$, in which case we get what is called a *local coordinate
system* at p.

If $y^1, \ldots, y^n$ is another local coordinate system at p, we may
write

$$y^i = y^i(x^1, \ldots, x^n),$$

and these functions will be infinitely differentiable where defined,
according to Definition (4.2).

A real function ϕ on **M** is *infinitely differentiable*, or C^∞, if it is an infinitely differentiable function of the local coordinates $x^1, \ldots, x^n$ on each such **U**.

Similarly, we define a C^∞ mapping

$$\phi: \mathbf{M} \to \mathbf{N}$$

from one manifold to another.

Almost every modern text on differential geometry covers the basics of C^∞ manifolds. An interesting treatment of the real analytic case is found in [**Ch1**, pp. 68, ff].

We now come to the definition of a differential form on a manifold **M**. To this end, let **M** be a manifold of dimension n with atlas $\mathscr{A}$.

DEFINITION (4.4). A *differential form* of degree p on **M** is defined by assigning to each chart $(\mathbf{U}, f)$ of $\mathscr{A}$ a p-form

$$\omega_{(f, \mathbf{U})} \quad \text{on} \quad f(\mathbf{U})$$

so that whenever $(f, \mathbf{U})$ and $(g, \mathbf{V})$ are two charts with $\mathbf{U} \cap \mathbf{V} \neq \varnothing$, and

$$g \circ f^{-1} \colon f(\mathbf{U} \cap \mathbf{V}) \to g(\mathbf{U} \cap \mathbf{V}),$$

then

$$\left(g \circ f^{-1} \right)^* \omega_{(g, \mathbf{V})} = \omega_{(f, \mathbf{U})} \quad \text{on} \quad g(\mathbf{U} \cap \mathbf{V}).$$

In other words, we take a p-form in each coordinate system and insist that these "pieces of a form" fit together properly. Once we have said it right, all of our previous work is applicable. In particular,

$$\omega_1 + \omega_2, \qquad \omega \wedge \eta, \qquad d\omega$$

are defined. The Poincaré lemma, $d(d\omega) = 0$, holds (but not necessarily its converse).

If

$$\mathbf{M} \xrightarrow{\phi} \mathbf{N}$$

then $\phi^*\omega$ is a p-form on $\mathbf{M}$ for each p-form ω on $\mathbf{N}$, and

$$\begin{cases} \phi^*(\omega_1 + \omega_2) = \phi^*\omega_1 + \phi^*\omega_2, \\ \phi^*(\omega \wedge \eta) = (\phi^*\omega) \wedge (\phi^*\eta), \\ \phi^*(d\omega) = d(\phi^*\omega). \end{cases} \tag{4.5}$$

If

$$\mathbf{M} \xrightarrow{\phi} \mathbf{N}$$
$$\psi \circ \phi \searrow \quad \downarrow \psi$$
$$\mathbf{P}$$

then

$$(\psi \circ \phi)^* = \phi^* \circ \psi^*. \tag{4.6}$$

All of these statements are proved by (what seem to the experienced to be) routine applications of the definitions and the corresponding statements in Section 2. However, the beginner must work through the proofs himself to be sure that he understands the subject.

Differential forms on a manifold can also be defined in a more sophisticated way as cross sections of suitable fiber bundles. See [**St1**, pp. 77–96] for instance.

NOTE. *Requirement* (3) *of Definition* (4.2) *seems puzzling at first. It is included only for the technical reason that without it we must define equivalence classes of atlases. For practical applications, an atlas is completely determined by any subset which contains enough charts* $(\mathbf{U}, f)$ *so that the collection of* $\mathbf{U}$'s *covers* $\mathbf{M}$. *To define a p-form, it is enough to give the* $\omega_{(f, \mathbf{U})}$ *for these charts of the subset, subject to the condition of fitting together, of course.*

5. INTEGRATION

Let us begin with a review of line and surface integrals in $\mathbf{E}^3$. For a line integral, we are given a curve $\mathbf{c}^1$ in $\mathbf{E}^3$ and a 1-form ω defined in a domain which includes the curve. The problem is to define

$$\int_{\mathbf{c}^1} \omega.$$

The usual procedure is to break the curve into several pieces, each of which has a parameterization, to give the integral by an explicit formula for each piece, and to sum the results.

The simplest kind of parametric curve will be given by a one-to-one smooth mapping

$$\phi: t \to \left(x^1(t), \ldots, x^3(t)\right) \qquad (5.1)$$

on $0 \leqslant t \leqslant 1$ into $\mathbf{E}^3$, such that the velocity vector

$$\left(\frac{dx^1}{dt}, \ldots, \frac{dx^3}{dt} \right)$$

never vanishes. We then substitute into

$$\omega = A_1 \, dx^1 + A_2 \, dx^2 + A_3 \, dx^3$$

the expressions for $x^1, \ldots, x^3$ in terms of t given by (5.1), and we set

$$\int_{c^1} \omega = \int_0^1 \left[A_1\left(x^1(t), \ldots, x^3(t)\right) \frac{dx^1}{dt} + \cdots \right] dt. \qquad (5.2)$$

In differential form language,

$$\int_{c^1} \omega = \int_0^1 \phi^* \omega. \qquad (5.3)$$

Let us look at it another way. We denote by $\mathbf{e}^1$ the unit interval on the t-axis, a very special parameterized curve. Its image under ϕ is $\mathbf{c}^1$, and we write

$$\phi_* \mathbf{e}^1 = \mathbf{c}^1 \qquad (5.4)$$

to emphasize this relation. Then (5.3) becomes

$$\int_{\phi_* \mathbf{e}^1} \omega = \int_{\mathbf{e}^1} \phi^* \omega, \qquad (5.5)$$

a suggestive formulation.

Next, let us examine a surface integral. We are given a 2-form,

$$\omega = P_1 \, dx^2 \, dx^3 + P_2 \, dx^3 \, dx^1 + P_3 \, dx^1 \, dx^2$$

on an open set U of E^3 (again, omitting $\wedge$). We are also given an oriented surface c^2, and we seek

$$\int_{c^2} \omega. \tag{5.6}$$

The surface c^2 is broken up into parameterized pieces and the integrals for these pieces are summed. Thus, let us suppose c^2 is one of these pieces—that is, c^2 is given as the image of a smooth one-one mapping,

$$\phi: (u^1, u^2) \to x(u^1, u^2) = (x^1(u^1, u^2), \ldots, x^3(u^1, u^2)). \tag{5.7}$$

We assume (u^1, u^2) runs over an integration domain $e^2 \subset E^2$ and the mapping is regular in the sense that the two velocity vectors,

$$\frac{\partial x}{\partial u^1} \quad \text{and} \quad \frac{\partial x}{\partial u^2}, \tag{5.8}$$

which correspond to motion along the parameter curves, are linearly independent at each point. The standard definition is

$$\int_{c^2} \omega = \int_{c^2} \left[P_1(x^1(u^1, u^2), \ldots) \frac{\partial(x^2, x^3)}{\partial(u^1, u^2)} + \cdots \right.$$
$$\left. + P_3(x^1(u^1, u^2), \ldots) \frac{\partial(x^1, x^2)}{\partial(u^1, u^2)} \right] du^1 \, du^2. \tag{5.9}$$

But, for example,

$$\phi^*(dx^2 \, dx^3) = \left(\frac{\partial x^2}{\partial u^1} du^1 + \frac{\partial x^2}{\partial u^2} du^2 \right) \left(\frac{\partial x^3}{\partial u^1} du^1 + \frac{\partial x^3}{\partial u^2} du^2 \right)$$

$$= \left(\frac{\partial x^2}{\partial u^1} \frac{\partial x^3}{\partial u^2} - \frac{\partial x^2}{\partial u^2} \frac{\partial x^3}{\partial u^1} \right) du^1 \, du^2$$

$$= \frac{\partial(x^2, x^3)}{\partial(u^1, u^2)} du^1 \, du^2, \quad \text{etc.,}$$

so that (5.9) is simply

$$\int_{c^2} \omega = \int_{e^2} \phi^* \omega. \tag{5.10}$$

Finally, if we formally recognize that c^2 is the image of e^2 under ϕ by writing

$$\phi_* e^2 = c^2, \tag{5.11}$$

we have

$$\int_{\phi_* e^2} \omega = \int_{e^2} \phi^* \omega.$$

These low dimensional examples motivate the formal theory of integrals of differential forms over chains. For a complete treatment of this subject, see [Wh1]. Treatments adequate for most of differential geometry appear in [St1, pp. 104–111] and [Fl1, pp. 57–66]. The result of this theory is that on a manifold **M** there is a class of objects called p-chains over which one can integrate p-forms. Each $(p+1)$-chain, c^{p+1}, has a boundary, $c^p = \partial c^{p+1}$, which is a p-chain, and the general Stokes' theorem, (2.13), is valid.

It is technically convenient to allow chains that cross over onto themselves and bunch up. Therefore, for the preceding examples, the mappings ϕ of (5.1) and (5.7) do not have to be either one-one or regular. Because of this added freedom, it is not hard to establish the following. Let **M** and **N** be manifolds and let

$$\phi: \mathbf{M} \to \mathbf{N} \tag{5.12}$$

be a smooth mapping. Let c^p be a p-chain on **M** and ω a p-form on **N**. Then $\phi_* c^p$ is a p-chain on **N**, $\phi^* \omega$ is a p-form on **M**, and

$$\int_{c^p} \phi^* \omega = \int_{\phi_* c^p} \omega. \tag{5.13}$$

In the next section, we shall make some explicit computations to illustrate these matters.

6. EXAMPLES OF THE THEOREMS OF DE RHAM

In Section 3, we stated the converse to the Poincaré lemma, which says that on $\mathbf{E}^n$ a form is exact (equal to an exterior derivative) if and only if it is closed (its exterior derivative vanishes). This theorem is local and is not true as stated for any manifold. If $\mathbf{M}$ is a manifold, let $F^p = F^p(\mathbf{M})$ denote the real vector space of all p-forms on $\mathbf{M}$. Then

$$d: F^p \to F^{p+1} \tag{6.1}$$

The kernel of d, in each dimension, is the space Z^p of closed forms. We write,

$$Z^p = \left\{ \omega \in F^P \mid d\omega = 0 \right\}. \tag{6.2}$$

The image of d (from the previous dimension) is the space B^p of exact forms. We write

$$B^p = d\left(F^{p-1} \right). \tag{6.3}$$

Since $d(d\alpha) = 0$ for any form α, we have

$$B^p \subseteq Z^p. \tag{6.4}$$

The quotient space

$$H^p = Z^p / B^p \tag{6.5}$$

is the pth *de Rham cohomology group of* $\mathbf{M}$. We know that it vanishes for $\mathbf{E}^n$. The main *de Rham theorem* asserts that it is naturally isomorphic (for $p > 0$) to the topological cohomology group (with real coefficients) of $\mathbf{M}$, provided $\mathbf{M}$ is compact. We shall explore some examples.

1. First we take $\mathbf{M} = \mathbf{S}^1$, the circle. We may take the central angle θ (mod 2π) as parameter. A 1-form

$$\omega = f(\theta) \, d\theta \qquad (f(\theta + 2\pi) = f(\theta))$$

is exact if there is a periodic function g such that

$$f(\theta) = \frac{dg}{d\theta}.$$

Then

$$\int_{\mathbf{S}^1} \omega = \int_0^{2\pi} \frac{dg}{d\theta} = g(2\pi) - g(0) = 0,$$

suggesting that a 1-form ω on $\mathbf{S}^1$ is exact if and only if

$$\int_{\mathbf{S}^1} \omega = 0. \tag{6.6}$$

We have just seen that the condition is necessary. If, on the other hand, the integral vanishes, then we may set

$$g(\theta) = \int_0^\theta f(t)\, dt$$

and, (the important point) this relation is well-defined for $\theta \bmod 2\pi$. Clearly,

$$\frac{dg}{d\theta} = f(\theta), \qquad dg = \omega.$$

2.   Our second example is not a compact manifold, but its properties are useful to us. It is the *cylinder*

$$(-1,1) \times \mathbf{S}^1 = \{(t,\theta) \mid -1 < t < 1,\ \theta \bmod 2\pi\}. \tag{6.7}$$

We denote its equator by

$$\mathbf{c}^1 = \{0\} \times \mathbf{S}^1. \tag{6.8}$$

PROPOSITION.   *Let ω be a closed 1-form on the cylinder. Then ω is exact if and only if*

$$\int_{\mathbf{c}^1} \omega = 0. \tag{6.9}$$

Proof. One direction is immediate. Suppose f is a 0-form on the cylinder and $\omega = df$. Then

$$\int_{c^1} \omega = \int_{c^1} df = \int_{\partial c^1} f = 0.$$

We could prove the sufficiency by showing that the line integral

$$\int_{P_0}^{P_1} \omega$$

is independent of the path. Instead, we shall give a type of proof that will guide our later study of the projective plane. To write something definite, we think of S^1 as the unit circle in the (x, y) plane, so that $\theta \to (\cos\theta, \sin\theta)$ is the polar angle parameterization of S^1. We consider the mapping

$$\phi: (-1,1) \times E^1 \to (-1,1) \times S^1 \qquad (6.10)$$

given by

$$\phi(t, \theta) = (t, \cos\theta, \sin\theta), \qquad (6.11)$$

which is a covering of the cylinder by the infinite strip. Suppose we are given a closed 1-form ω on the cylinder that satisfies the relation

$$\int_{c^1} \omega = 0.$$

We note first that the integral of ω taken over any circle parallel to the equator c^1 also vanishes. For example, if $0 < t_0 < 1$, then the 2-chain

$$c^2 = [0, t_0] \times S^1$$

has boundary

$$\partial c^2 = \{t_0\} \times S^1 - c^1.$$

Hence

$$\int_{\{t_0\}\times \mathbf{S}^1}\omega = \int_{\{t_0\}\times \mathbf{S}^1}\omega - \int_{\mathbf{c}^1}\omega = \int_{\partial\mathbf{c}^2}\omega = \int_{\mathbf{c}^2}d\omega = \int_{\mathbf{c}^2}0 = 0.$$

Having this, we consider the form $\phi^*\omega$, a 1-form on the infinite strip—a space which, from the viewpoint of its differentiable structure alone, is indistinguishable from $\mathbf{E}^2$. We have

$$d(\phi^*\omega) = \phi^*(d\omega) = \phi^*(0) = 0,$$

so $\phi^*\omega$ is a closed 1-form on the strip. By the converse of the Poincaré lemma, there exists a function g on the strip such that

$$\phi^*\omega = dg.$$

The crucial question follows: Is there a function f on the cylinder such that $\phi^*f = g$? Clearly, for this to be true it is necessary and sufficient that g be periodic of period 2π in θ. But

$$g(t,\theta+2\pi) - g(t,\theta) = \int_{\theta}^{\theta+2\pi}\left[\frac{d}{ds}g(t,s)\right]ds = \int_{\{t\}\times[\theta,\theta+2\pi]}dg$$

$$= \int_{\{t\}\times[\theta,\theta+2\pi]}\phi^*\omega = \int_{\{t\}\times \mathbf{S}^1}\omega = 0.$$

Thus g has the required periodicity, so there is a function f on the cylinder satisfying $\phi^*f = g$. Hence $dg = d(\phi^*f) = \phi^*(df)$, and

$$\phi^*\omega = \phi^*(df).$$

The covering mapping ϕ is locally one-one with a smooth inverse; hence ϕ^* is one-one, so finally

$$\omega = df.$$

The reader should think about 2-forms on the cylinder.

3. Our next example is the 2-sphere, S^2, the locus of $x^2 + y^2 + z^2 = 1$ in E^3. We cover S^2 with the two coordinate neighborhoods,

$$U^+ = \left\{ z > -\tfrac{1}{2} \right\} \quad \text{and} \quad U^- = \left\{ z < \tfrac{1}{2} \right\},$$

which overlap in the equatorial zone,

$$U^+ \cap U^- = \left\{ -\tfrac{1}{2} < z < \tfrac{1}{2} \right\}.$$

The set U^+ is diffeomorphic (that is, differentially isomorphic) to E^2, say, by projecting it onto the equatorial plane from the point $(0, 0, -1/2)$; so is U^-. The set $U^+ \cap U^-$ is diffeomorphic to the cylinder, a result of the Mercator projection.

(6.12) PROPOSITION. *Each closed 1-form on S^2 is exact.*

Proof. Let α be a 1-form such that $d\alpha = 0$. By what we know of E^2, there exists a function g on U^+ such that $\alpha = dg$ on U^+ and a function h on U^- such that $\alpha = dh$ on U^-. Now

$$d(g - h) = dg - dh = \alpha - \alpha = 0 \quad \text{on} \quad U^+ \cap U^-.$$

Since the zone $U^+ \cap U^-$ is connected, we conclude that $g - h$ is constant on this set—that is, $g - h = c$ on $U^+ \cap U^-$.

We now change the function h on all of U^- to the function $h_1 = h + c$. Then $dg = \alpha$ on U^+, $dh_1 = d(h + c) = dh = \alpha$ on U^-, and $g = h_1$ on $U^+ \cap U^-$. We define a function f on S^2 by setting $f = g$ on U^+ and $f = h_1$ on U^-—valid precisely because $g = h_1$ on the overlap—and $\alpha = df$ as required.

PROPOSITION. *A 2-form ω on S^2 is exact if and only if*

$$\int_{S^2} \omega = 0. \tag{6.13}$$

Proof. We give S^2 the orientation of the outward normal and give its equator,

$$c^1 = \{ z = 0 \},$$

the orientation of counterclockwise rotation in the x, y-plane, with the understanding that x, y, z is a right-handed system. Then,

$$\mathbf{c}^1 = \partial \mathbf{H}^+ = -\partial \mathbf{H}^-,$$

where $\mathbf{H}^+ = \{z \geqslant 0\}$ and $\mathbf{H}^- = \{z \leqslant 0\}$ are the upper and lower hemispheres, respectively.

In one direction, the proof is easy. If $\omega = d\alpha$, then

$$\int_{\mathbf{S}^2} \omega = \int_{\mathbf{S}^2} d\alpha = \int_{\partial \mathbf{S}^2} \alpha = 0.$$

The proof of the other direction is harder. We assume ω is a 2-form on $\mathbf{S}^2$ satisfying (6.13). Because $\mathbf{U}^+$ and $\mathbf{U}^-$ are diffeomorphic to $\mathbf{E}^2$, there is a 1-form β on $\mathbf{U}^+$ such that $\omega = d\beta$ on $\mathbf{U}^+$ and a 1-form γ on $\mathbf{U}^-$ such that $\omega = d\gamma$ on $\mathbf{U}^-$. The 1-form $\beta - \gamma$, defined on the equatorial zone $\mathbf{U}^+ \cap \mathbf{U}^-$, is closed, because

$$d(\beta - \gamma) = d\beta - d\gamma = \omega - \omega = 0.$$

As a result of Equation (6.9), this form, $\beta - \gamma$, will be exact on the zone if its integral over $\mathbf{c}^1$ vanishes. Now

$$\int_{\mathbf{c}^1} \beta = \int_{\partial \mathbf{H}^+} \beta = \int_{\mathbf{H}^+} d\beta = \int_{\mathbf{H}^+} \omega,$$

$$\int_{\mathbf{c}^1} \gamma = -\int_{\partial \mathbf{H}^-} \gamma = -\int_{\mathbf{H}^-} d\gamma = -\int_{\mathbf{H}^-} \omega,$$

$$\int_{\mathbf{c}^1} (\beta - \gamma) = \int_{\mathbf{H}^+} \omega + \int_{\mathbf{H}^-} \omega = \int_{\mathbf{S}^2} \omega = 0$$

by (6.13). Hence there is a function h on $\mathbf{U}^+ \cap \mathbf{U}^-$ such that

$$\beta - \gamma = dh \qquad \text{on} \quad \mathbf{U}^+ \cap \mathbf{U}^-.$$

We now encounter technical difficulties. If we could prolong the function h to a function defined on all of $\mathbf{U}^-$, then we would replace γ by $\gamma + dh$ and quickly end the argument. Unfortunately we cannot, because h may get nasty as we approach the lower boundary $\{z = -1/2\}$ of the equatorial zone. However, if we cut the zone back, it is possible to prolong h from what remains.

To do so, we assume for the moment that we can find a (smooth) function g on $\mathbf{U}^-$ such that $g = 1$ on $\{0 < z < 1/2\}$ and $g = 0$ on $\{z < -1/4\}$. We then define h_1 on $\mathbf{U}^-$ by

$$\begin{cases} h_1 = gh & \text{on } \mathbf{U}^+ \cap \mathbf{U}^-, \\ h_1 = 0 & \text{on } \{z \leqslant -\tfrac{1}{4}\}. \end{cases}$$

Finally, we define α on $\mathbf{S}^2$ by

$$\begin{cases} \alpha = \beta & \text{on } \mathbf{H}^+, \\ \alpha = \gamma + dh_1 & \text{on } \mathbf{U}^-, \end{cases}$$

which is a definition because on the common part $\mathbf{H}^+ \cap \mathbf{U}^-$ of these sets we have

$$\gamma + dh_1 = \gamma + d(gh) = \gamma + dh = \beta.$$

Next,

$$\begin{cases} d\alpha = d\beta = \omega & \text{on } \mathbf{H}^+, \\ d\alpha = d(\gamma - dh_1) = d\gamma = \omega & \text{on } \mathbf{H}^-. \end{cases}$$

Therefore, $d\alpha = \omega$ on $\mathbf{S}^2$.

For the function g, we may take a function of the height z alone. We thus seek a function $g = g(z) \in C^\infty$ which vanishs for $z < -1/4$ and equals 1 for $z > 0$. The following lemma, with a linear change of variable, suffices.

LEMMA. *There exists a C^∞ function $f(x)$ defined on $\mathbb{R}$ such that $f(x) = 0$ for $x \leqslant 0$ and $f(x) = 1$ for $x \geqslant 1$.*

Proof. First set $g(x) = 0$ for $x \leqslant 0$, $g(x) = \exp(-1/x)$ for $x > 0$. Next set $h(x) = g(x)g(1-x)$. Then $h(x)$ is C^∞, $h(x) = 0$ for $x \leqslant 0$ or $x \geqslant 1$, and $h(x) > 0$ for $0 < x < 1$. Finally set

$$f(x) = \int_0^x h(t)\, dt \Big/ \int_{-\infty}^\infty h(t)\, dt.$$

4.   Our next example is the *projective plane* $\mathbf{P}^2$, obtained from $\mathbf{S}^2$ by identifying antipodal points. Thus $\mathbf{S}^2$ is a 2-sheeted covering surface of $\mathbf{P}^2$; we let π denote the covering mapping. We also let α denote the mapping,

$$\alpha(x) = -x,$$

on $\mathbf{S}^2$ that maps each point x of $\mathbf{S}^2$ to its antipode. Then $\pi \circ \alpha = \pi$, that is the diagram

$$\mathbf{S}^2 \xrightarrow{\quad \alpha \quad} \mathbf{S}^2 \qquad (6.14)$$
$$\searrow_{\pi} \ \mathbf{P}^2 \ \swarrow_{\pi}$$

is commutative. If ω is a p-form on $\mathbf{P}^2$ and $\pi^*\omega = 0$, then $\omega = 0$. This statement is true because locally π is a diffeomorphism (indeed, this is the method by which we define the manifold structure on $\mathbf{P}^2$); hence π^* is a monomorphism. Our basic tool for constructing forms on $\mathbf{P}^2$ follows.

PROPOSITION.   *Let η be a p-form on* $\mathbf{S}^2$ *($p = 0, 1, 2$). Then there is a p-form ω on* $\mathbf{P}^2$ *such that*

$$\pi^*\omega = \eta, \qquad (6.15)$$

if and only if

$$\alpha^*\eta = \eta. \qquad (6.16)$$

Proof.   Suppose $\eta = \pi^*\omega$. Then

$$\alpha^*\eta = \alpha^*\pi^*\omega = (\pi \circ \alpha)^*\omega = \pi^*\omega = \eta.$$

Conversely, let η be a form on $\mathbf{S}^2$ which satisfies $\alpha^*\eta = \eta$. We selected a covering of $\mathbf{S}^2$ by open sets $\mathbf{U}$, so small that $\alpha\mathbf{U} \cap \mathbf{U} = \varnothing$. The sets $\pi(\mathbf{U})$ then cover $\mathbf{P}^2$, and for each $\mathbf{U}$,

$$\pi: \mathbf{U} \to \pi(\mathbf{U})$$

is a diffeomorphism. Let $\lambda_{\mathbf{U}}$ denote its inverse:

$$\lambda_{\mathbf{U}}: \pi(\mathbf{U}) \to \mathbf{U}.$$

We define

$$\omega_{\pi(U)} = \lambda_U^*(\eta) \qquad \text{on} \quad \pi(U)$$

so that

$$\pi^*(\omega_{\pi(U)}) = \pi^*\lambda_U^*(\eta) = (\lambda_U \circ \pi)^*\eta = (\text{identity})^*\eta = \eta \qquad \text{on} \quad U.$$

We have defined a p-form on each open set $\pi(U)$ of the covering of $\mathbf{P}^2$. Do these fit together? Suppose $\pi(U) \cap \pi(V) \neq \varnothing$. Then either U meets V or U meets αV. If U meets V, then

$$\pi(U \cap V) = \pi(U) \cap \pi(V),$$

$$\lambda_U = \lambda_V \qquad \text{on} \quad \pi(U) \cap \pi(V),$$

$$\omega_{\pi(U)} = \lambda_U^*(\eta) = \lambda_V^*(\eta) = \omega_{\pi(V)} \qquad \text{on} \quad \pi(U) \cap \pi(V).$$

Otherwise, U meets αV, and then

$$\pi(U \cap \alpha V) = \pi(U) \cap \pi(V),$$

$$\lambda_U = \alpha \circ \lambda_V \qquad \text{on} \quad \pi(U) \cap \pi(V),$$

$$\omega_{\pi(U)} = \lambda_U^*(\eta) = \lambda_V^*\alpha^*\eta = \lambda_V^*(\eta) = \omega_{\pi(V)} \text{ on } \pi(U) \cap \pi(V),$$

which shows us that the forms we have defined do fit into a single form ω on $\mathbf{P}^2$ such that $\pi^*\omega = \eta$.

PROPOSITION. *If ω is a closed 1-form on $\mathbf{P}^2$, then ω is exact.*

Proof. The form $\pi^*\omega$ is a closed 1-form on $\mathbf{S}^2$. By (6.12), there is a function (0-form) f on $\mathbf{S}^2$ such that $\pi^*\omega = df$. We have

$$d(\alpha^*f) = \alpha^*(df) = \alpha^*\pi^*\omega = (\pi \circ \alpha)^*\omega = \pi^*\omega = df.$$

Hence,

$$d(\alpha^*\eta - \eta) = 0 \qquad \text{on} \quad \mathbf{S}^2.$$

It follows that $\alpha^*f - f = c$, a constant, on S^2, that is,

$$f(-x) - f(x) = c \qquad \text{for each} \quad x \in S^2.$$

Apply this relation to $-x$:

$$f(x) - f(-x) = c.$$

We deduce that $c = 0$, so

$$\alpha^*f = f.$$

By (6.15), there is a function g on $\mathbf{P}^2$ satisfying $\pi^*g = f$; hence,

$$\pi^*dg = d\pi^*g = df = \pi^*\omega, \qquad dg = \omega.$$

This result, together with the following proposition, shows that the de Rham groups of $\mathbf{P}^2$ are trivial for $p = 1$ and $p = 2$.

PROPOSITION.　*Each 2-form on $\mathbf{P}^2$ is exact.*

(Note that no integral condition is required as in (6.13), because the surface $\mathbf{P}^2$ is nonorientable and consequently has no two-dimensional real homology.)

Proof.　Let σ be a 2-form on $\mathbf{P}^2$. Then $\tau = \pi^*\sigma$ is a 2-form on S^2 that satisfies $\alpha^*\tau = \tau$. Hence,

$$\int_{S^2} \tau = \int_{S^2} \alpha^*\tau = \int_{\alpha_* S^2} \tau = \int_{-S^2} \tau = -\int_{S^2} \tau, \qquad \int_{S^2} \tau = 0.$$

[The antipodal mapping α reverses the orientation on S^2.] We now may apply (6.13): There is a 1-form η on S^2 such that $\tau = d\eta$. Now

$$d(\alpha^*\eta) = \alpha^*d\eta = \alpha^*\tau = \alpha^*(\pi^*\sigma) = (\pi \circ \alpha)^*\sigma = \pi^*\sigma = \tau = d\eta.$$

Hence

$$d(\alpha^*\eta - \eta) = 0 \qquad \text{on} \quad S^2.$$

By (6.12), there is a function f on $\mathbf{S}^2$ satisfying the equation

$$\alpha^*\eta - \eta = df.$$

We apply α^* to this relation to find

$$\eta - \alpha^*\eta = \alpha^*df.$$

Hence

$$\alpha^*df = -df.$$

We now set

$$\eta_1 = \eta + \tfrac{1}{2}df.$$

Then

$$d\eta_1 = d\eta = \tau = \pi^*\sigma,$$

and

$$\alpha^*\eta_1 = \alpha^*\eta + \tfrac{1}{2}\alpha^*df = \eta + df - \tfrac{1}{2}df = \eta_1.$$

By (6.16), there is a 1-form ω on $\mathbf{P}^2$ such that $\pi^*\omega = \eta_1$. Hence

$$\pi^*d\omega = d\pi^*\omega = d\eta_1 = \pi^*\sigma, \qquad d\omega = \sigma.$$

See what you can do with $\mathbf{P}^3$ and $\mathbf{P}^n$. Another interesting exercise is to work out the situation first for the ordinary torus, $\mathbf{T}^2 = \mathbf{S}^1 \times \mathbf{S}^1$, and then for the n-dimensional torus.

7. THE BROUWER FIXED-POINT THEOREM

The following remarkable proof [Ka1] shows the power of differential form methods. The main step in the proof is the following lemma, which says that the unit ball cannot be retracted to its boundary.

LEMMA. *Suppose* $\mathbf{U}$ *is an open neighborhood of the unit ball* $\mathbf{B}^{n+1} = \{\mathbf{x} \in \mathbf{E}^{n+1} \mid |\mathbf{x}| \leqslant 1\}$ *in* $\mathbf{E}^{n+1}$. *Then there does not exist a*

smooth mapping

$$\phi: \mathbf{U} \to \mathbf{S}^n \tag{7.1}$$

such that $\phi(\mathbf{x}) = \mathbf{x}$ *for all* $\mathbf{x} \in \mathbf{S}^n$.

The Brouwer theorem says that if

$$f: \mathbf{B}^{n+1} \to \mathbf{B}^{n+1}$$

is a mapping (continuous function), then f has a fixed point. This follows from the lemma by a routine approximation argument plus a topological device—the details are not germane here, however—if f is fixed-point free, then f can be approximated uniformly by a polynomial p. By possibly shrinking slightly, one can assume that p maps a small neighborhood of $\mathbf{B}^{n+1}$ into $\mathbf{B}^{n+1}$, and by taking the approximation close enough, one can guarantee that p is also free of fixed points. Thus f continuous can be replaced by f smooth. Then define ϕ by letting $\phi(\mathbf{x})$ be the unique point where the ray from $f(\mathbf{x})$ through $\mathbf{x}$ meets $\mathbf{S}^n$, etc.

Proof of the Lemma.   The proof hinges largely on careful notation! We do best by thinking of $\mathbf{U}$ and $\mathbf{S}^n$ as independent spaces, related by an injection mapping. Thus we interpret the statement that $\mathbf{S}^n$ is a submanifold of $\mathbf{U}$ to mean that there is a smooth mapping (injection)

$$j: \mathbf{S}^n \to \mathbf{U}. \tag{7.2}$$

The relation of $\mathbf{S}^n$ to the subset $\mathbf{B}^{n+1}$ of $\mathbf{U}$ is expressed by

$$j(\mathbf{S}^n) = \partial \mathbf{B}^{n+1}. \tag{7.3}$$

We assume a smooth mapping ϕ of (7.1) exists and satisfies $\phi(\mathbf{x}) = \mathbf{x}$ on $\mathbf{S}^n$. More precisely, we assume a commutative diagram

$$\tag{7.4}$$

where i is the identity mapping.

We set

$$\omega = x^1 \, dx^2 \cdots dx^{n+1}, \qquad (7.5)$$

an n-form on $\mathbf{U}$. Then

$$d\omega = dx^1 \, dx^2 \cdots dx^{n+1}$$

is the element of volume on $\mathbf{U}$, so

$$\int_{\mathbf{B}^{n+1}} d\omega > 0. \qquad (7.6)$$

We set

$$\alpha = j^*\omega, \qquad (7.7)$$

an n-form on $\mathbf{S}^n$. We have

$$\int_{\mathbf{S}^n} \alpha = \int_{\mathbf{S}^n} j^*\omega = \int_{j(\mathbf{S}^n)} \omega = \int_{\partial \mathbf{B}^{n+1}} \omega = \int_{\mathbf{B}^{n+1}} d\omega$$

by (7.3) and Stokes' theorem. Thus

$$\int_{\mathbf{S}^n} \alpha > 0 \qquad (7.8)$$

by (7.6). We can compute this integral another way. From (7.4) we have $i = \phi \circ j$ so that $i^* = j^* \circ \phi^*$ and

$$\alpha = i^*\alpha = j^*\phi^*\alpha.$$

Therefore

$$\int_{\mathbf{S}^n} \alpha = \int_{\mathbf{S}^n} j^*\phi^*\alpha = \int_{j(\mathbf{S}^n)} \phi^*\alpha$$

$$= \int_{\partial(\mathbf{B}^{n+1})} \phi^*\alpha = \int_{\mathbf{B}^{n+1}} d(\phi^*\alpha) = \int_{\mathbf{B}^{n+1}} \phi^*(d\alpha). \qquad (7.9)$$

But $d\alpha = 0$ because α has maximal degree n on the n-dimensional manifold $\mathbf{S}^n$. Therefore (7.9) implies

$$\int_{\mathbf{S}^n} \alpha = 0,$$

which contradicts (7.8).

8. AN EXAMPLE OF A MOVING FRAME

Suppose we have vectors $\mathbf{v}_1, \ldots, \mathbf{v}_n$ in $\mathbf{E}^n$, each of which is a smooth function $\mathbf{v}_i = \mathbf{v}_i(u^1, \ldots, u^r)$ of $(u^1, \ldots, u^r)$, which varies over a domain in r-space. We assume that for each point in u-space, the vectors $\mathbf{v}_1, \ldots, \mathbf{v}_n$ are linearly independent in $\mathbf{E}^n$ and consequently form a linear basis of $\mathbf{E}^n$. If we let $\mathbf{e}_1, \ldots, \mathbf{e}_n$ denote the standard basis of $\mathbf{E}^n$ [$\mathbf{e}_1 = (1, 0, \ldots, 0)$, $\mathbf{e}_2 = (0, 1, 0, \ldots, 0)$, etc.] then we may write

$$\mathbf{v}_i = \left(a_i^1, a_i^2, \ldots, a_i^n \right) = \Sigma a_i^j \mathbf{e}_j. \qquad (8.1)$$

The $a_i^j = a_i^j(u^1, \ldots, u^n)$ make up an $n \times n$ matrix,

$$A = \|a_i^j\|, \qquad (8.2)$$

which is nonsingular at each point of the u-space by our assumption that $\mathbf{v}_1, \ldots, \mathbf{v}_n$ are linearly independent.

We differentiate (8.1) as follows:

$$d\mathbf{v}_i = \left(da_i^1, \ldots, da_i^n \right) = \Sigma \left(da_i^j \right) \mathbf{e}_j = \Sigma \left(\frac{\partial a_i^j}{\partial u^k} \right) (du^k) \mathbf{e}_j. \quad (8.3)$$

These equations give us the components of the differential of each $\mathbf{v}_i$ in terms of the standard basis $\mathbf{e}_1, \ldots, \mathbf{e}_n$. The moving frame idea, as applied to this situation, is to express the differentials of the vectors $\mathbf{v}_i$ *in terms of the* $\mathbf{v}_1, \ldots, \mathbf{v}_n$ *themselves*. We accomplish this by first inverting equation (8.1) (possible because A is nonsingular) to express the $\mathbf{e}_i$ as linear combinations of the $\mathbf{v}_j$, and then substituting those expressions into (8.3). What results is

$$d\mathbf{v}_i = \Sigma \omega_i^j \mathbf{v}_j \qquad (8.4)$$

where the ω_i^j are 1-forms.

Precisely, if we write

$$\mathbf{v} = \begin{pmatrix} \mathbf{v}_1 \\ \vdots \\ \mathbf{v}_n \end{pmatrix}, \qquad \mathbf{e} = \begin{pmatrix} \mathbf{e}_1 \\ \vdots \\ \mathbf{e}_n \end{pmatrix}, \qquad \Omega = \| \omega_i^j \|, \qquad (8.5)$$

then

$$\mathbf{v} = A\mathbf{e}, \qquad \mathbf{e} = A^{-1}\mathbf{v}, \qquad d\mathbf{v} = (dA)\mathbf{e} = (dA)A^{-1}\mathbf{v} = \Omega\mathbf{v},$$
$$\Omega = (dA)A^{-1}. \qquad (8.6)$$

As an application, we pose the following question: What is the differential of $(\det A)$? We recall that we may consider the determinant of a matrix to be an alternating multilinear functional of the rows of the matrix,

$$\det A = \Delta(\mathbf{v}_1, \ldots, \mathbf{v}_n). \qquad (8.7)$$

Therefore, the rule for differentiating a product is applicable; for example,

$$\frac{\partial}{\partial u^k}(\det A) = \Delta\left(\frac{\partial \mathbf{v}_1}{\partial u^k}, \mathbf{v}_2, \ldots, \mathbf{v}_n\right) + \Delta\left(\mathbf{v}_1, \frac{\partial \mathbf{v}_2}{\partial u^k}, \mathbf{v}_3, \ldots, \mathbf{v}_n\right)$$
$$+ \cdots + \Delta\left(\mathbf{v}_1, \ldots, \mathbf{v}_{n-1}, \frac{\partial \mathbf{v}_n}{\partial u^k}\right).$$

Multiplying by du^k and summing, we have

$$d(\det A) = \Delta(d\mathbf{v}_1, \mathbf{v}_2, \ldots, \mathbf{v}_n) + \Delta(\mathbf{v}_1, d\mathbf{v}_2, \mathbf{v}_3, \ldots, \mathbf{v}_n)$$
$$+ \cdots + \Delta(\mathbf{v}_1, \ldots, \mathbf{v}_{n-1}, d\mathbf{v}_n). \qquad (8.8)$$

Now,

$$\Delta(d\mathbf{v}_1, \mathbf{v}_2, \ldots, \mathbf{v}_n) = \Delta\left(\Sigma\omega_1^j\mathbf{v}_j, \mathbf{v}_2, \ldots, \mathbf{v}_n\right)$$
$$= \Sigma\omega_1^j\Delta(\mathbf{v}_j, \mathbf{v}_2, \ldots, \mathbf{v}_n)$$
$$= \omega_1\Delta(\mathbf{v}_1, \ldots, \mathbf{v}_n)$$
$$= \omega_1^1(\det A),$$

because determinants with two equal rows vanish. We come to

$$d(\det A) = \omega_1^1(\det A) + \omega_2^2(\det A) + \cdots + \omega_n^n(\det A)$$

$$= \left(\omega_1^1 + \omega_2^2 + \cdots + \omega_n^n \right)(\det A). \tag{8.9}$$

In view of Equation (8.6), we may write this relation as

$$\frac{d(\det A)}{(\det A)} = \text{trace}\left[(dA) A^{-1} \right]. \tag{8.10}$$

This formula is useful in differential geometry.

9.   THE GAUSS-BONNET THEOREM FOR SURFACES

In this section, **M** is a compact oriented manifold of dimension two which has a Riemannian structure. This last condition means that each tangent plane at each point of **M** has a Euclidean structure—inner products are defined—and if **v** and **w** are smooth vector fields over a region on **M**, then **v** · **w** is a smooth scalar over that region. (The case to keep in mind is the classical one in which **M** is a closed surface in $\mathbf{E}^3$. The tangent planes at the various points of **M** inherit their euclidean structures from that of the ambient space, $\mathbf{E}^3$. From topology, we know that a closed surface in $\mathbf{E}^3$ divides $\mathbf{E}^3$ into inside and outside regions. Taking the outward normal imposes an orientation on **M** by the right-hand rule.)

Over a local coordinate neighborhood **U** on **M**, we may find vector fields $\mathbf{e}_1$ and $\mathbf{e}_2$ that make up a right-handed orthonormal frame in the tangent space of each point of **U**. Letting σ_1 and σ_2 be the dual basis of 1-forms,[†] we may write, symbolically,

$$dx = \sigma_1 \mathbf{e}_1 + \sigma_2 \mathbf{e}_2. \tag{9.1}$$

[†]The duality between vector fields and 1-forms is most simply explained by the natural bases associated with a local coordinate system $u^1, \ldots, u^n$ (in arbitrary dimension). Here $\mathbf{e}_1 = \partial/\partial u^1, \ldots, \mathbf{e}_n = \partial/\partial u^n$ is a basis of the tangent space, and

In the case of an imbedded surface in $\mathbf{E}^3$ this relation is precise. On an abstract surface, we use it in order to motivate our steps—in particular, to "see" the transformation rule for the σ's. We may consider (9.1) to mean that an infinitesimal displacement of the point $\mathbf{x}$ on the surface will have components σ_1 in the $\mathbf{e}_1$ direction and σ_2 in the $\mathbf{e}_2$ direction.

We may compute σ_1 and σ_2 explicitly as follows. We start with a local coordinate system, u^1 and u^2. Then

$$\frac{\partial}{\partial u^i} \cdot \frac{\partial}{\partial u^j} = g_{ij}(u^1, u^2)$$

gives us a positive definite matrix $\|g_{ij}\|$. Indeed, the Riemannian structure on $\mathbf{M}$ is usually given in advance by means of such matrices, one for each chart of a covering with appropriate relations on overlapping charts. By some process of orthonormalization we construct the orthonormal frame $\mathbf{e}_1$ and $\mathbf{e}_2$, and we have

$$\begin{cases} \mathbf{e}_1 = a_{11}\dfrac{\partial}{\partial u^1} + a_{12}\dfrac{\partial}{\partial u^2} \\[2mm] \mathbf{e}_2 = a_{21}\dfrac{\partial}{\partial u^1} + a_{22}\dfrac{\partial}{\partial u^2}. \end{cases} \tag{9.2}$$

Substitution of these expressions into

$$dx = \sigma_1\mathbf{e}_1 + \sigma_2\mathbf{e}_2 = du^1\frac{\partial}{\partial u^1} + du^2\frac{\partial}{\partial u^2}$$

gives us

$$\begin{cases} a_{11}\sigma_1 + a_{21}\sigma_2 = du^1 \\[2mm] a_{12}\sigma_1 + a_{22}\sigma_2 = du^2, \end{cases} \tag{9.3}$$

$du^1, \ldots, du^n$ is a basis of the form space. The dual pairing between the two is

$$\left(du^i, \frac{\partial}{\partial u^j} \right) = \delta^i_j,$$

which is easily seen to be independent of local coordinates. Details are found in [St1, p. 72] and [Ko1, p. 6].

which we may solve for σ_1 and σ_2. We note that the relations $\mathbf{e}_i \cdot \mathbf{e}_j = \delta_{ij}$ imply that

$$\delta_{ij} = \Sigma a_{ik} g_{kl} a_{jl}. \tag{9.4}$$

The *element of area* on **M** is the 2-form

$$\sigma_1 \sigma_2 = \sqrt{g}\, du^1\, du^2 \tag{9.5}$$

where

$$g = \begin{vmatrix} g_{11} & g_{12} \\ g_{21} & g_{22} \end{vmatrix}. \tag{9.6}$$

This relation follows from (9.3), which implies $[\det(a_{ij})]\sigma_1\sigma_2 = du^1\, du^2$, and from (9.4), which implies $[\det(a_{ij})]^2 g = 1$. (Note that $\det(a_{ij}) > 0$, because both frames agree with the orientation.)

We next claim that there exists a unique 1-form ϖ on our neighborhood **U**, such that

$$\begin{cases} d\sigma_1 = \varpi \sigma_2 \\ d\sigma_2 = -\varpi \sigma_1, \end{cases} \tag{9.7}$$

which is true because $\sigma_1\sigma_2$ is a basis of 2-forms. Hence, $d\sigma_1 = a_1\sigma_1\sigma_2$, $d\sigma_2 = a_2\sigma_1\sigma_2$, and we are forced to the unique solution $\varpi = a_1\sigma_1 + a_2\sigma_2$.

Suppose we have a second coordinate neighborhood $\overline{\mathbf{U}}$ which overlaps **U**, and an orthonormal frame $\bar{\mathbf{e}}_1, \bar{\mathbf{e}}_2$ for $\overline{\mathbf{U}}$. Then,

$$dx = \bar{\sigma}_1 \bar{\mathbf{e}}_1 + \bar{\sigma}_2 \bar{\mathbf{e}}_2 \tag{9.8}$$

on $\overline{\mathbf{U}}$. On the intersection $\mathbf{U} \cap \overline{\mathbf{U}}$, we have

$$\begin{cases} \bar{\mathbf{e}}_1 = (\cos\alpha)\mathbf{e}_1 + (\sin\alpha)\mathbf{e}_2 \\ \bar{\mathbf{e}}_2 = -(\sin\alpha)\mathbf{e}_1 + (\cos\alpha)\mathbf{e}_2, \end{cases} \tag{9.9}$$

because $\mathbf{e}_1, \mathbf{e}_2$ and $\bar{\mathbf{e}}_1, \bar{\mathbf{e}}_2$ are both right-handed orthonormal systems. Here α is a scalar on $\mathbf{U} \cap \overline{\mathbf{U}}$. Because (9.1) and (9.8) are both valid on $U \cap \bar{U}$, we have

$$\begin{cases} \sigma_1 = (\cos \alpha)\,\bar{\sigma}_1 - (\sin \alpha)\,\bar{\sigma}_2 \\ \sigma_2 = (\sin \alpha)\,\bar{\sigma}_1 + (\cos \alpha)\,\bar{\sigma}_2. \end{cases} \qquad (9.10)$$

On $\bar{U}$, we have

$$\begin{cases} d\bar{\sigma}_1 = \bar{\varpi}\,\bar{\sigma}_2 \\ d\bar{\sigma}_2 = -\bar{\varpi}\,\bar{\sigma}_1. \end{cases} \qquad (9.11)$$

To find the relation between ϖ and $\bar{\varpi}$ on $U \cap \bar{U}$, we differentiate the relations in (9.10):

$$d\sigma_1 = -(\sin \alpha)(d\alpha)\,\bar{\sigma}_1 - (\cos \alpha)(d\alpha)\,\bar{\sigma}_2$$
$$+ (\cos \alpha)\,\bar{\varpi}\,\bar{\sigma}_2 + (\sin \alpha)\,\bar{\varpi}\,\bar{\sigma}_1$$
$$= (\bar{\varpi} - d\alpha)\sigma_2;$$
$$d\sigma_2 = (\cos \alpha)(d\alpha)\,\bar{\sigma}_1 - (\sin \alpha)(d\alpha)\,\bar{\sigma}_2$$
$$+ (\sin \alpha)\,\bar{\varpi}\,\bar{\sigma}_2 - (\cos \alpha)\,\bar{\varpi}\,\bar{\sigma}_1$$
$$= -(\bar{\varpi} - d\alpha)\sigma_1.$$

It follows that

$$\varpi = \bar{\varpi} - d\alpha. \qquad (9.12)$$

The important consequence is that the 2-form $d\varpi = d\bar{\varpi}$ is defined on all of M, independent of the moving frame. We set

$$d\varpi + K\sigma_1\sigma_2 = 0, \qquad (9.13)$$

defining the *Gaussian* or *total curvature K* of M. We shall prove the Gauss-Bonnet theorem:

$$\int_M K\sigma_1\sigma_2 = 2\pi\chi(M) \qquad (9.14)$$

where $\chi(M)$ is the *Euler-Poincaré characteristic* of M.

We begin by introducing the *unit tangent bundle* of **M**, the space **B** of all unit tangent vectors **v** at all points of **M**. This space is three-dimensional, and there exists a natural mapping (projection)

$$p: \mathbf{B} \to \mathbf{M} \tag{9.15}$$

which sends each tangent vector **v** to the (base) point $p(\mathbf{v})$ at which it is a tangent vector. If **U** is the preceding local coordinate neighborhood, and **x** is any point of **U**, the typical unit tangent vector at **x** is

$$\mathbf{v} = (\cos \phi)\mathbf{e}_1\big|_{\mathbf{x}} + (\sin \phi)\mathbf{e}_2\big|_{\mathbf{x}}. \tag{9.16}$$

Since **U** is parameterized by u^1, u^2, we see that

$$u^1, \qquad u^2, \qquad \phi$$

parameterize the set $p^{-1}(\mathbf{U})$ in **B**. In this way, we turn **B** into a differentiable manifold. (You should recognize **B** as a fiber bundle with **M** as the base space, p as projection, the fiber as the unit circle, and the group as the rotation group.) The differential forms,

$$du^1, \qquad du^2, \qquad d\phi,$$

or, equivalently,

$$\sigma_1, \qquad \sigma_2, \qquad d\phi, \tag{9.17}$$

form a basis for 1-forms on that part $p^{-1}(\mathbf{U})$ of **B** over **U**.

We can, and should, be more precise. The forms σ_1, σ_2 are defined on **U**, and we have

$$p: p^{-1}(\mathbf{U}) \to \mathbf{U}.$$

Thus, what we are really discussing are

$$p^*\sigma_1, \qquad p^*\sigma_2, \qquad d\phi,$$

which form a basis of 1-forms on $p^{-1}(\mathbf{U})$.

If we consider what happens on the common part $U \cap \overline{U}$ of two coordinate neighborhoods, we have, using the notation of (9.8), etc.,

$$\mathbf{v} = (\cos \overline{\phi}) \overline{\mathbf{e}}_1 + (\sin \overline{\phi}) \overline{\mathbf{e}}_2$$

$$= (\cos \overline{\phi} \cos \alpha - \sin \overline{\phi} \sin \alpha) \mathbf{e}_1 + (\cos \overline{\phi} \sin \alpha + \sin \overline{\phi} \cos \alpha) \mathbf{e}_2$$

$$= [\cos(\overline{\phi} + \alpha)] \mathbf{e}_1 + [\sin(\overline{\phi} + \alpha)] \mathbf{e}_2.$$

Hence, comparison with (8.16) yields

$$\phi = \overline{\phi} + \alpha. \tag{9.18}$$

We shall now define three 1-forms on $\mathbf{B}$; here we mean all of $\mathbf{B}$, not just a neighborhood. Each of these forms, which we call Σ_1, Σ_2, and Π, respectively, we first define on each $p^{-1}(U)$. Then we show that the "pieces of a form" coincide on the intersections $p^{-1}(U) \cap p^{-1}(\overline{U}) = p^{-1}(U \cap \overline{U})$.

First we define Σ_1 and Σ_2 on $p^{-1}(U)$ by the equations

$$\begin{cases} p^* \sigma_1 = (\cos \phi) \Sigma_1 - (\sin \phi) \Sigma_2 \\ p^* \sigma_2 = (\sin \phi) \Sigma_1 + (\cos \phi) \Sigma_2. \end{cases} \tag{9.19}$$

On U, the corresponding forms Σ_1, Σ_2 are defined by the analogous equations,

$$\begin{cases} p^* \overline{\sigma}_1 = (\cos \overline{\phi}) \overline{\Sigma}_1 - (\sin \overline{\phi}) \overline{\Sigma}_2 \\ p^* \overline{\sigma}_2 = (\sin \overline{\phi}) \overline{\Sigma}_1 + (\cos \overline{\phi}) \overline{\Sigma}_2. \end{cases} \tag{9.19'}$$

We consider the overlap $p^{-1}(U) \cap p^{-1}(\overline{U}) = p^{-1}(U \cap \overline{U})$, where (9.10) and (9.18) connect the various quantities [first applying p^* to (9.10)]. We wish to conclude that on this overlap, $\overline{\Sigma}_1 = \Sigma_1$ and $\overline{\Sigma}_2 = \Sigma_2$. We accomplish this most easily by expressing (9.19),

(9.19′), and (9.10) with p^* applied in matrix notation:

$$(p^*\sigma_1, p^*\sigma_2) = (\Sigma_1, \Sigma_2)\begin{pmatrix} \cos\phi & \sin\phi \\ -\sin\phi & \cos\phi \end{pmatrix};\qquad(9.19)$$

$$(p^*\bar\sigma_1, p^*\bar\sigma_2) = (\bar\Sigma_1, \bar\Sigma_2)\begin{pmatrix} \cos\bar\phi & \sin\bar\phi \\ -\sin\bar\phi & \cos\bar\phi \end{pmatrix};\qquad(9.19')$$

$$(p^*\sigma_1, p^*\sigma_2) = (p^*\bar\sigma_1, p^*\bar\sigma_2)\begin{pmatrix} \cos\alpha & \sin\alpha \\ -\sin\alpha & \cos\alpha \end{pmatrix}.\qquad(9.10')$$

From these equations, we readily deduce

$$(\bar\Sigma_1, \bar\Sigma_2)\begin{pmatrix} \cos\bar\phi & \sin\bar\phi \\ -\sin\bar\phi & \cos\bar\phi \end{pmatrix}\begin{pmatrix} \cos\alpha & \sin\alpha \\ -\sin\alpha & \cos\alpha \end{pmatrix}$$

$$= (\Sigma_1, \Sigma_2)\begin{pmatrix} \cos\phi & \sin\phi \\ -\sin\phi & \cos\phi \end{pmatrix}.$$

But relation (9.18) tells us precisely that the product of the two rotation matrices on the left-hand side equals the rotation matrix on the right-hand side; therefore,

$$(\bar\Sigma_1, \bar\Sigma_2) = (\Sigma_1, \Sigma_2),\qquad \bar\Sigma_1 = \Sigma_1,\qquad \bar\Sigma_2 = \Sigma_2.$$

Thus we have defined 1-forms Σ_1, Σ_2 on all of **B**. The remaining form Π is defined on $p^{-1}(U)$ by

$$p^*\varpi = \Pi - d\phi.\qquad(9.20)$$

On $p^{-1}(\bar U)$, the analogous form is defined by

$$p^*\bar\varpi = \bar\Pi - d\bar\phi.\qquad(9.20')$$

But (9.12) gives us

$$p^*\varpi = p^*\bar\varpi - p^*d\alpha\qquad(9.12')$$

On $p^{-1}(\mathbf{U} \cap \mathbf{U})$,[†] and (9.18) gives us

$$d\phi = d\bar{\phi} + p^*d\alpha. \tag{9.18'}$$

Combining these last four equations gives us $\Pi = \bar{\Pi}$. We have therefore succeeded in defining our third 1-form, Π, on all of $\mathbf{B}$.

From (9.19) we have (all exterior products)

$$\Sigma_1\Sigma_2 = p^*(\sigma_1\sigma_2) = (p^*\sigma_1)(p^*\sigma_2). \tag{9.21}$$

From this equation and from (9.20),

$$\Sigma_1\Sigma_2\Pi = (p^*\sigma_1)(p^*\sigma_2)\, d\phi, \tag{9.22}$$

because

$$(p^*\sigma_1)(p^*\sigma_2)(p^*\bar{\omega}) = p^*(\sigma_1\sigma_2\bar{\omega}) = 0.$$

From this equation, we deduce that Σ_1, Σ_2, and Π form a basis of the space of 1-forms on $\mathbf{B}$. Hence, any 2-form on $\mathbf{B}$ is a linear combination of $\Sigma_1\Sigma_2$, $\Pi\Sigma_1$, and $\Pi\Sigma_2$. We may work it out for $d\Sigma_1$, $d\Sigma_2$, and $d\Pi$. It suffices to work over one neighborhood of $p^{-1}(\mathbf{U})$.

First we invert (9.19):

$$\begin{cases} \Sigma_1 = (\cos\phi)p^*\sigma_1 + (\sin\phi)p^*\sigma_2 \\ \Sigma_2 = -(\sin\phi)p^*\sigma_1 + (\cos\phi)p^*\sigma_2. \end{cases}$$

Next we differentiate the first of these and use (9.7), with p^* applied:

$$\begin{aligned} d\Sigma_1 &= -d\phi(\sin\phi)p^*\sigma_1 + d\phi(\cos\phi)p^*\sigma_2 \\ &\quad + (\cos\phi)p^*\varpi p^*\sigma_2 - (\sin\phi)p^*\varpi p^*\sigma_1 \\ &= d\phi\Sigma_2 + p^*\varpi\Sigma_2 = \Pi\Sigma_2. \end{aligned}$$

[†]Note the slight inaccuracy in notation. We have identified the function (scalar) $p^*\alpha$ on $p^{-1}(\mathbf{U})$ with the scalar α on $\mathbf{U}$, just as we call the function x on the x-axis and the x-coordinate of the point (x, y) by the same letter.

Similarly we find $d\Sigma_2 = -\Pi\Sigma_1$. Hence

$$\begin{cases} d\Sigma_1 = \Pi\Sigma_2 \\ d\Sigma_2 = -\Pi\Sigma_1. \end{cases} \tag{9.23}$$

Next, from (9.20), (9.13), and (9.21), we have

$$d\Pi = p^*(d\varpi) = -K(p^*\sigma_1)(p^*\sigma_2) = -K\Sigma_1\Sigma_2,$$
$$d\Pi + K\Sigma_1\Sigma_2 = 0. \tag{9.24}$$

To illustrate the applicability of this relation, we begin with a special case of the Gauss-Bonnet theorem.

THEOREM. *If there exists a smooth unit tangent vector field on* **M**, *then*

$$\int_{\mathbf{M}} K\sigma_1\sigma_2 = 0. \tag{9.25}$$

Proof. A unit field means a smooth mapping r which assigns to each point **x** of **M** a tangent vector at **x**. In other words,

$$\begin{cases} r: \mathbf{M} \to \mathbf{B} \\ p \circ r = i = (\text{identity on } \mathbf{M}). \end{cases} \tag{9.26}$$

Since $\Sigma_1\Sigma_2 = p^*(\sigma_1\sigma_2)$, we have

$$r^*(\Sigma_1\Sigma_2) = (r^* \circ p^*)(\sigma_1\sigma_2) = (p \circ r)^*\sigma_1\sigma_2 = \sigma_1\sigma_2.$$

Hence, from (9.24),

$$d[r^*(\Pi)] + K\sigma_1\sigma_2 = 0.$$

The form $r^*\Pi$ is a 1-form on all of **M**. Therefore,

$$\int_{\mathbf{M}} K\sigma_1\sigma_2 = -\int_{\mathbf{M}} d(r^*\Pi) = -\int_{\partial(\mathbf{M})} r^*\Pi = 0,$$

because **M** is a compact surface and hence has no boundary.

For a surface to have a smooth vector field it is necessary and sufficient that its Euler characteristic vanish. This results from the following more general considerations.

If $\mathbf{M}$ is a compact oriented surface of Euler characteristic χ, then there exists a vector field on $\mathbf{M}$ with a finite number of points deleted. The sum of the indices of the field at these singular points is exactly χ. It is even possible to find a field with only one singularity and, if $\chi = 0$, with no singularities. These are standard facts in surface topology which we shall take for granted.

Suppose, then, we have a smooth unit tangent vector field defined on

$$\mathbf{M}' = \mathbf{M} - \{\mathbf{x}_1, \ldots, \mathbf{x}_n\},$$

which means we have a commutative diagram of smooth mappings

where j is the natural injection of $\mathbf{M}'$ into $\mathbf{M}$ (submanifold). As before, we now have

$$d(r^*\Pi) + K\sigma_1\sigma_2 = 0 \qquad \text{on} \quad \mathbf{M}'.$$

We take a small local coordinate neighborhood $\mathbf{U}_i$ centered at $\mathbf{x}_i$, so small that none of these $\mathbf{U}_i$ intersect and each $\mathbf{U}_i \cup \partial \mathbf{U}_i$ is diffeomorphic to the closed unit ball. Then

$$\int_{\mathbf{M} - \cup_1^n \mathbf{U}_i} K\sigma_1\sigma_2 = -\int_{\mathbf{M} - \cup_1^n \mathbf{U}_i} d(r^*\Pi) = -\int_{\partial(\mathbf{M} - \cup_1^n \mathbf{U}_i)} r^*\Pi$$

$$= \sum_{i=1}^n \int_{\partial \mathbf{U}_i} r^*\Pi.$$

Let us examine one of these summands,

$$\int_{\partial U_1} r^*\Pi.$$

We take a moving frame e_1, e_2 on U_1, so that (9.20) applied on $p^{-1}(U_1)$ gives

$$\Pi = p^*\varpi + d\phi.$$

Then

$$r^*\Pi = \varpi + r^*d\phi \qquad \text{on} \quad U_1 - \{x_1\}. \qquad (9.27)$$

Now

$$\int_{\partial U_1} \varpi = \int_{U_1} d\varpi = -\int_{U_1} K\sigma_1\sigma_2, \qquad (9.28)$$

so that it remains to evaluate

$$\int_{\partial U_1} r^*d\phi.$$

We recall the meaning of ϕ given by (9.16). At each point x of $U_1 - \{x_1\}$, we have

$$r(x) = [\cos r^*(\phi)]e_1 + [\sin r^*(\phi)]e_2.$$

The angle $r^*(\phi) = \phi \circ r$ is only determined $\mod 2\pi$, but its differential $d[r^*\phi] = r^*d\phi$ is completely determined. The boundary ∂U_1 is like a circle; it is a simple closed curve extending once around x_1. The integral of $r^*d\phi = d(r^*\phi)$ taken along it gives the variation of the angle $r^*\phi$ in following $r(x)$ over one circuit of this curve—that is,

$$\int_{\partial U_1} r^*d\phi = 2\pi\left[\text{ind}_{x_1}(r)\right].$$

Combining (9.27), (9.28), and (9.29), we arrive at the Gauss-Bonnet theorem,

$$\int_{M} K\sigma_1\sigma_2 = 2\pi \sum_{1}^{n} \mathrm{ind}\,_{\pi_i}(r) = 2\pi\chi,$$

which also proves that any two vector fields on **M**, each with a finite number of singularities, have the same index sum.

REFERENCES

Au1. Auslander, L. and MacKenzie, R. E., *Introduction to Differentiable Manifolds*, McGraw-Hill, 1963.

Bi1. Bishop, R. L. and Crittenden, R. J., *Geometry of Manifolds*, Academic Press, 1964.

Bl1. Blaschke, W., *Vorlesungen über Differentialgeometrie* I, Springer, 1924.

Bo1. Bott, R. and Tu, L. W., *Differential Forms in Algebraic Topology*, Springer-Verlag, 1986.

Bl2. ____, *Einführung in die Differentialgeometrie*, Springer, 1950.

Ca1. Cartan, E., *Leçons sur la Géométrie des Espaces de Riemann*, Gauthier-Villars, 1951.

Ca2. ____, *Les Systèmes Différentiels Extérieurs et Leurs Applications Géométriques*, Hermann, 1945.

Ca3. ____, *Leçons sur les Invariant Intégraux*, Hermann, 1922, 1958.

Ca4. ____, *Théorie des Groupes Finis et Continus et la Géométrie Différentielle*, Gauthier-Villars, 1951.

Ca5. ____, *Oeuvres Complètes*, 6 vols, Gauthier-Villars, 1952–1955.

Ca6. Cartan, H., *Differential Forms*, Hermann, 1970.

Ch1. Chevalley, C., *Theory of Lie Groups*, Princeton University Press, 1946.

Ch2. Choquet-Bruhat, Y., Witt-Morette, C. de and Dillard-Bleick, M., *Analysis, Manifolds, and Physics*, North-Holland, 1977.

Da1. Darboux, G., *Théorie des Surfaces* I-IV, Gauthier-Villars, 1887–1896.

Ed1. Edelen, D. G. B., *Applied Exterior Calculus*, John Wiley & Sons, 1985.

Fl1. Flanders, H., *Differential Forms with Applications to the Physical Sciences*, Academic Press, 1963.

Go1. Goldberg, S. I., *Curvature and Homology*, Academic Press, New York, 1962.

Gr1. Greub, W., Halpern, S. and Vanstone, R., *Connections, Curvature, and Cohomology* I, Academic Press, 1972, esp. section 3.4, ff.

Gr2. Grunsky, H., *The General Stokes' Theorem*, Pitman Pub. Co., 1983.

Gu1. Guggenheimer, H. W., *Differential Geometry*, McGraw-Hill, 1963.

He1. Helgason, S., *Differential Geometry, Lie groups, and Symmetric Spaces*, Academic Press, 1978.

Ka1. Kannai, Y., An elementary proof of the no-retraction theorem, *Amer. Math. Monthly*, 88 (1981) 264–268.

Kl1. Klingenberg, W., *A Course in Differential Geometry*, Springer-Verlag, 1978, esp. Section 5.6, ff.

Ko1. Kobayashi, S. and Nomizu, K., *Foundations of Differential Geometry* I, John Wiley & Sons, 1963.

Ko2. Kock, A., *Synthetic Differential Geometry*, Cambridge University Press, 1981, esp. Chapters 14–15.

La1. Lang, S., *Introduction to Differentiable Manifolds*, John Wiley & Sons, 1962.

Li1. Lichnerowicz, A., "Théorie globale des connexions et des groupes d'holonomie," *Consiglio Nazionale delle Ricerche*, Ed. Cremonese, Rome, 1955.

Li2. ____, *Géométrie des groupes de transformations*, Dunod, 1958.

Lo1. Lovelock, D. and Rund, H., *Tensors, Differential Forms, and Variational Principles*, John Wiley & Sons, 1975, esp. Chapter 5.

M1. Matsushima, Y., *Differentiable Manifolds*, Marcel Dekker Pub. Co., 1972, esp. Chapter 3, ff.

Rh1. Rham, G. de, *Differentiable Manifolds*, Springer-Verlag, 1984, esp. Chapter 2.

Sc1. Schreiber, M., *Differential Forms, a Heuristic Approach*, Springer-Verlag, 1977.

Sl1. Ślebodziński, W., *Exterior Forms and their Applications*, PWN-Polish Scientific Publishers, 1970, esp. Chapter 2, ff.

Sp1. Spivak, M., *A Comprehensive Introduction to Differential Geometry*, Vol. 1, Publish or Perish, 1970, esp. Chapter 7.

St1. Sternberg, S., *Lectures on Differential Geometry*, Prentice-Hall, 1964.

Sv1. Švec, A., *Global Differential Geometry of Surfaces*, D. Reidel Pub. Co., 1981, esp. Sect. 2.3, ff.

Va1. Vaisman, I., *Cohomology and Differential Forms*, Marcel Dekker Pub. Co., 1973, esp. Sect. 4.3, ff.

We1. Weil, A., *Introduction a l'Étude des Variétés Kählériennes*, Act. Sci. et Ind. 1267, Hermann, 1958.

We2. Westenholz, C. von, *Differential Forms in Mathematical Physics*, North-Holland, 1981.

Wh1. Whitney, H., Self-intersection of a manifold, *Annals of Math.* (2) 45 (1944), 220–246.

Wh2. ____, *Geometric Integration Theory*, Princeton University Press, 1957.

Wi1. Willmore, T. J., *Introduction to Differential Geometry*, Oxford University Press, 1959.

Y1. Yano, K. and Bochner, S., *Curvature and Betti Numbers*, Princeton University Press, 1953.

MINIMAL SURFACES IN $\mathbf{R}^3$

Robert Osserman

1. THE MINIMAL SURFACE EQUATION

Let D be a domain in the x, y-plane, and let S be a surface defined as a graph $z = f(x, y)$ over the domain D. S is a *minimal surface* if it satisfies the equation

$$\left(1 + f_y^2\right)f_{xx} - 2f_x f_y f_{xy} + \left(1 + f_x^2\right)f_{yy} = 0, \qquad (1.1)$$

called the *minimal surface equation*. The reason for the terminology is the following.

THEOREM 1.1. *If $f(x, y)$ satisfies the minimal surface equation in D and if f extends continuously to the closure of D, then the area of the surface S defined by f is less than the area of any other surface $\tilde{S}$ defined by a function $\tilde{f}(x, y)$ in D having the same values as f on the boundary of D.*

Proof. In the domain $D \times \mathbf{R}$ of $\mathbf{R}^3$, consider the unit vector field $v(x, y, z)$ defined by

$$v = \left(-\frac{f_x}{W}, -\frac{f_y}{W}, \frac{1}{W} \right) \qquad (1.2)$$

73

where

$$W = \sqrt{1 + f_x^2 + f_y^2}. \tag{1.3}$$

Note that v is independent of z, so that

$$v_z \equiv 0, \tag{1.4}$$

while a calculation shows that

$$(v_1)_x + (v_2)_y = \left[\left(1 + f_y^2\right)f_{xx} + \left(1 + f_x^2\right)f_{yy} - 2f_x f_y f_{xy}\right]\Big/ W^3. \tag{1.5}$$

Since f satisfies equation (1.1), it follows from (1.4) and (1.5) that

$$\operatorname{div} v \equiv 0 \qquad \text{in } D \times \mathbf{R}. \tag{1.6}$$

The surfaces S and $\tilde{S}$ have the same boundary, and therefore $S - \tilde{S}$ may be considered as the oriented boundary of a signed open set Δ in $D \times \mathbf{R}$. By the divergence theorem,

$$0 = \int_\Delta \operatorname{div} v = \int_{S - \tilde{S}} v \cdot N \, dA \tag{1.7}$$

where N is the unit normal corresponding to the orientation on $S - \tilde{S}$. But from the definition (1.2) of v, it follows that

$$v \equiv N \qquad \text{on } S. \tag{1.8}$$

Hence, from (1.7),

$$\text{Area of } S = \int_S v \cdot N \, dA = \int_{\tilde{S}} v \cdot N \, dA \leqslant \int_{\tilde{S}} 1 \, dA = \text{Area of } \tilde{S}, \tag{1.9}$$

since v and N are both unit vector fields. Furthermore, the inequality is strict unless $v \cdot N = 1$, which would mean that f and $\tilde{f}$ have the same gradient at each point, so that $\tilde{S}$ would be a translate of S, and having the same boundary values, would have to coincide with S. This proves the theorem.

REMARK.   This proof is elementary if the surfaces S and $\tilde{S}$ do not intersect except at their common boundary. If one wishes to avoid the problem of complicated domains with multiplicity, one can use the fact that since $\operatorname{div} v \equiv 0$ in $D \times \mathbf{R}$, there must exist a vector field Y with $\operatorname{curl} Y = v$. The result then follows by applying Stokes' Theorem to S and $\tilde{S}$. Equivalently, we can identify v with a 2-form ω in $D \times \mathbf{R}$, which is closed, by (1.6), hence exact, and once again, we may apply Stokes' Theorem.

The geometric significance of the minimal surface equation comes from the fact that the right-hand side of equation (1.5) is independent of coordinates, and equals the sum of the principal curvatures of the surface S. That is, the *mean curvature H* of a surface S represented as a graph $z = f(x, y)$ is given by

$$H = \frac{\left(1 + f_y^2\right)f_{xx} - 2f_x f_y f_{xy} + \left(1 + f_x^2\right)f_{yy}}{2W^3}$$

(1.10)

$$= \frac{1}{2}\left[\frac{\partial}{\partial x}\left(\frac{f_x}{W}\right) + \frac{\partial}{\partial y}\left(\frac{f_y}{W}\right)\right].$$

Thus, equation (1.1) signifies the vanishing of the mean curvature, and the following result may be viewed as a kind of converse to Theorem 1.1.

THEOREM 1.2.   *Let S be an arbitrary surface immersed in $\mathbf{R}^3$, not necessarily given as the graph of a function. If S has least area among all surfaces $\tilde{S}$ that differ from S on a compact set, then S has mean curvature identically zero.*

Proof.   Since every immersed surface is locally a graph, for any point p of S we may choose a neighborhood S_0 of p and coordinates in $\mathbf{R}^3$ so that S_0 is represented by $z = f(x, y)$ over a domain D in the x, y-plane. The area of S_0 is then given by

$$A = \int_D \int W \, dx \, dy$$

(1.11)

Robert Osserman

where W is expressed by (1.3). If g is a smooth function that vanishes on the boundary of D, then setting

$$z = f(x, y) + \epsilon g(x, y) \tag{1.12}$$

defines a surface S_ϵ over D for every value of ϵ, with area $A(\epsilon)$ given by

$$A(\epsilon) = \int_D \int W(\epsilon) \, dx \, dy; \tag{1.13}$$

$W(\epsilon)$ is defined by (1.3) using the function $f + \epsilon g$ in place of f. Then

$$
\begin{aligned}
A'(0) &= \int_D \int W'(0) \, dx \, dy \\
&= \int_D \int \left(\frac{f_x}{W} g_x + \frac{f_y}{W} g_y \right) dx \, dy \\
&= -\int_D \int g \left[\frac{\partial}{\partial x} \left(\frac{f_x}{W} \right) + \frac{\partial}{\partial y} \left(\frac{f_y}{W} \right) \right] dx \, dy \\
&= -\int_D \int 2gH \, dx \, dy
\end{aligned}
\tag{1.14}
$$

by (1.10). The usual argument shows that if $H(p) \neq 0$, then by choosing a neighborhood of p in which H is not zero, and choosing g to be nonnegative in that neighborhood, positive at p, and identically zero outside the neighborhood we obtain from (1.14) that $A'(0) \neq 0$. Hence, for small values of ϵ, $A(\epsilon)$ takes on values both larger and smaller than $A = A(0)$. Thus S_0 does not have least area among surfaces with the same boundary. This proves the theorem.

One may give a direct proof of Theorem 2, without writing the surface locally as a graph, by expressing the surface in parametric form and using an arbitrary variation. For the record, we state the general formula:

$$A'(0) = \int_{S_0} \left(-2h \cdot V + \operatorname{div} V^T \right) dA \tag{1.15}$$

where S_0 is any compact portion of S, V is the *variation vector field*: the tangent vector to the curve along which each point of S_0 is moved to obtain a nearby surface S_t, $A(t)$ is the area of S_t, V^T is the projection of V into the tangent plane of S_0 at the corresponding point, div V^T refers to the divergence of V^T as a vector field on S_0, and h is the *mean curvature vector*; for a neighborhood of S_0 expressed in graph form $z = f(x, y)$, we have $h = Hv$, in terms of (1.2) and (1.10). In fact, h is independent of local coordinates, and is a purely geometric quantity. Note that (1.15) reduces to

$$A'(0) = \int_{S_0} -2h \cdot V \, dA \qquad (1.16)$$

if V is everywhere normal to S_0 (in which case $V^T \equiv 0$) or even if V is normal to S_0 along the boundary of S_0, since the divergence theorem on surfaces yields

$$\int_{S_0} (\operatorname{div} V^T) \, dA = \int_{\partial S_0} V^T \cdot Z \, ds \qquad (1.17)$$

where Z is a unit normal field along the boundary ∂S_0. In particular, (1.16) holds whenever the comparison surfaces S_t have the same boundary as S_0, since in that case $V \equiv 0$ on ∂S_0.

Note that (1.14) is a special case of (1.16), where

$$V = (0, 0, g(x, y))$$

$$h = Hv = H\left(-\frac{f_x}{W}, -\frac{f_y}{W}, \frac{1}{W}\right)$$

$$dA = W \, dx \, dy.$$

In view of (1.16), minimal surfaces may be thought of as surfaces that are *stationary* for area; that is, the first variation of area is zero for one-parameter variations keeping the boundary fixed. They may or may not actually minimize area among all surfaces with the same boundary. In order to study that question (which we do not) it is useful to examine the *second variation* of area: $A''(0)$. Again for the record we state the result. Let S be a minimal surface, S_0 a compact portion, S_t a one-parameter family of comparison surfaces having the same boundary as S_0, and $A(t)$ the area of S_t. Assume that the variation field V is normal at each point of S_0:

$$V = \varphi v$$

where v is a unit normal field along S_0. Then

$$A''(0) = \int_{S_0} \left(|\nabla \varphi|^2 + 2K\varphi^2 \right) dA$$

where K is the Gauss curvature of S_0. A minimal surface is called *stable* if for every variation of a compact part S_0 of S, $A''(0) \geqslant 0$. Stability is thus an infinitesimal version of area minimization. By virtue of Theorem 1.1, every minimal surface that can be represented as a graph is stable. Thus, when trying to get at the underlying geometric content of theorems about solutions to the minimal surface equation, one natural approach is to study stable minimal surfaces. Another is to impose a restriction on the Gauss map of the surface. As an example, consider the famous theorem of Bernstein: *the only solutions $f(x, y)$ of equation* (1.1) *defined on the whole x, y-plane are the trivial ones:* $f(x, y) = ax + by + c$. (For a proof, see this volume.) One generalization is the following theorem (do Carmo/Peng [3]), (Fischer-Colbrie/Schoen [4]). *A complete stable minimal surface in $\mathbf{R}^3$ must be a plane.*

We will not prove that theorem here, but in Section 4 below, we give two generalizations of Bernstein's theorem involving restrictions on the Gauss map.

2. THE ISOPERIMETRIC INEQUALITY

Let C be a simple closed curve in the plane, bounding a domain D. Let the length of C be L and the area of D be A. Then the classical isoperimetric inequality states that

$$L^2 \geqslant 4\pi A \tag{2.1}$$

with equality if and only if C is a circle.

Next, let C be a simple closed curve in $\mathbf{R}^3$ bounding a surface S. If S minimizes area among surfaces bounded by C, then it may not seem surprising that (2.1) should continue to hold. However, as we have seen in Section 1, if S is an arbitrary minimal surface bounded by C, then S need not have minimal area, and hence it is not at all clear that the isoperimetric inequality should still hold. But the fact is that it does.

THEOREM 2.1. *Inequality* (2.1) *holds for the area* A *of any minimal surface* S *bounded by a smooth simple closed curve* C *of length* L.

Proof. Let the curve C be represented by

$$X(s) = (x_1(s), x_2(s), x_3(s)), \qquad 0 \leqslant s \leqslant L, \qquad (2.2)$$

where s is the element of arc length on C. By a translation, we may assume that

$$\int_0^L x_k(s)\, ds = 0, \qquad k = 1, 2, 3. \qquad (2.3)$$

An elementary and classical formula for the area A of a surface S is

$$A = \frac{1}{2} \int_C X \cdot Z\, ds - \int_S \int X \cdot h\, dA \qquad (2.4)$$

where h is the mean curvature vector of S, and Z is the "outer normal" along X; that is, Z is the unit vector tangent to S and normal to C, pointing away from S. One proof of (2.4) is obtained by using the first variation formula (1.15), (1.17) applied to the radial vector field $V = X$, since the comparison surface S_t is given by $X + tV = (1 + t)X$ and therefore has area $A(t) = (1 + t)^2 A$, so that $A'(0) = 2A$. For a minimal surface, (2.4) takes the form

$$2A = \int_C X \cdot Z\, ds. \qquad (2.5)$$

We now introduce a vector field along C:

$$W = X - \frac{L}{2\pi} Z \qquad (2.6)$$

and observe that

$$
\begin{aligned}
0 \leqslant \frac{2\pi^2}{L} &\int_C |W|^2 \, ds \\
&= \frac{2\pi^2}{L} \left[\int_C |X|^2 \, ds - \frac{L}{\pi} \int_C X \cdot Z \, ds + \frac{L^2}{4\pi^2} \int_C |Z|^2 \, ds \right] \\
&= \frac{2\pi^2}{L} \left[\int_C |X|^2 \, ds - \frac{2L}{\pi} A + \frac{L^3}{4\pi^2} \right] \\
&= \frac{2\pi^2}{L} \int_C |X|^2 \, ds - 4\pi A + \frac{L^2}{2}
\end{aligned}
\qquad (2.7)
$$

using (2.5) and the fact that $|Z|^2 \equiv 1$ on C.

We now use an elementary inequality known as "Wirtinger's inequality" (see [9], and this volume):

$$\int_0^{2\pi} f(t)^2 \, dt \leqslant \int_0^{2\pi} f'(t)^2 \, dt \qquad (2.8)$$

for any smooth 2π-periodic function satisfying

$$\int_0^{2\pi} f(t) \, dt = 0.$$

By virtue of (2.3), we may apply (2.8) to each of the functions

$$y_k(t) = x_k\left(\frac{L}{2\pi} t\right), \qquad 0 \leqslant t \leqslant 2\pi,$$

and we find

$$\int_0^L x_k^2 \, ds \leqslant \frac{L^2}{4\pi^2} \int_0^L \left(\frac{dx_k}{ds}\right)^2 \, ds.$$

Summing over k, we obtain

$$\int_0^L |X|^2 \, ds \leqslant \frac{L^2}{4\pi^2} \int_0^L \left| \frac{dX}{ds} \right|^2 ds = \frac{L^3}{4\pi^2}.$$

Inserting this into (2.7) gives the desired inequality.

This proof is due to Chakerian [2]. (It holds, without change, for minimal surfaces in $\mathbf{R}^n$.) A special case of Theorem 2.1, where S is assumed to be simply-connected, was originally proved by Carleman [1] in 1921. It led to a whole series of further results, including the isoperimetric inequality of André Weil [18]: *inequality* (1) *holds for the area A of any simply-connected surface S of Gauss curvature $K \leqslant 0$, bounded by a simple closed curve C of length L*.

The question arose: what happens if the boundary of S has more than one component? Partial results led to the following.

CONJECTURE. *Inequality (2.1) holds for the area A of any minimal surface S whose boundary is the union of a finite number of simple closed curves, where L is the total length of the boundary curves.*

The truth of this conjecture is not yet known, except in special cases. For example, it was shown for doubly-connected surfaces [16] and later for arbitrary surfaces with no more than two boundary components [8]. It also holds for surfaces S that are not only minimal, but *area-minimizing*. The reason is that each separate boundary component is known to bound an immersed minimal surface to which we may apply Theorem 2.1. Then adding all the inequalities we obtain one for the sum of the area of the individual surfaces, and by assumption, it is at least as great as the area of our original surface.

3. THE WEIERSTRASS REPRESENTATION

One of the most useful tools for the global study of minimal surfaces is the "Weierstrass representation." It allows many geometric questions on minimal surfaces to be examined using complex functions. One formulation is the following.

THEOREM 3.1.   *Let D be a simply-connected domain in the complex plane, let $g(\zeta)$ be any meromorphic function in D and $f(\zeta)$ any holomorphic function in D, with the condition that the zeros of f coincide with the poles of g, and if g has a pole of order n at a point, then f has a zero of order 2n at that point. Choose any point ζ_0 in D and set*

$$x = \mathscr{R}\left\{ \int \frac{1}{2} f(1 - g^2) \, d\zeta \right\}$$

$$y = \mathscr{R}\left\{ \int \frac{i}{2} f(1 + g^2) \, d\zeta \right\} \qquad (3.1)$$

$$z = \mathscr{R}\left\{ \int fg \, d\zeta \right\}$$

where the integrals are along an arbitrary path in D joining the fixed point ζ_0 to a variable point. Then equations (3.1) define a conformal immersion of D onto a minimal surface S in $\mathbf{R}^3$. Conversely, any simply-connected minimal surface immersed in $\mathbf{R}^3$ can be represented in the form (3.1), where f and g satisfy the prescribed conditions, and D may be chosen to be either the unit disk or the entire plane.

For a proof of this theorem, see [15], p. 64.

We note the following expressions for fundamental geometric quantities in terms of the representation functions f, g:

the *element of arc length* on S: $ds = \lambda |d\zeta|, \quad \lambda = \frac{1}{2}|f|(1 + |g|^2)$

$$(3.2)$$

Gauss curvature: $K = -\left[4|g'| \Big/ |f|(1 + |g|^2)^2 \right]^2 \qquad (3.3)$

the *unit normal*: $N = \left(\dfrac{2\mathscr{R}\{g\}}{|g|^2 + 1}, \dfrac{2\mathscr{I}\{g\}}{|g|^2 + 1}, \dfrac{|g|^2 - 1}{|g|^2 + 1} \right). \qquad (3.4)$

From (3.4) we see that the unit normal depends only on g and not on f, and that g is in turn determined by N; in fact, (3.4) tells us that if the unit normal at a point is considered to be a point p on the unit sphere, then g is obtained by stereographic projection of p from the North pole. In particular, the poles of g correspond simply to points on S whose unit normal points vertically upward: $N = (0, 0, 1)$.

To observe the representation theorem in action, consider Bernstein's theorem, mentioned in Section 1. Let $F(x, y)$ be a solution of the minimal surface equation in the whole x, y-plane. Then the surface $z = F(x, y)$ is a simply-connected minimal surface S, and we wish to show that S is a plane. We may use the representation (3.1) for S, and since S is a plane if and only if the normal N is constant, we wish to show (according to (3.4)) that g is constant. We may orient S by choosing either the upward or downward normal at each point. We choose the latter—the negative of (1.2)—so that the third component, $-1/W$, is strictly negative. It follows from (3.4) that $|g| < 1$.

Now recall that there were two possibilities for the domain D: either the unit disk or the whole plane. In the latter case, g would be a bounded holomorphic function in the whole plane, hence constant. Thus S would be a plane. It only remains to show that the other alternative—D a unit disk—cannot arise. We shall do that in the next section. We note here that a key element in the proof is Liouville's theorem for the function g. Analyzing the proof, Nirenberg was led to conjecture that a more general and more geometric result might hold. In fact Nirenberg made two conjectures, depending on whether Liouville's theorem or the much stronger Picard theorem was used.

NIRENBERG'S FIRST CONJECTURE. *A complete simply-connected minimal surface whose Gauss map omits a neighborhood of some point must be a plane.*

NIRENBERG'S SECOND CONJECTURE. *A complete simply-connected minimal surface whose Gauss map omits three points must be a plane.*

In both cases, if we could exclude the case where D is the unit disk, we would be done, since if D is the entire plane, then the function g would be constant, by Picard's theorem in Conjecture 2, and by Liouville's theorem in Conjecture 1 (after, if necessary, a rotation of $\mathbf{R}^3$ so that the omitted neighborhood on the unit sphere contains the North pole). We shall show in the next section how to rule out the case that D is a disk when g is bounded, thus completing the proof of Nirenberg's First Conjecture. On the other hand, it turns out that D *can* be a disk when g omits three points, and thus Nirenberg's second conjecture is false. The construction of specific counterexamples provides a good illustration of the use of the Weierstrass representation theorem.

Choose any finite set of points on the unit sphere. We wish to construct a minimal surface whose Gauss map omits precisely that set of points. By a preliminary rotation, we may assume that one of the points is the North pole. Denote the image of the others under stereographic projection by $\alpha_1, \ldots, \alpha_n$. Let D be the domain consisting of the whole ζ-plane with the points $\alpha_1, \ldots, \alpha_n$ deleted. It then follows that inserting $g(\zeta) = \zeta$ and *any* holomorphic function $f(\zeta)$ that does not vanish in D (say $f(\zeta) \equiv 1$) into the equations (3.1) will produce a minimal surface S immersed in $\mathbf{R}^3$ whose Gauss map covers the entire unit sphere except for points corresponding to $\alpha_1, \ldots, \alpha_n, \infty$ under stereographic projection. Note that since D is not simply connected, the integrals may depend on the choice of path, so that the representation really defines a map from the universal covering surface $\tilde{D}$ of D into $\mathbf{R}^3$, but the image is still the sphere with the given finite set of points removed.

The only question that remains is whether or not there exists a choice of the function $f(\zeta)$ for which the resulting surface will be complete. To achieve that, the obvious choice is to have a pole of f at each of the α_k. Thus, we set

$$f(\zeta) = 1 \bigg/ \prod_{k=1}^{n} (\zeta - \alpha_k). \tag{3.5}$$

According to (3.2), if C is any curve in D, then the length of the corresponding curve on the surface is given by

$$\int_C \frac{1}{2} |f| (1 + |g|^2) |d\zeta| = \frac{1}{2} \int_C \frac{1 + |\zeta|^2}{\prod_{k=1}^{n} |\zeta - \alpha_k|} |d\zeta|. \tag{3.6}$$

Recall that S is *complete* if the length of every divergent curve on S is infinite, where a *divergent curve* is a map $\gamma(t)$ of the half-line $t \geqslant 0$ into S such that for every compact set E on S there exists $t_0 > 0$ such that for $t > t_0$, $\gamma(t)$ lies outside of E. Suppose then that C is a curve in D corresponding to a divergent curve Γ on S. We must show that the length of Γ, given by (3.6), is infinite.

Consider first the case that C lies in a bounded domain of the ζ-plane. Then the integrand in (3.6) is bounded below by a positive constant. It follows that if C has infinite length, then the integral in (3.6) diverges, so that the length of Γ is also infinite. If, on the other hand, C has finite length, then it must converge to a point ζ_0 in the plane. Since Γ is divergent, ζ_0 cannot be a point of D. Hence ζ_0 must be one of the points $\alpha_1, \ldots, \alpha_n$. But if C tends to one of the α_k, then the integral in (3.6) clearly diverges, and Γ again has infinite length.

The only remaining case is where C does not lie in any bounded domain. There are again two possibilities: either C tends to infinity, or else there is a subsequence of points on C remaining bounded while another subsequence tends to infinity. In the latter case, the curve C must cross some fixed annulus an infinite number of times, and since the integrand of (3.6) is again bounded below on that annulus, the integrand must diverge, and Γ would have infinite length. In the former case, where C tends to infinity, we see that the integral in (3.6) diverges if and only if $n \leqslant 3$. We have thus proved the following result.

THEOREM 3.2. *Let S be the surface obtained by inserting in the Weierstrass formulas* (3.1) *the function $f(\zeta)$ of* (3.5) *and $g(\zeta) = \zeta$, where the domain D is the ζ-plane with the points $\alpha_1, \ldots, \alpha_n$ deleted. Then S is complete if and only if $n \leqslant 3$.*

Since, as we have noted, the Gauss map of the surface covers the whole unit sphere except for the $n + 1$ points corresponding to $\alpha_1, \ldots, \alpha_n, \infty$, we have:

COROLLARY. *Given any set of four or fewer points on the unit sphere, there exists a complete minimal surface whose Gauss map omits precisely that set of points.*

This corollary is due to Voss [17]. Note that if we want the omitted set to consist of a single point, then we may choose $f(\zeta) \equiv 1$.

4. THE GAUSS MAP OF COMPLETE MINIMAL SURFACES

THEOREM 4.1. *If S is a complete minimal surface, not a plane, then the image of S under the Gauss map is everywhere dense on the unit sphere.*

This theorem is just a restatement of Nirenberg's First Conjecture. It is no loss of generality to assume the surface to be simply connected, since otherwise we can pass to the universal covering surface, whose image under the Gauss map is the same. As we observed in the previous section, the proof reduces to showing that the domain D cannot be the unit disk.

Suppose then that D *is* the unit disk, and that the image of S under the Gauss map omits a full neighborhood of some point. We shall show that S cannot be complete. After a preliminary rotation, so that the omitted neighborhood contains the North pole, we have from (3.4) that the function g in the Weierstrass representation is bounded: say $|g| < M$. Then from (3.2), the length L of a curve Γ on S corresponding to a curve C in D, satisfies

$$L = \int_C \frac{1}{2}|f|(1 + |g|^2)|d\zeta| \leqslant \frac{1 + M^2}{2} \int_C |f| |d\zeta|. \qquad (4.1)$$

The curve Γ is divergent on S if and only if C is divergent in the unit disk D; that is, C tends to the boundary, $|\zeta| = 1$. To show that S is not complete, we must prove that there exists such a curve C for which the value of L in (4.1) is finite. Making use of the fact that f has no zeros, since g has no poles, we see that the theorem follows from (4.1) and the following function-theoretic fact.

LEMMA 4.2. *Let f be an analytic function in the unit disk that is never zero. Then there is a path C to the boundary such that*

$$\int_C |f(\zeta)| |d\zeta| < \infty. \qquad (4.2)$$

Proof. Denote by D the unit disk in the ζ-plane, and define the function $F(\zeta)$ in D by

$$F(\zeta) = \int f(\zeta) \, d\zeta,$$

where we may assume that the integral starts at $\zeta = 0$, so that $F(0) = 0$. Note that $F'(\zeta) = f(\zeta) \neq 0$, so that $F(\zeta)$ has a well-defined holomorphic inverse function $G(w)$ near $w = 0$, with $G(0) = 0$. Let R be the radius of convergence of the power series for $G(w)$ at the origin. R is finite since otherwise G would be an entire function with $|G(w)| < 1$, and hence G would be constant, which it is not. Thus there is a point w_0 with $|w_0| = R$ such that G cannot be extended from $|w| < R$ into any neighborhood of w_0. Let C' be the line segment from 0 to w_0, and let C be the image of C' under G. Then

$$\int_C |f(\zeta)| \, |d\zeta| = \int_C |F'(\zeta)| \, |d\zeta| = \int_{C'} |dw| = R < \infty \qquad (4.3)$$

so that (4.2) is satisfied. We need only show that C is divergent in D. Suppose not. Then there would be a sequence of points ζ_k on C such that $F(\zeta_k)$ tends to w_0, but all the ζ_k lie in a compact subset of D. There would then be a subsequence of the ζ_k converging to ζ_0 in D, and by continuity, $F(\zeta_0) = w_0$. But since $F'(\zeta_0) = f(\zeta_0) \neq 0$, $F(\zeta)$ defines a conformal diffeomorphism of a neighborhood of ζ_0 onto a neighborhood of w_0, whose inverse is defined in a neighborhood of w_0 and agrees with G at a sequence of points tending to w_0, hence it provides an extension of G to a neighborhood of w_0. But this contradicts the defining property of w_0, and hence shows that C must be divergent. Thus the lemma is proved.

A closer examination of the proof shows that not only does it prove Nirenberg's first conjecture, but it leads easily to a sharper, more quantitative form of the result. Namely, if the Gauss map of a minimal surface S omits a fixed neighborhood on the unit sphere, then not only can S not be complete, but for any point Γ on S, we can give a concrete upper bound on the distance from p to the boundary. To see that, we may assume that S is simply connected and represent S in the unit disk D, where we may assume that the point p corresponds to the origin. (If S is not simply connected, we may work with the universal covering surface $\tilde{S}$ of S; any point $\tilde{p}$ covering p has the same distance to the boundary of $\tilde{S}$ as p does in S.) Let C be the curve obtained in Lemma 4.2, and Γ the image of C on S. Then combining (4.1), and (4.3), the length L of Γ satisfies

$$L \leqslant \frac{R}{2}(1 + M^2). \qquad (4.4)$$

We may further estimate R by Schwarz' Lemma, since G maps $|w| < R$ into $|\zeta| < 1$:

$$|G'(0)| \leqslant \frac{1}{R}$$

or

$$R \leqslant \frac{1}{|G'(0)|} = |F'(0)| = |f(0)|. \qquad (4.5)$$

Now the distance d from p to the boundary of S is by definition the *infimum* of the lengths of all divergent paths on S starting at p. Hence, combining (4.4) and (4.5),

$$d \leqslant L \leqslant \frac{|f(0)|}{2}(1 + M^2) \qquad (4.6)$$

where

$$|g(w)| < M \quad \text{in} \quad D. \qquad (4.7)$$

To get a purely geometric bound on d, we recall the expression (3.3) for the Gauss curvature of S:

$$K(p) = -\left[4|g'(0)|\Big/|f(0)|\left(1 + |g(0)|^2\right)^2\right]^2. \qquad (4.8)$$

Combining this with (4.6) yields

$$\sqrt{|K(p)|}\,d \leqslant \frac{2|g'(0)|}{\left(1+|g(0)|^2\right)^2}(1+M^2). \qquad (4.9)$$

In view of (4.7) we may apply Schwarz' Lemma to the function g to obtain

$$|g'(0)| \leqslant \frac{M^2-|g(0)|^2}{M},$$

which inserted into (4.9) yields

$$d \leqslant \frac{2}{\sqrt{|K(p)|}}\,\frac{(1+M^2)\left(M^2-|g(0)|^2\right)}{M\left(1+|g(0)|^2\right)^2}. \qquad (4.10)$$

Note that in this estimate for the distance d of p to the boundary, the quantities on the right all have direct geometric interpretations: The constant M, an upper bound for g, is easily expressed in terms of the size of the neighborhood omitted by the normals; $g(0)$ corresponds to the direction of the normal at p, and $K(p)$ is the Gauss curvature of S at p. Using $|g(0)| \geqslant 0$ gives a simpler estimate independent of the normal at p:

$$d \leqslant \frac{2M(1+M^2)}{\sqrt{|K(p)|}}. \qquad (4.11)$$

Inequalities (4.10) and (4.11) do not give direct information on d at points where the Gauss curvature $K(p)$ vanishes. However, by (4.8), such points are isolated if the surface is not a plane, and we may always choose a point q where $K(q) \neq 0$, apply (4.10) or (4.11) at q, and then add the distance from p to q to get the desired bound. In any case, it is often preferable to look at these inequalities in the other direction as providing an upper bound for the Gauss curvature at any point p of a domain on a surface whose

Gauss map omits some neighborhood on the sphere

$$|K(p)| \leqslant \frac{c}{d^2} \tag{4.12}$$

where c depends only on the size of the neighborhood omitted and d is the distance from p to the boundary of the domain. If the surface is complete, then for any point p, d may be made arbitrarily large, and $K(p)$ must be zero. But if $K \equiv 0$ on a minimal surface S, then S must lie on a plane, and Nirenberg's first conjecture follows as a limiting case of (4.12).

Finally, if the domain is representable in the form $z = f(x, y)$, then we have $M \leqslant 1$ in (4.11), and hence

$$|K(p)| \leqslant \frac{16}{d^2}. \tag{4.13}$$

Again, Bernstein's theorem is an immediate consequence of (4.13).

The first inequality on the Gauss curvature, giving a quantitative version of Bernstein's Theorem, was obtained by Erhard Heinz [7]. Theorem 4.1 was proved in 1959 [12]. A sharper version, proved by Ahlfors and Osserman [13] in 1961, shows that the Gauss map must not only be everywhere dense, but that the omitted set must have logarithmic capacity zero. The proof follows the same general lines as the one given above. There was then a gap of twenty years before further progress was made. Using an entirely different method, Xavier [19] proved in 1981 that the Gauss map can omit at most six points. Finally, returning to an elaboration of the original method of proof, Fujimoto [5] obtained in 1987 the optimal result.

THEOREM 4.3. *Let S be a complete minimal surface in* $\mathbf{R}^3$, *not a plane. Then the Gauss map of S can omit at most four points.*

Proof. Let S be a minimal surface whose Gauss map omits five points on the unit sphere. By a rotation, one of the points may be chosen to be the North pole. By passing to the universal covering surface, we may assume that S is simply connected. We may then use the representation (3.1) for S, where g will omit certain values

α_1, α_2, α_3, α_4 and $\alpha_5 = \infty$. Thus the domain D cannot be the whole plane. We may therefore assume that D is the unit disk $|\zeta| < 1$, and that g is holomorphic in D. The function f is then also holomorphic in D and never vanishes.

The proof of the theorem falls into two parts. The first is purely function-theoretic. It consists in obtaining a constraint on g from the fact that it omits four points. The second involves defining an auxiliary map $F(\zeta)$ used to get an estimate of the distance to the boundary of S. We start with Part I.

Denote by Δ the complex plane with the points $\alpha_1, \alpha_2, \alpha_3, \alpha_4$ deleted. Recall that the universal covering surface of Δ is conformally equivalent to the unit disk, and that the standard Poincaré metric on the disk pulls back to a conformal metric $d\sigma = \lambda(z)|dz|$ on Δ with constant Gauss curvature equal to -1.

LEMMA 4.4. *Let ϵ, ϵ' be any two numbers satisfying*

$$0 < \epsilon < 1, \qquad 0 < \epsilon' < \epsilon/4. \qquad (4.14)$$

Then the function

$$\frac{\left(1 + |z|^2\right)^{(3-\epsilon)/2}}{\lambda(z)\prod_{j=1}^4 |z - \alpha_j|^{1-\epsilon'}} \leqslant B < \infty \qquad \text{on} \quad \Delta. \qquad (4.15)$$

Proof. The asymptotic behavior of $\lambda(z)$ at each of the boundary points of Δ is well known. (See, for example [11] p. 250.) At each α_j, $j = 1, \ldots, 4$,

$$\lambda(z) \sim \frac{c_j}{|z - \alpha_j||\log|z - \alpha_j|}, \qquad c_j \neq 0,$$

and at $\alpha_5 = \infty$,

$$\lambda(z) \sim \frac{c_0}{|z||\log|z|}, \qquad c_0 \neq 0.$$

It follows that the left-hand side of (4.15) is a positive continuous function on Δ which by virtue of the constraints (4.14) on ϵ, ϵ', tends to zero at $\alpha_1, \ldots, \alpha_5$. Hence it has a positive maximum.

LEMMA 4.5.   *Let $h(w)$ be analytic in $|w| < R$ and omit the points $\alpha_1, \ldots, \alpha_4$. Let ϵ, ϵ' satisfy (4.14). Then*

$$\frac{\left(1 + |h(w)|^2\right)^{(3-\epsilon)/2} |h'(w)|}{\prod_{j=1}^4 |h(w) - \alpha_j|^{1-\epsilon'}} \leqslant B \frac{2R}{R^2 - |w|^2} \qquad (4.16)$$

where B is the bound in (4.15).

Proof.   By the Schwarz-Pick lemma, non-Euclidean lengths are reduced under the map h. That means

$$\lambda(z) |dz| \leqslant \frac{2R}{R^2 - |w|^2} |dw|$$

or

$$\lambda(h(w)) |h'(w)| \leqslant \frac{2R}{R^2 - |w|^2}. \qquad (4.17)$$

But by Lemma 4.4,

$$\frac{\left(1 + |h(w)|^2\right)^{(3-\epsilon)/2}}{\lambda(h(w)) \prod |h(w) - \alpha_j|^{1-\epsilon'}} \leqslant B. \qquad (4.18)$$

Combining (4.17) and (4.18) yields (4.16).

This completes the first part of the proof. Note that the constant B in (4.15) and (4.16) depends only on the values $\alpha_1, \ldots, \alpha_4$ and the choice of ϵ and ϵ'.

We now apply (4.16) to the problem at hand. We want to adapt the argument in the proof of Theorem 4.1, using Lemma 4.2. Specifically, we want to define a map of D:

$$w = F(\zeta) = \int \psi(\zeta) \, d\zeta, \qquad \psi(\zeta) \neq 0 \qquad (4.19)$$

where $\psi(\zeta)$ is a suitably defined function, and the integral is taken from $\zeta = 0$ to a variable point. For any such map, if $G(w)$ is the inverse of $F(\zeta)$ defined in a neighborhood of the origin, with $G(0) = 0$, then as in Lemma 4.2, there is a largest disk $|w| < R$ in which G is defined, and a point w_0 satisfying $|w_0| = R$ such that G cannot be extended to a neighborhood of w_0. Let C' be the line segment from 0 to w_0, let C be the image of C' under G, and let Γ be the corresponding curve on S. Then the length L of Γ is given by

$$L = \frac{1}{2} \int_C |f|(1 + |g|^2) \, |d\zeta| = \int_{C'} |f \circ G|(1 + |g \circ G|^2) \left| \frac{d\zeta}{dw} \right| |dw|.$$

Also

$$\frac{d\zeta}{dw} = \frac{1}{dw/d\zeta} = \frac{1}{\psi \circ G}. \tag{4.20}$$

If we choose ψ to be of the form

$$\psi = f\varphi \tag{4.21}$$

then the term involving f in the integrand cancels out, and we have

$$L = \int_{C'} \frac{1 + |h(w)|^2}{|\varphi \circ G|} |dw| \tag{4.22}$$

where

$$h = g \circ G. \tag{4.23}$$

So h is analytic in $|w| < R$ and omits the values $\alpha_1, \ldots, \alpha_4$. We may therefore apply Lemma 4.5 and conclude that inequality (4.16) holds. What we *want* is to choose φ (or equivalently, ψ) to make L finite. In view of (4.22), it would suffice if

$$\frac{1 + |h(w)|^2}{|\varphi \circ G|} \leq \frac{C}{(R^2 - |w|^2)^p}, \qquad 0 < p < 1. \tag{4.24}$$

By virtue of (4.16), we can achieve that if

$$\frac{1+|h(w)|^2}{|\varphi \circ G|} = \left[\frac{\left(1+|h(w)|^2\right)^{(3-\epsilon)/2}|h'(w)|}{\Pi_{j=1}^4|h(w)-\alpha_j|^{1-\epsilon'}}\right]^p. \quad (4.25)$$

So

a) choose $p = 2/(3-\epsilon)$; then $0 < \epsilon < 1 \Rightarrow 2/3 < p < 1$;

b) make

$$\varphi \circ G = \frac{\Pi\big(h(w)-\alpha_j\big)^{p(1-\epsilon')}}{h'(w)^p}. \quad (4.26)$$

The problem is that G is defined in terms of φ, so that we cannot use this equation as a definition of φ. However, we can express everything as a function of ζ:

$$\frac{dw}{d\zeta} = \psi = f\varphi = \frac{f(\zeta)\Pi\big(g(\zeta)-\alpha_j\big)^{p(1-\epsilon')}}{g'(\zeta)^p\Big/\left(\dfrac{dw}{d\zeta}\right)^p}$$

or

$$\left(\frac{dw}{d\zeta}\right)^{1-p} = f(\zeta)\prod\big(g(\zeta)-\alpha_j\big)^{p(1-\epsilon')}\Big/g'(\zeta)^p$$

or finally

$$\psi(\zeta) = \frac{dw}{d\zeta} = \frac{f(\zeta)^{1/(1-p)}\Pi\big(g(\zeta)-\alpha_j\big)^{(p/(1-p))(1-\epsilon')}}{g'(\zeta)^{p/(1-p)}}. \quad (4.27)$$

CASE 1. $g'(\zeta) \neq 0$ in D. Then we can use (4.27) to define ψ and (4.19) to define F. Then working backwards through (4.26) and (4.25) we find that (4.24) holds, and hence by (4.22), the length L of Γ is finite. It follows that either Γ tends to a point P of S or else Γ is divergent. But the former is not possible, since it would imply that the image C of Γ would tend to a point ζ_0 in D, and then, as in the proof of Lemma 4.2, G would be extendable over w_0, contradicting the definition of w_0. Thus Γ must be a divergent curve on S of finite length, and S cannot be complete.

CASE 2. $g'(\zeta)$ vanishes on a nonempty set E in D. In this case, since S is not a plane, the set E is either finite or else consists of a countable set of points tending to the boundary. Let $\tilde{D}$ be the universal covering surface of $D \backslash E$. We may again define $\psi(\zeta)$ by (4.27), but only in $D \backslash E$. The definition (4.19) of $F(\zeta)$ may lead to a multivalued function in $D \backslash E$, but it can be lifted to a single-valued function $\tilde{F}$ in $\tilde{D}$. Since $\tilde{D}$ is again conformally the unit disk, the same argument as before produces a largest disk $|w| < R$ in which the inverse $\tilde{G}$ of $\tilde{F}$ is defined and a boundary point w_0 over which G cannot be extended. Let G be the map of $|w| < R$ into $D \backslash E$ obtained by composing $\tilde{G}$ with the projection of $\tilde{D}$ onto $D \backslash E$. If C' and C are defined as before, then the same arguments show that the image Γ of C on S has finite length. The theorem is proved if we can show that Γ is a divergent curve. But suppose it were not. Then C is not divergent in D. That would mean that for some sequence of points w_n on C' with $w_n \to w_0$, the corresponding points ζ_n on C converge to an interior point ζ_0 of D. If ζ_0 is in $D \backslash E$, then the same argument as in Case 1 leads to a contradiction. If, on the other hand, ζ_0 is a point of E, then $g'(\zeta_0) = 0$ and $g'(\zeta) \sim a(\zeta - \zeta_0)^m$ for some $m \geqslant 1$. Then

$$g'(\zeta)^{(p/(1-p))} \sim b(\zeta - \zeta_0)^{m(p/(1-p))},$$

where

$$p = \frac{2}{3 - \epsilon}, \qquad \frac{p}{1 - p} = \frac{2}{1 - \epsilon} > 2.$$

Now if C tends to ζ_0 as w tends to w_0, then

$$R = \int_{C'} |dw| = \int_C |\psi(\zeta)| |d\zeta| > c \int_C \frac{|d\zeta|}{|\zeta - \zeta_0|^2} = \infty$$

which is a contradiction. Finally, if C does not tend to ζ_0, then it has some other accumulation point on $D \setminus E$, leading to a contradiction as before. We conclude that C must diverge in D, hence Γ is a divergent path of finite length on S, and S cannot be complete. This proves the theorem.

ELABORATIONS AND EXTENSIONS

1. Just as in the case of Nirenberg's conjecture, there is a quantitative, finite version of Fujimoto's theorem, obtainable by a refinement of the above argument.

THEOREM 4.4. *Let S be a minimal surface in* $\mathbf{R}^3$ *whose Gauss map omits five or more points. Let p be any point of S, K the Gauss curvature of S at p, and d the distance from p to the boundary of S. Then*

$$|K| d^2 \leqslant c,$$

where c is an absolute constant depending not on the particular surface S, but only on the omitted values and the choice of ϵ, ϵ' *satisfying* (4.14).

For details of the proof, see Fujimoto [5].

2. Recall that in Theorem 3.2 we have shown how to construct complete minimal surfaces whose Gauss map omits any four points on the unit sphere. A closer examination reveals that every other point of the sphere is covered infinitely often. That turns out not to be an artifact of the construction, but a general property. In fact, one may give two different but related strengthenings of Fujimoto's Theorem.

THEOREM 4.5. *Let S be a complete minimal surface in* $\mathbf{R}^3$, *not a plane. If the Gauss map of S omits four points, then it must cover all other points of the sphere infinitely often.*

THEOREM 4.6. *Let S be a complete minimal surface in $\mathbf{R}^3$. If there are five points on the sphere covered only finitely often, then S has finite total curvature.*

Note that for complete minimal surfaces of finite total curvature the Gauss map can be described quite explicitly, since either the surface is a plane and its Gauss map is constant, or else the image under the Gauss map is a finite-sheeted branched covering surface over the sphere with a finite number of points removed; every point of the sphere is covered a fixed finite number of times, except for a finite set covered less often, and at most three omitted altogether [13].

The proof of Theorems 4.5 and 4.6 combines the ideas of Fujimoto's proof of Theorem 4.3 with earlier results on surfaces of finite and infinite total curvature [14]. For details, see the paper of Mo and Osserman [10].

3. In subsequent work [6], Fujimoto has given defect relations, providing extensions of his theorem in precise analogy to Nevanlinna's extensions of Picard's Theorem.

4. Two further papers based on Xavier's method were written before Fujimoto's paper [5] appeared. They are

R. S. Earp and H. Rosenberg, "On the values of the Gauss map of complete minimal surfaces in $\mathbf{R}^3$," *Comment. Math. Helv.* 63 (1988), 579–86.

F. J. Lopez and A. Ros, "On the values of the Gauss map of minimal surfaces", preprint.

The first of these papers proves a somewhat weaker form of Theorems 4.5 and 4.6 above. The second one adapts ideas of Fischer-Colbrie and Schoen [4] to improve Xavier's result, showing that the Gauss map cannot omit more than five values, unless the surface is a plane.

REFERENCES

[1] T. Carleman, "Zur Theorie der Minimalflächen," *Math. Zeitschrift* 9 (1921), 154–160.

[2] G. D. Chakerian, "The isoperimetric theorem for curves on minimal surfaces," *Proc. Amer. Math. Soc.* 69 (1978), 312–313.

[3] M. do Carmo and C. K. Peng, "Stable complete minimal surfaces in $\mathbb{R}^3$ are planes," *Bull. Amer. Math. Soc.* 1 (1979), 903–906.

[4] D. Fischer-Colbrie and R. Schoen, "The structure of complete stable minimal surfaces in 3-manifolds of nonnegative scalar curvature," *Comm. Pure Appl. Math.* 33 (1980), 199–211.

[5] H. Fujimoto, "On the number of exceptional values of the Gauss map of minimal surfaces," *J. Math. Soc. Japan* 40 (1988), 237–249.

[6] ___, "Modified defect relations for the Gauss map of minimal surfaces," *J. Diff. Geometry* (to appear).

[7] E. Heinz, "Über die Lösungen der Minimalflächengleichung," *Nachr. Akad. Wiss., Göttingen Math., Phys. Kl. II* (1952), 51–56.

[8] P. Li, R. Schoen, and S.-T. Yau, "On the isoperimetric inequality for minimal surfaces," *Annali della Scuola Normale Sup. di Pisa,* Cl. Scienze Ser. IV Vol. XI, 2 (1984), 237–244.

[9] D. S. Mitrinović, *Analytic Inequalities,* Springer, Berlin, 1970.

[10] X. Mo and R. Osserman, "On the Gauss map and total curvature of complete minimal surfaces and an extension of Fujimoto's Theorem," *J. Diff. Geometry* (to appear).

[11] R. Nevanlinna, *Analytic Functions,* Grundlehren der Mathematischen Wissenchaften, Vol. 162, Springer-Verlag, New York, 1970.

[12] R. Osserman, "Proof of a conjecture of Nirenberg," *Comm. Pure Appl. Math.* 12 (1959), 229–232.

[13] ___, "Minimal surfaces in the large," *Comment. Math. Helv.* 35 (1961), 65–76.

[14] ___, "Global properties of minimal surfaces in E^3 and E^n," *Ann. of Math.* 80 (1964), 340–364.

[15] ___, *A Survey of Minimal Surfaces,* 2nd edition, Dover, New York 1986.

[16] R. Osserman and M. Schiffer, "Doubly-connected minimal surfaces," *Arch. Rat'l. Mech. Anal.* 58 (1975), 285–307.

[17] K. Voss, "Über vollständige Minimalflächen," *L'Enseignement Math.* 10 (1964), 316–317.

[18] A. Weil, "Sur les surfaces à courbure négative," *C. R. Acad. Sci.* 182 (1926), 1069–1071. (*Collected Papers,* Vol. I, p. 1; commentary p. 522.)

[19] F. Xavier, "The Gauss map of a complete non-flat minimal surface cannot omit 7 points on the sphere," *Ann. Math.* 113 (1981), 211–214; Erratum: *Ann. Math.* 115 (1982), p. 667.

CURVES AND SURFACES
IN EUCLIDEAN SPACE

S. S. Chern

1. INTRODUCTION

This article contains a treatment of some of the most elementary theorems in differential geometry in the large. They are the seeds for further developments and the subject should have a promising future. We shall consider the simplest cases, where the geometrical ideas are most clear.

1. THEOREM OF TURNING TANGENTS

Let E be the euclidean plane, which is oriented so that there is a prescribed sense of rotation. We define a smooth curve by expressing its position vector $X = (x_1, x_2)$ as a function of its arc length s. We suppose the function $X(s)$—that is, the functions $x_1(s)$, $x_2(s)$—to be twice continuously differentiable and the vector $X'(s)$ to be nowhere 0. The latter allows the definition of the unit tangent vector $e_1(s)$, which is the unit vector in the direction of $X'(s)$ and, since E is oriented, the unit normal vector $e_2(s)$, so that the rotation from e_1 to e_2 is positive. The vectors $X(s)$, $e_1(s)$, $e_2(s)$ are related by the so-called Frenet formulas

$$(1) \qquad \frac{dX}{ds} = e_1, \qquad \frac{de_1}{ds} = ke_2, \qquad \frac{de_2}{ds} = -ke_1.$$

The function $k(s)$ is called the *curvature*. It is defined together with its sign and changes its sign if the orientation of the curve or of the plane is reversed.

The curve C is called *closed*, if $X(s)$ is periodic of period L, L being the length of C. It is called *simple* if $X(s_1) \neq X(s_2)$, when $0 < s_1 - s_2 < L$. It is said to be *convex* if it lies in one side of every tangent line.

Let C be an oriented closed curve of length L, with the position vector $X(s)$ as a function of the arc length s. Let O be a fixed point in the plane, which we take as the origin of our coordinate system. Denote by Γ the unit circle about O. We define the tangential mapping $T: C \to \Gamma$ as the one which maps a point P of C to the endpoint of the unit vector through O parallel to the tangent vector to C at P. Obviously T is a continuous mapping. It is intuitively clear that when a point goes around C once its image point goes around Γ a number of times. This number will be called the rotation index of C. The theorem of turning tangents asserts that if C is simple, the rotation index is ± 1. We begin by giving a rigorous definition of the rotation index.

We choose a fixed vector through O, say Ox, and denote by $\tau(s)$ the angle which Ox makes with the vector $e_1(s)$. We assume that $0 \leqq \tau(s) < 2\pi$, so that $\tau(s)$ is uniquely determined. This function $\tau(s)$ is, however, not continuous, for in every neighborhood of s_0 at which $\tau(s_0) = 0$ there may be values of $\tau(s)$ differing from 2π by an arbitrarily small quantity. There exists nevertheless a continuous function $\bar{\tau}(s)$ closely related to $\tau(s)$, as given by the following lemma.

LEMMA: *There exists a continuous function $\bar{\tau}(s)$ such that $\bar{\tau}(s) \equiv \tau(s)$, mod 2π.*

Proof: To prove the lemma, we observe that the mapping T, being continuous, is uniformly continuous. Therefore, there exists a number $\delta > 0$, such that, for $|s_1 - s_2| < \delta$, $T(s_1)$ and $T(s_2)$ lie in the same open half-plane. From our conditions on $\bar{\tau}(s)$, it follows that, if $\bar{\tau}(s_1)$ is known, $\bar{\tau}(s_2)$ is completely determined. We divide the interval $0 \leqq s \leqq L$ by the points $s_0 \,(= 0) < s_1 < \cdots < s_m \,(= L)$ such that $|s_i - s_{i-1}| < \delta$, $i = 1, \cdots, m$. To define $\bar{\tau}(s)$, we assign to $\bar{\tau}(s_0)$ the value $\tau(s_0)$. Then it is determined in the subinterval $s_0 \leqq s \leqq s_1$, in particular at s_1, which determines it in the second subinterval, etc. The function $\bar{\tau}(s)$ so defined clearly satisfies the conditions of the lemma.

The difference $\bar{\tau}(L) - \bar{\tau}(0)$ is an integral multiple of 2π, say, $= \gamma 2\pi$. We assert that the integer γ is independent of the choice of the function $\bar{\tau}(s)$. In fact, let $\bar{\tau}'(s)$ be a function satisfying the same conditions. Then we have

$$\bar{\tau}'(s) - \bar{\tau}(s) = n(s) \cdot 2\pi,$$

where $n(s)$ is an integer. Since $n(s)$ is continuous in s, it must be a constant. It follows that

$$\bar{\tau}'(L) - \bar{\tau}'(0) = \bar{\tau}(L) - \bar{\tau}(0),$$

which proves the independence of γ from the choice of $\bar{\tau}(s)$. We define γ to be the rotation index of C. The *theorem of turning tangents* follows.

THEOREM: *The rotation index of a simple closed curve is* ± 1.

Proof: To prove this theorem, we consider the mapping Σ which carries an ordered pair of points of C, $X(s_1)$, $X(s_2)$, $0 \leqq s_1 \leqq s_2 \leqq L$, into the endpoint of the unit vector through O parallel to the secant joining $X(s_1)$ to $X(s_2)$. These ordered pairs of points can be represented as a triangle Δ in the (s_1, s_2)-plane defined by $0 \leqq s_1 \leqq s_2 \leqq L$. The mapping Σ of Δ into Γ is continuous. We also observe that its restriction to the side $s_1 = s_2$ is the tangential mapping T.

To a point $p \in \Delta$, let $\tau(p)$ be the angle which Ox makes with $O\Sigma(p)$, such that $0 \leqq \tau(p) < 2\pi$. Again this function need not be continuous. We shall, however, prove that there exists a continuous function $\bar{\tau}(p)$, $p \in \Delta$, such that $\bar{\tau}(p) \equiv \tau(p)$ mod 2π.

In fact, let m be an interior point of Δ. We cover Δ by the radii through m. By the arguments used in the proof of the preceding lemma, we can define a function $\bar{\tau}(p)$, $p \in \Delta$, such that $\bar{\tau}(p) \equiv \tau(p)$, mod 2π, and such that it is continuous along every radius through m. It remains to prove that it is continuous in Δ. For this purpose, let p_0 be a point of Δ. Since Σ is continuous, it follows from the compactness of the segment mp_0 that there exists a number $\eta = \eta(p_0) > 0$, such that, for $q_0 \in mp_0$, and for any point of $q \in \Delta$ for which the distance $d(q, q_0) < \eta$, the points $\Sigma(q)$ and $\Sigma(q_0)$ are never antipodal. The latter condition is equivalent to the relation

$$(2) \qquad \bar{\tau}(q) - \bar{\tau}(q_0) \not\equiv 0, \quad \text{mod } \pi.$$

Now let $\epsilon > 0$, $\epsilon < \pi/2$, be given. We choose a neighborhood U of p_0, such that U is contained in the η-neighborhood of p_0, and such that, for $p \in U$, the angle between $O\Sigma(p_0)$ and $O\Sigma(p)$ is less than ϵ. This is possible, because the mapping Σ is continuous. The last condition can be expressed in the form

$$(3) \qquad \bar{\tau}(p) - \bar{\tau}(p_0) = \epsilon' + 2k(p)\pi, \quad |\epsilon'| < \epsilon,$$

where $k(p)$ is an integer. Let q_0 be any point on the segment mp_0. Draw the segment q_0q parallel to p_0p, with q on mp. The function $\bar{\tau}(q) - \bar{\tau}(q_0)$ is continuous in q along mp and is zero when q coincides with m. Since $d(q, q_0)$ is less than η, it follows from Equation (2) that $|\bar{\tau}(q) - \bar{\tau}(q_0)| < \pi$. In particular, for $\dot{q}_0 = p_0$, $|\bar{\tau}(p) - \bar{\tau}(p_0)| < \pi$. Combining this result with Equation (3), we get $k(p) = 0$, which proves that $\bar{\tau}(p)$ is continuous in Δ. Since $\bar{\tau}(p) \equiv \tau(p)$, mod 2π, it is easy to see that $\bar{\tau}(p)$ is differentiable.

Now let $A(0, 0)$, $B(0, L)$, and $D(L, L)$ be the vertices of Δ. The rotation index γ of C is defined by the line integral

$$2\pi\gamma = \int_{AD} d\bar{\tau}.$$

Since $\bar{\tau}(p)$ is defined in Δ, we have

$$\int_{AD} d\bar{\tau} = \int_{AB} d\bar{\tau} + \int_{BD} d\bar{\tau}.$$

To evaluate the line integrals on the right-hand side, we make use of a suitable coordinate system. We can suppose $X(0)$ to be the "lowest" point of C—that is, the point when the vertical coordinate is a minimum, and we choose $X(0)$ to be the origin O. The tangent vector to C at $X(0)$ is horizontal, and we call it Ox. The curve C then lies in the upper half-plane bounded by Ox, and the line integral $\int_{AB} d\bar{\tau}$ is equal to the angle rotated by OP as P traverses once along C. Since OP never points downward, this angle is $\epsilon\pi$, with $\epsilon = \pm 1$. Similarly, the integral $\int_{BD} d\bar{\tau}$ is the angle rotated by PO as P goes once along C. Its value is also equal to $\epsilon\pi$. Hence, the sum of the two integrals is $\epsilon 2\pi$ and the rotation index of C is ± 1, which completes our proof.

We can also define the rotation index by an integral formula. In fact, using the function $\bar{\tau}(s)$ in our lemma, we can express the components of the unit tangent and normal vectors as follows:

$$e_1 = (\cos \bar{\tau}(s), \sin \bar{\tau}(s)), \qquad e_2 = (-\sin \bar{\tau}(s), \cos \bar{\tau}(s)).$$

It follows that

$$d\bar{\tau}(s) = de_1 \cdot e_2 = k\,ds.$$

From this equation, we derive the following formula for the rotation index:

$$(4) \qquad\qquad 2\pi\gamma = \int_C k\,ds.$$

This formula holds for closed curves which are not necessarily simple.

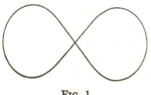

The accompanying figure gives an example of a closed curve with rotation index zero.

Many interesting theorems in differential geometry are valid for a more general class of curves, the so-called *sectionally smooth curves*. Such a curve is the union of a finite number of smooth arcs A_0A_1, A_1A_2, $\cdots$, $A_{m-1}A_m$, where the tangents of the two arcs through a common vertex A_i, $i = 1, \cdots, m - 1$, may be different. The curve is called *closed*, if $A_0 = A_m$. The simplest example of a closed sectionally smooth curve is a rectilinear polygon.

The notion of rotation index and the theorem of turning tangents can be extended to closed sectionally smooth curves; we summarize, without proof, the result as follows. Let s_i, $i = 1, \cdots, m$, be the arc length measured from A_0 to A_i, so that $s_m = L$ is the length of the curve. The curve supposedly being oriented, the tangential mapping is defined at all points different from A_i. At a vertex A_i there are two unit vectors, tangent respectively to $A_{i-1}A_i$ and A_iA_{i+1}. (We define $A_{m+1} = A_1$.) The corresponding points on Γ we denote by $T(A_i)^-$ and $T(A_i)^+$. Let φ_i be the angle from $T(A_i)^-$ to $T(A_i)^+$, with $0 < \varphi_i < \pi$, briefly the exterior angle from the tangent to $A_{i-1}A_i$ to the tangent to A_iA_{i+1}. For each arc $A_{i-1}A_i$, a continuous function $\bar\tau(s)$ can be defined which is one of the determinations of the angle from Ox to the tangent at $X(s)$. The number γ defined by the equation

$$(5) \qquad 2\pi\gamma = \sum_{i=1}^{m} \{\bar\tau(s_i) - \bar\tau(s_{i-1})\} + \sum_{i=1}^{m} \varphi_i$$

is an integer, which will be called the *rotation index* of the curve. The theorem of turning tangents is again valid.

THEOREM. *If a sectionally smooth curve is simple, the rotation index is equal to ± 1.*

As an application of the theorem of turning tangents, we wish to give the following characterization of a simple closed convex curve.

REMARK: *A simple closed curve is convex, if and only if it can be so oriented that its curvature is greater than, or equal to, 0.*

FIG. 2

Let us first remark that the theorem is not true without the assumption that the curve is simple. In fact, the accompanying figure gives a nonconvex curve with $k > 0$.

Proof: To prove the theorem, we let $\bar\tau(s)$ be the function constructed, so that we have $k = d\bar\tau/ds$. The condition $k \geqq 0$ is equivalent to the assertion that $\bar\tau(s)$ is a monotone nondecreasing function. Because C is simple, we can suppose that $\bar{\tau}(s)$, $0 \leqq s \leqq L$, increases from 0 to 2π. It follows that if the tangents at $X(s_1)$ and $X(s_2)$, $0 \leqq s_1 <$ $s_2 < L$, are parallel in the same sense, the arc of C from $X(s_1)$ to $X(s_2)$ is a straight line segment and these tangents must coincide.

Suppose $\bar{\tau}(s)$, $0 \leqq s \leqq L$, is monotone nondecreasing and C is not convex. There is a point $A = X(s_0)$ on C such that there are points of C at both sides of the tangent t to C at A. Choose a positive side of t and consider the oriented perpendicular distance from a point $X(s)$ of C to t. This is a continuous function in s and attains a maximum and a minimum at the points M and N of C, respectively. Clearly M and N are not on t and the tangents to C at M and N are parallel to t. Among these two tangents and t itself, there are two tangents parallel in the same sense, which, according to the preceding remark, is impossible.

Next we let C be convex. To prove that $\bar{\tau}(s)$ is monotone, we suppose $\bar{\tau}(s_1) = \bar{\tau}(s_2)$, $s_1 < s_2$. Then the tangents at $X(s_1)$ and $X(s_2)$ are parallel in the same sense. But there exists a tangent parallel to them in the opposite sense. From the convexity of C it follows that two of them coincide.

We are thus led to the consideration of a line t tangent to C at two distinct points, A and B. We claim that the segment AB must be a part of C. In fact, suppose this is not the case and let D be a point of AB not on C. Draw through D a perpendicular u to t in the half-plane which contains C. Then u intersects C in at least two points. Among these points of intersection, let F be the farthest from t and G the nearest, so that $F \neq G$. Then G is an interior point of the triangle ABF. The tangent to C at G must have points of C in both sides, which contradicts the convexity of C.

It follows that, under the hypothesis of the last paragraph, the segment AB is a part of C and that the tangents at A and B are parallel in the same sense. This proves that the segment joining $X(s_1)$ to $X(s_2)$ belongs to C. The latter implies that $\bar{\tau}(s)$ remains constant in the interval $s_1 \leqq s \leqq s_2$. Hence, the function $\bar{\tau}(s)$ is monotone, and our theorem is proved.

The first half of the theorem can also be stated as follows.

REMARK: *A closed curve with $k(s) \geqq 0$ and rotation index equal to 1 is convex.*

The theorem of turning tangents was essentially known to Riemann. The above proof was given by H. Hopf, *Compositio Mathematica* 2 (1935), pp. 50–62. For further reading, see:

1. H. Whitney, "On regular closed curves in the plane," *Compositio Mathematica* 4 (1937), pp. 276–84.

2. S. Smale, "Regular curves on a Riemannian manifold," *Transactions of the American Mathematical Society* 87 (1958), pp. 492–511.

3. S. Smale, "A classification of immersions of the two-sphere," *Transactions of the American Mathematical Society* 90 (1959), pp. 281–90.

2. THE FOUR-VERTEX THEOREM

An interesting theorem on closed plane curves is the so-called "four-vertex theorem." By a *vertex* of an oriented closed plane curve we mean a point at which the curvature has a relative extremum. Since the curve forms a compact point set, it has at least two vertices, corresponding respectively to the absolute minimum and maximum of the curvature. Our theorem says that there are at least four.

THEOREM: *A simple closed convex curve has at least four vertices.*

This theorem was first presented by Mukhopadhyaya (1909); the proof we shall give was the work of G. Herglotz. It is also true for nonconvex curves, but the proof is more difficult. The theorem cannot be improved, because an ellipse with unequal axes has exactly four vertices, which are its points of intersection with the axes.

Proof: We suppose that the curve C has only two vertices, M and N, and we shall show that this leads to a contradiction. The line MN does not meet C in any other point, for if it does, the tangent line to C at the middle point must contain the other two points. By the last section, this condition is possible only when the segment MN is a part of C. It would follow that the curvature vanishes at M and N, which is not possible, since they are the points where the curvature takes the absolute maximum and minimum respectively.

We denote by 0 and s_0 the parameters of M and N respectively and take MN to be the x_1-axis. Then we can suppose

$$x_2(s) < 0, \qquad 0 < s < s_0,$$
$$x_2(s) > 0, \qquad s_0 < s < L,$$

where L is the length of C. Let $(x_1(s), x_2(s))$ be the position vector of a point of C with the parameter s. Then the unit tangent and normal vectors have the components

$$e_1 = (x_1', x_2'), \qquad e_2 = (-x_2', x_1'),$$

where primes denote differentiations with respect to s. From the Frenet formulas we get

(6) $$x_1'' = -kx_2', \qquad x_2'' = kx_1'.$$

It follows that

$$\int_0^L kx_2' \, ds = -x_1' \Big|_0^L = 0.$$

The integral in the left-hand side can be written as a sum:

$$\int_0^L kx_2' \, ds = \int_0^{s_0} kx_2' \, ds + \int_{s_0}^L kx_2' \, ds.$$

To each summand we apply the second mean value theorem, which is stated as follows. Let $f(x)$, $g(x)$, $a \leq x \leq b$, be two functions in x such that $f(x)$ and $g'(x)$ are continuous and $g(x)$ is monotone. Then there exists ξ, $a < \xi < b$, satisfying the equation,

$$\int_a^b f(x)g(x) \, dx = g(a) \int_a^\xi f(x) \, dx + g(b) \int_\xi^b f(x) \, dx.$$

Since $k(s)$ is monotone in each of the intervals $0 \leq s \leq s_0$, $s_0 \leq s \leq L$, we get

$$\int_0^{s_0} kx_2' \, ds = k(0) \int_0^{\xi_1} x_2' \, ds + k(s_0) \int_{\xi_1}^{s_0} x_2' \, ds$$
$$= x_2(\xi_1)(k(0) - k(s_0)), \qquad\qquad 0 < \xi_1 < s_0$$

$$\int_{s_0}^{L} kx_2' \, ds = k(s_0) \int_{s_0}^{\xi_2} x_2' \, ds + k(L) \int_{\xi_1}^{L} x_2' \, ds$$

$$= x_2(\xi_2)(k(s_0) - k(0)), \qquad\qquad s_0 < \xi_2 < L.$$

Since the sum of the left-hand members is zero, these equations give

$$(x_2(\xi_1) - x_2(\xi_2))(k(0) - k(s_0)) = 0,$$

which is a contradiction, because

$$x_2(\xi_1) - x_2(\xi_2) < 0, \qquad k(0) - k(s_0) > 0.$$

It follows that there is at least one more vertex on C. Since the relative extrema occur in pairs, there are at least four vertices and the theorem is proved.

At a vertex we have $k' = 0$. Hence, we can also say that on a simple closed convex curve there are at least four points at which $k' = 0$.

The four-vertex theorem is also true for simple closed nonconvex plane curves; see:

1. S. B. Jackson, "Vertices for plane curves," *Bulletin of the American Mathematical Society* 50 (1944), pp. 564–578.

2. L. Vietoris, "Ein einfacher Beweis des Vierscheitelsatzes der ebenen Kurven," *Archiv der Mathematik* 3 (1952), pp. 304–306.

For further reading, see:

1. P. Scherk, "The four-vertex theorem," *Proceedings of the First Canadian Mathematical Congress.* Montreal: 1945, pp. 97–102.

3. ISOPERIMETRIC INEQUALITY
FOR PLANE CURVES

The theorem can be stated as follows.

THEOREM: *Among all simple closed curves having a given length the circle bounds the largest area. In other words, if L is the length of a simple closed curve C, and A is the area it bounds, then*

$$(7) \qquad\qquad L^2 - 4\pi A \geq 0.$$

Moreover, the equality sign holds only when C is a circle.

Many proofs have been given of this theorem, differing in degree of elegance and in the range of curves under consideration—that

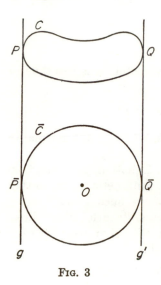

FIG. 3

is, whether differentiability or convexity is supposed. We shall give two proofs, the work of E. Schmidt (1939) and A. Hurwitz (1902), respectively.

Schmidt's Proof: We enclose C between two parallel lines, g and g', such that C lies between g and g' and is tangent to them at the points P and Q, respectively. We let $s = 0$, s_0 being the parameters of P and Q, and construct a circle $\overline{C}$ tangent to g and g' at $\overline{P}$ and $\overline{Q}$, respectively. Denote its radius by r and take its center to be the origin of a coordinate system. Let $X(s) = (x_1(s), x_2(s))$ be the position vector of C, so that $(x_1(0), x_2(0)) = (x_1(L), x_2(L))$. As the position vector of $\overline{C}$ we take $(\overline{x}_1(s), x_2(s))$, such that

$$\overline{x}_1(s) = x_1(s),$$

(8) $$\overline{x}_2(s) = -\sqrt{r^2 - x_1^2(s)}, \quad 0 \leqq s \leqq s_0$$

$$= +\sqrt{r^2 - x_1^2(s)}, \quad s_0 \leqq s \leqq L.$$

Denote by $\overline{A}$ the area bounded by $\overline{C}$. Now the area bounded by a closed curve can be expressed by the line integral

$$A = \int_0^L x_1 x_2' \, ds = -\int_0^L x_2 x_1' \, ds = \tfrac{1}{2} \int_0^L (x_1 x_2' - x_2 x_1') \, ds.$$

Applying this to our two curves C and $\overline{C}$, we get

$$A = \int_0^L x_1 x_2' \, ds$$

$$\overline{A} = \pi r^2 = -\int_0^L \overline{x}_2 \overline{x}_1' \, ds = -\int_0^L \overline{x}_2 x_1' \, ds.$$

Adding these two equations, we have

$$A + \pi r^2 = \int_0^L (x_1 x_2' - \bar{x}_2 x_1') \, ds \leqq \int_0^L \sqrt{(x_1 x_2' - \bar{x}_2 x_1')^2} \, ds$$

$$(9) \qquad\qquad \leqq \int_0^L \sqrt{(x_1^2 + \bar{x}_2^2)(x_1'^2 + x_2'^2)} \, ds$$

$$= \int_0^L \sqrt{x_1^2 + \bar{x}_2^2} \, ds = Lr.$$

Since the geometric mean of two positive numbers is less than or equal to their arithmetic mean, it follows that

$$\sqrt{A} \, \sqrt{\pi r^2} \leqq \tfrac{1}{2}(A + \pi r^2) \leqq \tfrac{1}{2} Lr,$$

which gives, after squaring and cancellation of r^2, the inequality in Equation (7).

Suppose now that the equality sign in Equation (7) holds; then A and πr^2 have the same geometric and arithmetic mean, so that $A = \pi r^2$ and $L = 2\pi r$. The direction of the lines g and g' being arbitrary, this means that C has the same "width" in all directions. Moreover, we must have the equality sign everywhere in Equation (9). It follows, in particular, that

$$(x_1 x_2' - \bar{x}_2 x_1')^2 = (x_1^2 + \bar{x}_2^2)(x_1'^2 + x_2''^2),$$

which gives

$$\frac{x_1}{x_2'} = \frac{-\bar{x}_2}{x_1'} = \frac{\sqrt{x_1^2 + \bar{x}_2^2}}{\sqrt{x_1'^2 + x_2'^2}} = \pm r.$$

From the first equality in Equation (9), the factor of proportionality is seen to be r, that is,

$$x_1 = r x_2', \qquad \bar{x}_2 = -r x_1',$$

which remains true when we interchange x_1 and x_2, so that

$$x_2 = r x_1'.$$

Therefore, we have

$$x_1^2 + x_2^2 = r^2,$$

which means that C is a circle.

Hurwitz's proof makes use of the theory of Fourier series. We shall first prove the lemma of Wirtinger.

LEMMA: *Let $f(t)$ be a continuous periodic function of period 2π, possessing a continuous derivative $f'(t)$. If $\int_0^{2\pi} f(t)\, dt = 0$, then*

(10)
$$\int_0^{2\pi} f'(t)^2\, dt \geqq \int_0^{2\pi} f(t)^2\, dt.$$

Moreover, the equality sign holds if and only if

(11)
$$f(t) = a \cos t + b \sin t.$$

Proof: To prove the lemma, we let the Fourier series expansion of $f(t)$ be

$$f(t) \sim \frac{a_0}{2} + \sum_{n=1}^{\infty} (a_n \cos nt + b_n \sin nt).$$

Since $f'(t)$ is continuous, its Fourier series can be obtained by differentiation term by term, and we have

$$f'(t) \sim \sum_{n=1}^{\infty} (nb_n \cos nt - na_n \sin nt).$$

Since

$$\int_0^{2\pi} f(t)\, dt = \pi a_0,$$

it follows from our hypothesis that $a_0 = 0$. By Parseval's formula, we get

$$\int_0^{2\pi} f(t)^2\, dt = \sum_{n=1}^{\infty} (a_n^2 + b_n^2),$$

$$\int_0^{2\pi} f'(t)^2\, dt = \sum_{n=1}^{\infty} n^2(a_n^2 + b_n^2).$$

Hence,

$$\int_0^{2\pi} f'(t)^2\, dt - \int_0^{2\pi} f(t)^2\, dt = \sum_{n=1}^{\infty} (n^2 - 1)(a_n^2 + b_n^2),$$

which is greater than, or equal to, 0. It is equal to zero, only if $a_n = b_n = 0$ for all $n > 1$. Therefore, $f(t) = a_1 \cos t + b_1 \sin t$, which proves the lemma.

Hurwitz's Proof: In order to prove the inequality in Equation (7), we assume, for simplicity, that $L = 2\pi$, and that

$$\int_0^{2\pi} x_1(s)\, ds = 0.$$

The latter means that the center of gravity lies on the x_1-axis, a condition which can always be achieved by a proper choice of the coordinate system. The length and the area are given by the integrals,

$$2\pi = \int_0^{2\pi} (x_1'^2 + x_2'^2)\, ds, \quad \text{and} \quad A = \int_0^{2\pi} x_1 x_2'\, ds.$$

From these two equations we get

$$2(\pi - A) = \int_0^{2\pi} (x_1'^2 - x_1^2)\, ds + \int_0^{2\pi} (x_1 - x_2')^2\, ds.$$

The first integral is greater than, or equal to, 0 by our lemma and the second integral is clearly greater than, or equal to, 0. Hence, $A \leqq \pi$, which is our isoperimetric inequality.

The equality sign holds only when

$$x_1 = a \cos s + b \sin s, \qquad x_2' = x_1,$$

which gives

$$x_1 = a \cos s + b \sin s, \qquad x_2 = a \sin s - b \cos s + c.$$

Thus, C is a circle.

For further reading, see:

1. E. Schmidt, "Beweis der isoperimetrischen Eigenschaft der Kugel im hyperbolischen und sphärischen Raum jeder Dimensionenzahl," *Math. Zeit.* 49 (1943), pp. 1–109.

4. TOTAL CURVATURE OF A SPACE CURVE

The *total curvature* of a closed space curve C of length L is defined by the integral

$$(12) \qquad \mu = \int_0^L |k(s)|\, ds,$$

where $k(s)$ is the curvature. For a space curve, only $|k(s)|$ is defined.

Suppose C is oriented. Through the origin O of our space we draw vectors of length 1 parallel to the tangent vectors of C. Their end-points describe a closed curve Γ on tne unit sphere, to be called the *tangent indicatrix* of C. A point of Γ is singular (that is, with either no tangent or a tangent of higher contact) if it is the image of a point of zero curvature of C. Clearly the total curvature of C is equal to the length of Γ.

Fenchel's theorem concerns the total curvature.

THEOREM: *The total curvature of a closed space curve C is greater than, or equal to, 2π. It is equal to 2π if and only if C is a plane convex curve.*

The following proof of this theorem was found independently by B. Segre (*Bolletino della Unione Matematica Italiana* 13 (1934), 279–283), and by H. Rutishauser and H. Samelson (*Comptes Rendus Hebdomadaires des Séances de l'Académie des Sciences* 227 (1948), 755–757). See also W. Fenchel, *Bulletin of the American Mathematical Society* 57 (1951), 44–54. The proof depends on the following lemma:

LEMMA: *Let Γ be a closed rectifiable curve on the unit sphere, with length $L < 2\pi$. There exists a point m on the sphere such that the spherical distance $\overline{mx} \leqq L/4$ for all points x of Γ. If Γ is of length 2π but is not the union of two great semicircular arcs, there exists a point m such that $\overline{mx} < \pi/2$ for all x of Γ.*

We use the notation $\overline{ab}$ to denote the spherical distance of two points, a and b. If $\overline{ab} < \pi$, their *midpoint* m is the point defined by the conditions $\overline{am} = \overline{bm} = \frac{1}{2}\overline{ab}$. Let x be a point such that $\overline{mx} \leqq \frac{1}{2}\pi$. Then $2\overline{mx} \leqq \overline{ax} + \overline{bx}$. In fact, let x' be the symmetry of x relative to m. Then,

$$\overline{x'a} = \overline{xb}, \qquad \overline{x'x} = \overline{x'm} + \overline{mx} = 2\overline{mx}.$$

If we use the triangle inequality, it follows that

$$(13) \qquad 2\overline{mx} = \overline{x'x} \leqq \overline{x'a} + \overline{ax} = \overline{ax} + \overline{bx},$$

as to be proved.

Lemma Proof: To prove the first part of the lemma, we take two points, a and b, on Γ which divide the curve into two equal arcs. Then $\overline{ab} < \pi$, and we denote the midpoint by m. Let x be a point of Γ such that $2\overline{mx} < \pi$. Such points exist—for example, the point a. Then we have

$$\overline{ax} \leqq \widehat{ax}, \qquad \overline{bx} \leqq \widehat{bx},$$

where $\widehat{ax}$ and $\widehat{bx}$ are respectively the arc lengths along Γ. From Equation (13), it follows that

$$2\overline{mx} \leqq \widehat{ax} + \widehat{bx} = \widehat{ab} = \frac{L}{2}.$$

Hence, the function $f(x) = \overline{mx}$, $x \in \Gamma$, is either $\geqq \pi/2$ or $\leqq L/4 < \pi/2$. Since Γ is connected and $f(x)$ is a continuous function in Γ, the range of the function $f(x)$ is connected in the interval $(0, \pi)$. Therefore, we have $f(x) = \overline{mx} \leqq L/4$.

Consider next the case that Γ is of length 2π. If Γ contains a pair of antipodal points, then, being of length 2π, it must be the union of two great semicircular arcs. Suppose that there is a pair of points, a and b, which bisect Γ such that

$$\overline{ax} + \overline{bx} < \pi$$

for all $x \in \Gamma$. Again, let m denote the midpoint of a and b. If $f(x) = \overline{mx} \leqq \frac{1}{2}\pi$, we have, from Equation (13),

$$2\overline{mx} \leqq \overline{ax} + \overline{bx} < \pi,$$

which means that $f(x)$ omits the value $\pi/2$. Since its range is connected and since $f(a) < \pi/2$, we have $f(x) < \pi/2$ for all $x \in \Gamma$. Thus the lemma is true in this case.

It remains to consider the case that Γ contains no pair of antipodal points, and that for any pair of points a and b which bisect Γ, there is a point $x \in \Gamma$ with

$$\overline{ax} + \overline{bx} = \pi.$$

An elementary geometrical argument, which we leave to the reader, will show that this is impossible. Thus, the lemma is proved.

Theorem Proof: To prove Fenchel's theorem, we take a fixed unit vector A and put

$$g(s) = AX(s),$$

where the right-hand side denotes the scalar product of the vectors A and $X(s)$. The function $g(s)$ is continuous on C and hence must have a maximum and a minimum. Since $g'(s)$ exists, we have, at such an extremum s_0,

$$g'(s_0) = AX'(s_0) = 0.$$

Thus A, as a point on the unit sphere, has a distance $\pi/2$ from at least two points of the tangent indicatrix. Since A is arbitrary, the tangent indicatrix is met by every great circle. It follows from the lemma that its length is greater than, or equal to, 2π.

Suppose next that the tangent indicatrix Γ is of length 2π. By our lemma, it must be the union of two great semicircular arcs. It follows that C itself is the union of two plane arcs. Since C has a tangent everywhere, it must be a plane curve. Suppose C be so oriented that its rotation index

$$\frac{1}{2\pi} \int_0^L k \, ds \geq 0.$$

Then we have

$$0 \leq \int_0^L \{|k| - k\} \, ds = 2\pi - \int_0^L k \, ds$$

so that the rotation index is either 0 or 1. To a given vector in the plane there is parallel to it a tangent t of C such that C lies to the left of t. Then t is parallel to the vector in the same sense, and at its point of contact we have $k \geq 0$, implying that $\int_{k>0} k \, ds \geq 2\pi$. Since $\int_C |k| \, ds = 2\pi$, there is no point with $k < 0$, and $\int k \, ds = 2\pi$. From the remark at the end of Section 1, we conclude that C is convex.

As a corollary we have the following theorem.

COROLLARY: *If $|k(s)| \leq 1/R$ for a closed space curve C, C has a length $L \geq 2\pi R$.*

We have

$$L = \int_0^L ds \geq \int_0^L R|k| \, ds = R \int_0^L |k| \, ds \geq 2\pi R.$$

Fenchel's theorem holds also for sectionally smooth curves. As the total curvature of such a curve we define

$$(14) \qquad\qquad \mu = \int_0^L |k| \, ds + \sum_i a_i$$

where the a_i are the angles at the vertices. In other words, in this case the tangent indicatrix consists of a number of arcs each corresponding to a smooth arc of C; we join successive vertices by the shortest great circular arc on the unit sphere. The length of the curve so obtained is the total curvature of C. It can be proved that for a closed sectionally smooth curve we have also $\mu \geqq 2\pi$.

We wish to give another proof of Fenchel's theorem and a related theorem of Fary-Milnor on the total curvature of a knot.† The basis is Crofton's theorem on the measure of great circles which cut an arc on the unit sphere. Every oriented great circle determines uniquely a "pole," the endpoint of the unit vector normal to the plane of the circle. By the measure of a set of great circles on the unit sphere is meant the area of the domain of their poles. Then Crofton's theorem is stated as follows.

THEOREM: *Let Γ be a smooth arc on the unit sphere Σ_0. The measure of the oriented great circles of Σ_0 which meet Γ, each counted a number of times equal to the number of its common points with Γ, is equal to four times the length of Γ.*

Proof: We suppose Γ is defined by a unit vector $e_1(s)$ expressed as a function of its arc length s. Locally (that is, in a certain neighborhood of s), let $e_2(s)$ and $e_3(s)$ be unit vectors depending smoothly on s, such that the scalar products

(15) $$e_i \cdot e_j = \delta_{ij}, \quad 1 \leqq i, j \leqq 3$$

and

(16) $$\det (e_1, e_2, e_3) = +1.$$

Then we have

(17)
$$
\begin{cases}
\dfrac{de_1}{ds} = a_2e_2 + a_3e_3, \\[2ex]
\dfrac{de_2}{ds} = -a_2e_1 + a_1e_3, \\[2ex]
\dfrac{de_3}{ds} = -a_3e_1 - a_1e_2.
\end{cases}
$$

---

† I. Fary (*Bulletin de la Société Mathématique de France*, 77 (1949), pp. 128–138), and J. Milnor (*Annals of Mathematics*, 52 (1950), pp. 248–257).

The skew-symmetry of the matrix of the coefficients in the above system of equations follows from differentiation of Equations (15). Since s is the arc length of Γ, we have

$$(18) \qquad a_2^2 + a_3^2 = 1,$$

and we put

$$(19) \qquad a_2 = \cos \tau(s), \qquad a_3 = \sin \tau(s).$$

If an oriented great circle meets Γ at the point $e_1(s)$, its pole is of the form $Y = \cos \theta \, e_2(s) + \sin \theta \cdot e_3(s)$, and vice versa. Thus (s, θ) serve as local coordinates in the domain of these poles; we wish to find an expression for the element of area of this domain.

For this purpose, we write

$$dY = (-\sin \theta \, e_2 + \cos \theta \, e_3)(d\theta + a_1 \, ds) - e_1(a_2 \cos \theta + a_3 \sin \theta) \, ds.$$

Since $-\sin \theta \, e_2 + \cos \theta \, e_3$ and e_1 are two unit vectors orthogonal to Y, the element of area of Y is

$$(20) \quad |dA| = |a_2 \cos \theta + a_3 \sin \theta| \, d\theta \, ds = |\cos (\tau - \theta)| \, d\theta \, ds,$$

where the absolute value at the left-hand side means that the area is calculated in the measure-theoretic sense, with no regard to orientation. To the point Y let $Y^\perp$ be the oriented great circle with Y as its pole, and let $n(Y^\perp)$ be the (arithmetic) number of points common to $Y^\perp$ and Γ. Then the measure μ in our theorem is given by

$$\mu = \int n(Y^\perp)|dA| = \int_0^\lambda ds \int_0^{2\pi} |\cos (\tau - \theta)| \, d\theta,$$

where λ is the length of Γ. As θ ranges from 0 to 2π, the variation of $|\cos (\tau - \theta)|$, for a fixed s, is 4. Hence, we get $\mu = 4\lambda$, which proves Crofton's theorem.

By applying the theorem to each subarc and adding, we see that the theorem remains true when Γ is a sectionally smooth curve on the unit sphere. Actually, the theorem is true for any rectifiable arc on the sphere, but the proof is much longer.

For a closed space curve the tangent indicatrix of which fulfills the conditions of Crofton's theorem, Fenchel's theorem is an easy consequence. In fact, the proof of Fenchel's theorem shows us that the tangent indicatrix of a closed space curve meets every great circle in at least two points—that is, $n(Y^\perp) \geqq 2$. It follows that its length is

$$\lambda = \int |k|\, ds = \tfrac{1}{4} \int n(Y^\perp)|dA| \geqq 2\pi,$$

because the total area of the unit sphere is 4π.

Crofton's theorem also leads to the following theorem of Fary and Milnor, which gives a necessary condition on the total curvature of a knot.

THEOREM: *The total curvature of a knot is greater than, or equal to, 4π.*

Since $n(Y^\perp)$ is the number of relative maxima or minima of the "height function," $Y \cdot X(s)$, it is even. Suppose that the total curvature of a closed space curve C is $< 4\pi$. There exists $Y \in \Sigma_0$, such that $n(Y^\perp) = 2$. By a rotation, suppose Y is the point $(0, 0, 1)$. Then the function $x_3(s)$ has only one maximum and one minimum. These points divide C into two arcs, such that x_3 increases on the one and decreases on the other. Every horizontal plane between the two extremal horizontal planes meets C in exactly two points. If we join them by a segment, all these segments will form a surface which is homeomorphic to a circular disk, which proves that C is not knotted.

For further reading, see:

1. S. S. Chern and R. K. Lashof, "On the total curvature of immersed manifolds," I, *American Journal of Mathematics* 79 (1957), pp. 302–18, and II, *Michigan Mathematical Journal* 5 (1958), pp. 5–12.

2. N. H. Kuiper, "Convex immersions of closed surfaces in E^5," *Comm. Math. Helv.* 35 (1961), pp. 85–92.

On integral geometry compare the article of Santalo in this volume.

5. DEFORMATION OF A SPACE CURVE

It is well-known that a one-one correspondence between two curves under which the arc lengths, the curvatures (when not equal to 0), and the torsions are respectively equal, can only be established by a proper motion. It is natural to study the correspondences under which only s and k are equal. We shall call such a correspondence a deformation of the space curve (in German, *Verwindung*). The most notable result in this direction is a theorem of A. Schur, which formulates the geometrical fact that if an arc is "stretched," the distance between its endpoints becomes longer. Using the name curvature to mean here always its absolute value, we state Schur's theorem as follows.

THEOREM: *Let C be a plane arc with the curvature $k(s)$ which forms a convex curve with its chord, AB. Let C^* be an arc of the same length referred to the same parameter s such that its curvature $k^*(s) \leq k(s)$. If d^* and d denote the lengths of the chords joining their endpoints, then $d \leq d^*$. Moreover, the equality sign holds when and only when C and C^* are congruent.*

Proof: Let Γ and Γ^* be the tangent indicatrices of C and C^* respectively, P_1 and P_2 two points on Γ, and P_1^* and P_2^* their corresponding points on Γ^*. We denote by $\widehat{P_1P_2}$ and $\widehat{P_1^*P_2^*}$ their arc lengths and by $\overline{P_1P_2}$ and $\overline{P_1^*P_2^*}$ their spherical distances. Then we have

$$\overline{P_1P_2} \leq \widehat{P_1P_2}, \qquad \overline{P_1^*P_2^*} \leq \widehat{P_1^*P_2^*}.$$

The inequality on the curvature implies

$$(21) \qquad \widehat{P_1^*P_2^*} \leq \widehat{P_1P_2}.$$

Since C is convex, Γ lies on a great circle, and we have

$$\overline{P_1P_2} = \widehat{P_1P_2},$$

provided that $\overline{P_1P_2} \leq \pi$. Now let Q be a point on C at which the tangent is parallel to the chord. Denote by P_0 its image point on Γ. Then the condition $\overline{P_0P} \leq \pi$ is satisfied by any point P on Γ, and if P_0^* denotes the point on Γ^* corresponding to P_0, we have

$$(22) \qquad \overline{P_0^*P^*} \leq \overline{P_0P},$$

from which it follows that

$$(23) \qquad \cos \overline{P_0^*P^*} \geq \cos \overline{P_0P},$$

since the cosine function is a monotone decreasing function of its argument when the latter lies between 0 and π.

Because C is convex, d is equal to the projection of C on its chord:

$$(24) \qquad\qquad d = \int_0^L \cos \overline{P_0 P} \, ds.$$

On the other hand, we have

$$(25) \qquad\qquad d^* \geqq \int_0^L \cos \overline{P_0^* P^*} \, ds,$$

for the integral on the right-hand side is equal to the projection of C^*, and hence of the chord joining its endpoints, on the tangent at the point Q^* corresponding to Q. Combining Equations (23), (24), and (25), we get $d^* \geqq d$.

Suppose that $d = d^*$. Then the inequalities in Equations (22), (23), and (25) become equalities, and the chord joining the endpoints A^* and B^* of C^* must be parallel to the tangent at Q^*. In particular, we have

$$\overline{P_0^* P^*} = \overline{P_0 P},$$

which implies that the arcs $A^* Q^*$ and $B^* Q^*$ are plane arcs. On the other hand, we have, by using Equation (21),

$$\overline{P_0^* P^*} \leqq \widehat{P_0^* P^*} \leqq \widehat{P_0 P} = \overline{P_0 P},$$

or

$$\widehat{P_0^* P^*} = \widehat{P_0 P}.$$

Hence, the arcs $A^* Q^*$ and $B^* Q^*$ have the same curvature as AQ and BQ at corresponding points and are therefore respectively congruent.

It remains to prove that the arcs $A^* Q^*$ and $B^* Q^*$ lie in the same plane. Suppose the contrary. They must be tangent at Q^* to the line of intersection of the two distinct planes on which they lie. Since this line is parallel to $A^* B^*$, the only possibility is that it contains A^* and B^*; however, then the tangent to C at Q must also contain the endpoints A and B, which is a contradiction. Hence, C^* is a plane arc and is congruent to C.

Schur's theorem has many applications. For example, it gives a solution of the following minimum problem: Determine the shortest closed curve with a curvature $k(s) \leq 1/R$, R being a constant. The answer is, of course, a circle.

REMARK: *The shortest closed curve with curvature $k(s) \leq 1/R$, R being a constant, is a circle of radius R.*

By the corollary to Fenchel's theorem, such a curve has length $2\pi R$. Comparing it with a circle of radius R, we conclude from Schur's theorem (with $d^* = d = 0$) that it must itself be a circle.

As a second application of Schur's theorem, we shall derive a theorem of Schwarz. It is concerned with the lengths of arcs joining two given points having a curvature bounded from the above by a fixed constant. The statement of Schwarz's theorem is as follows:

THEOREM: *Let C be an arc joining two given points A and B, with curvature $k(s) \leq 1/R$, such that $R \geq \frac{1}{2}d$, where $d = \overline{AB}$. Let S be a circle of radius R through A and B. Then the length of C is either less than, or equal to, the shorter arc AB or greater than, or equal to, the longer arc AB on S.*

Proof: We remark that the assumption $R \geq \frac{1}{2}d$ is necessary for the circle S to exist. To prove the theorem, we can assume that the length L of C is less than $2\pi R$; otherwise, there is nothing to prove. We then compare C with an arc of the same length on S having a chord of length d'. The conditions of Schur's theorem are satisfied and we get $d' \leq d$, d being the distance between A and B. Hence, L is either greater than, or equal to, the longer arc of S with the chord AB, or less than, or equal to, the shorter arc of S with the chord AB.

In particular, we can consider arcs joining A and B with curvature of $1/R$, $R \geq d/2$. The lengths of such arcs have no upper bound, as shown by the example of a helix. They have d as a lower bound, but can be as close to d as possible. Therefore, we have an example of a minimum problem which has no solution.

Finally, we remark that Schur's theorem can be generalized to sectionally smooth curves. We give here a statement of this generalization without proof.

REMARK: *Let C and C^* be two sectionally smooth curves of the same length, such that C forms a simple convex plane curve with its chord. Referred to the arc length s from one endpoint as parameter, let $k(s)$ be the curvature of C at a regular point and $a(s)$ the angle between the oriented tangents at a vertex; denote corresponding quantities for C^* by the same notations with asterisks. Let d and d^* be the distances between the endpoints of C and C^*, respectively. Then, if*

$$k^*(s) \leqq k(s) \quad \text{and} \quad a^*(s) \leqq a(s),$$

we have $d^* \geqq d$. The equality sign holds if and only if

$$k^*(s) = k(s) \quad \text{and} \quad a^*(s) = a(s).$$

The last set of conditions does not necessarily imply that C and C^* are congruent. In fact, there are simple rectilinear polygons in space which have equal sides and equal angles, but are not congruent.

6. THE GAUSS-BONNET FORMULA

We consider the intrinsic Riemannian geometry on a surface M. To simplify calculations and without loss of generality, we suppose the metric to be given in the isothermal parameters u and v:

$$(26) \qquad ds^2 = e^{2\lambda(u, v)}(du^2 + dv^2).$$

The element of area is then

$$(27) \qquad dA = e^{2\lambda} du\, dv$$

and the area of a domain D is given by the integral

$$(28) \qquad A = \iint_D e^{2\lambda} du\, dv.$$

Also, the Gaussian curvature of the surface is

$$(29) \qquad K = -e^{-2\lambda}(\lambda_{uu} + \lambda_{vv}).$$

It is well-known that the Riemannian metric defines the parallelism of Levi-Civita. To express it analytically, we write

$$(30) \qquad u^1 = u \quad \text{and} \quad u^2 = v$$

and

(31)
$$ds^2 = \Sigma \, g_{ij} \, du^i \, du^j.$$

In this last formula and throughout this paragraph, our small Latin indices will range from 1 to 2 and a summation sign will mean summation over all repeated indices. From g_{ij} we introduce the g^{ij}, according to the equation

(32)
$$\Sigma \, g_{ij} g^{jk} = \delta_i^k$$

and the Christoffel symbols

(33)
$$\left\{ \begin{array}{l} \Gamma_{ijk} = \dfrac{1}{2} \left(\dfrac{\partial g_{ij}}{\partial u^k} + \dfrac{\partial g_{jk}}{\partial u^i} - \dfrac{\partial g_{ik}}{\partial u^j} \right) \\[2mm] \Gamma_{lk}^j = \Sigma \, g^{jh} \Gamma_{ihk} \end{array} \right.$$

To a vector with the components ξ^i, the Levi-Civita parallelism defines the "covariant differential"

(34)
$$D\xi^i = d\xi^i + \Sigma \, \Gamma_{jk}^i \, du^k \, \xi^j.$$

All these equations are well-known in classical Riemannian geometry following the introduction of tensor analysis. The following is a new concept. Suppose the surface M is oriented. Consider the space B of all *unit* tangent vectors of M. This space B is a three-dimensional space, because the set of all unit tangent vectors with the same origin is one-dimensional. (It is called a *fiber space*, meaning that all the unit tangent vectors with origins in a neighborhood form a space which is topologically a product space.) To a unit tangent vector $\xi = (\xi^1, \xi^2)$, let $\eta = (\eta^1, \eta^2)$ be the uniquely determined unit tangent vector, orthogonal to ξ, such that ξ and η form a positive orientation. We introduce the linear differential form

(35)
$$\varphi = \sum_{1 \leq i,j \leq 2} g_{ij} D\xi^i \eta^j.$$

Then φ is well-defined in B and is usually called the *connection form*.

Because the vector ξ is a unit vector, we can write its components as follows:

(36)
$$\xi^1 = e^{-\lambda} \cos \theta \quad \text{and} \quad \xi^2 = e^{-\lambda} \sin \theta.$$

Then

(37) $\qquad \eta^1 = -e^{-\lambda} \sin \theta \quad \text{and} \quad \eta^2 = e^{-\lambda} \cos \theta.$

Routine calculation gives

(38) $\qquad \begin{aligned} \Gamma^1_{11} &= \Gamma^2_{12} = -\Gamma^1_{22} = \lambda_u, \\ \Gamma^1_{12} &= \Gamma^2_{22} = -\Gamma^2_{11} = \lambda_v, \end{aligned}$

whence the important relation

(39) $\qquad \varphi = d\theta - \lambda_v \, du + \lambda_u \, dv.$

Its exterior derivative is therefore

(40) $\qquad d\varphi = -K \, dA.$

Equation (40) is perhaps the most important formula in two-dimensional local Riemannian geometry.

The connection form φ is a differential form in B. We get from φ a differential form in a subset of M, when there is defined on it a field of unit tangent vectors. For example, let C be a smooth curve on M with the arc length s and let $\xi(s)$ be a smooth unit vector field along C. Then $\varphi = \sigma \, ds$, and σ is called the *variation* of ξ along C. The vectors ξ are said to be parallel along C, if $\sigma = 0$. If ξ is everywhere tangent to C, σ is called the geodesic curvature of C. C is a geodesic of M, if along C the unit tangent vectors are parallel, that is, if its geodesic curvature is 0.

Consider a domain D of M, such that there is a unit vector field defined over D, with an isolated singularity at an interior point $p_0 \in D$. Let γ_ϵ be a circle of geodesic radius ϵ about p_0. Then, from Equation (39), the limit

(41) $\qquad \dfrac{1}{2\pi} \lim_{\epsilon \to 0} \int_{\gamma_\epsilon} \varphi$

is an integer, to be called the *index* of the vector field at p_0.

Examples of vector fields with isolated singularities are shown in Figure 4. These singularities are, respectively, (a) a source or maximum, (b) a sink or minimum, (c) a center, (d) a simple saddle point, (e) a monkey saddle, and (f) a dipole. The indices are, respectively, 1, 1, 1, -1, -2, and 2.

The Gauss-Bonnet formula is the following theorem.

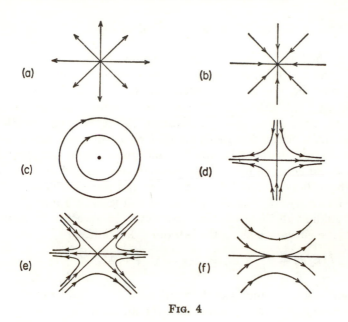

FIG. 4

THEOREM: *Let D be a compact oriented domain in M bounded by a sectionally smooth curve C. Then*

$$(42) \qquad \int_C k_g \, ds + \int_D K \, dA + \sum_i (\pi - \alpha_i) = 2\pi\chi,$$

where k_g is the geodesic curvature of C, $\pi - \alpha_i$ are the exterior angles at the vertices of C, and χ is the Euler characteristic of D.

Proof: Consider first the case that D belongs to a coordinate domain (u, v) and is bounded by a simple polygon C of n sides, C_i, $1 \leq i \leq n$, with the angles α_i at the vertices. Suppose D is positively oriented. To the points of the arcs C_i we associate the unit tangent vectors to C_i. Thus, to each vertex is associated two vectors at an angle $\pi - \alpha_i$. By the theorem of turning tangents (see Section 1), the total variation of θ as the C_i's are traversed once is $2\pi - \sum (\pi - \alpha_i)$. It follows that

$$\int_C k_g \, ds = 2\pi - \sum_i (\pi - \alpha_i) + \int_C -\lambda_v \, du + \lambda_u \, dv.$$

By Stokes theorem, the last integral is equal to $- \iint_D K \, dA$.

Thus, the formula is proved in this special case.

In the general case, suppose D is subdivided into a union of polygons D_λ, $\lambda = 1, \cdots, f$, such that (1) each D_λ lies in one coordinate neighborhood and (2) two D_λ have either no point, or one vertex, or a whole side, in common. Moreover, let the D_λ be coherently oriented with D, so that every interior side has different senses induced by the two polygons of which it is a side. Let v and e be the numbers of interior vertices and interior sides in this subdivision of D—i.e., vertices and sides which are not on the boundary, C. The above formula can then be applied to each D_λ. Adding all these relations, we have, because the integrals of geodesic curvature along the interior sides cancel,

$$\int_C k_g \, ds + \iint_D K \, dA = 2\pi f - \sum_{i,\lambda} (\pi - \alpha_{\lambda i}) - \sum_i (\pi - \alpha_i)$$

where α_i are the angles at the vertices of D, while the first sum in the right-hand side is extended over all interior vertices of the subdivision. Since each interior side is on exactly two D_λ and since the sum of interior angles about a vertex is 2π, this sum is equal to

$$-2\pi e + 2\pi v.$$

We call the integer

(43) $$\chi(D) = v - e + f$$

the Euler characteristic of D. Substituting, we get Equation (42). Equation (42) has the consequence that the integer χ is independent of the subdivision.

In particular, if C has no vertex, we have

(44) $$\int_C k_g \, ds + \iint_D K \, dA = 2\pi\chi.$$

Moreover, if D is the whole surface M, we get

(45) $$\iint_S K \, dA = 2\pi\chi.$$

It follows that if $K = 0$, the Euler characteristic of M is 0, and M is homeomorphic to a torus. If $K > 0$, then $\chi > 0$, and S is homeomorphic to a sphere.

The Euler characteristic plays an important role in the study of vector fields on a surface.

REMARK: *On a closed orientable surface M, the sum of the indices of a vector field with a finite number of singularities, is equal to the Euler characteristic, $\chi(M)$ of M.*

Proof: Let p_i, $1 \leqq i \leqq n$, be the singularities of the vector field. Let $\gamma_i(\epsilon)$ be a circle of radius ϵ about p_i, and let $\Delta_i(\epsilon)$ be the disk bounded by $\gamma_i(\epsilon)$. Integrating K of A over the domain $M - \bigcup_i \Delta_i(\epsilon)$ and using Equation (40), we get

$$\iint_{M - \bigcup_i \Delta_i(\epsilon)} K \, dA = \sum_i \int_{\gamma_i(\epsilon)} \varphi,$$

where $\gamma_i(\epsilon)$ is oriented so that it is the boundary of $\Delta_i(\epsilon)$. The theorem follows by letting $\epsilon \to 0$.

We wish to give two further applications of the Gauss-Bonnet formula. The first is a theorem of Jacobi. Let $X(s)$ be the coordinate vector of a closed space curve, with the arc length s. Let $T(s)$, $N(s)$, and $B(s)$ be the unit tangent, principal normal, and binormal vectors, respectively. In particular, the curve on the unit sphere with the coordinate vector $N(s)$ is the *principal normal indicatrix*. It has a tangent, wherever

$$(46) \qquad k^2 + w^2 \neq 0,$$

where k (when not equal to 0) and w are, respectively, the curvature and torsion of $X(s)$. Jacobi's theorem follows.

THEOREM: *If the principal normal indicatrix of a closed space curve has a tangent everywhere, it divides the unit sphere in two domains of the same area.*

Proof: To prove the theorem, we define τ by the equations

$$(47) \qquad k = \sqrt{k^2 + w^2} \cos \tau, \qquad w = \sqrt{k^2 + w^2} \sin \tau.$$

Then we have

$$d(-\cos \tau T + \sin \tau B) = (\sin \tau T + \cos \tau B) \, d\tau - \sqrt{k^2 + w^2} \, N \, ds.$$

Hence, if σ is the arc length of $N(s)$, $d\tau/d\sigma$ is the geodesic curvature of $N(s)$ on the unit sphere. Let D be one of the domains bounded by $N(s)$, and A its area. By the Gauss-Bonnet formula, we have, since $K = 1$,

$$\int_{N(s)} d\tau + \iint dA = 2\pi.$$

It follows that $A = 2\pi$, and the theorem is proved.

Our second application is Hadamard's theorem on convex surfaces.

THEOREM: *If the Gaussian curvature of a closed orientable surface in euclidean space is everywhere positive, the surface is convex (that is, it lies at one side of every tangent plane).*

We discussed a similar theorem for curves in Section 1. For surfaces, it is not necessary to suppose that it has no self-intersection.

Proof: It follows from the Gauss-Bonnet formula that the Euler characteristic $\chi(M)$ of the surface M is positive, so that $\chi(M) = 2$ and

$$\iint_M K \, dA = 4\pi.$$

Suppose M is oriented. We consider the Gauss mapping

(48) $g : M \rightarrow \Sigma_0$

(where Σ_0 is the unit sphere about a fixed point 0), which assigns to every point $p \in M$ the end of the unit vector through 0 parallel to the unit normal vector to M at p. The condition $K > 0$ implies that g has everywhere a nonzero functional determinant and is locally one-to-one. It follows that $g(M)$ is an open subset of Σ_0. Since M is compact, $g(M)$ is a compact subset of Σ_0, and hence is also closed. Therefore, g maps onto Σ_0.

Suppose that g is not one-to-one, that is, there exist points p and q of M, $p \neq q$, such that $g(p) = g(q)$. There is then a neighborhood U of q, such that $g(M - U) = \Sigma_0$. Since $\iint_{M-U} K \, dA$ is the area of $g(M - U)$, counted with multiplicities, we will have

$$\iint\limits_{M-U} K \, dA \geqq 4\pi.$$

But

$$\iint\limits_{U} K \, dA > 0,$$

so that

$$\iint\limits_{M} K \, dA = \iint\limits_{U} K \, dA + \iint\limits_{M-U} K \, dA > 4\pi,$$

which is a contradiction, and Hadamard's theorem is proved.

Hadamard's theorem is true under the weaker hypothesis $K \geqq 0$, but the proof is more difficult; see the article by Chern-Lashof mentioned in Section 4.

For further reading, see:

1. S. S. Chern, "On the curvatura integra in a Riemannian manifold," *Annals of Mathematics*, 46 (1945), pp. 674–84.

2. H. Flanders, "Development of an extended exterior differential calculus," *Transactions of the American Mathematical Society*, 75 (1953), pp. 311–26.

See also Section 8 of Flanders's article in this volume.

7. UNIQUENESS THEOREMS OF COHN-VOSSEN AND MINKOWSKI

The "rigidity" theorem of Cohn-Vossen can be stated as follows.

THEOREM: *An isometry between two closed convex surfaces is established either by a motion or by a motion and a reflection.*

In other words, such an isometry is always trivial, and the theorem is obviously not true locally. The following proof is the work of G. Herglotz.

Proof: We shall first discuss some notations on surface theory in euclidean space. Let the surface S be defined by expressing its position vector X as a function of two parameters, u and v. These functions are supposed to be continuously differentiable up to the second order. Suppose that X_u and X_v are everywhere linearly independent, and let ξ be the unit normal vector, so that S is oriented. As usual, let

$$(49) \quad \begin{aligned} \mathrm{I} &= dX \cdot dX = E \, du^2 + 2F \, du \, dv + G \, dv^2 \\ \mathrm{II} &= -dX \cdot d\xi = L \, du^2 + 2M \, du \, dv + N \, dv^2 \end{aligned}$$

be the first and second fundamental forms of the surface. Let H and K denote respectively the mean and Gaussian curvatures.

It is sufficient to prove that under the isometry, the second fundamental forms are equal. Assume the local coordinates are such that corresponding points have the same local coordinates. Then E, F, and G are equal for both surfaces, and the same is true of the Christoffel symbols. Let the second surface be S^*, and denote the quantities pertaining to S^* by the same symbols with asterisks. We introduce

$$(50) \qquad \lambda = \frac{L}{D}, \qquad \mu = \frac{M}{D}, \qquad \nu = \frac{N}{D},$$

where $D = \sqrt{EG - F^2}$. Then the Gaussian curvature is

$$(51) \qquad K = \lambda \nu - \mu^2 = \lambda^* \nu^* - \mu^{*2},$$

and is the same for both surfaces. The mean curvatures are

$$(52) \quad H = \frac{1}{2D} \left(G\lambda - 2F\mu + E\nu \right) \quad \text{and}$$

$$H^* = \frac{1}{2D} \left(G\lambda^* - 2F\mu^* + E\nu^* \right).$$

We introduce further

$$(53) \qquad J = \lambda \nu^* - 2\mu\mu^* + \nu\lambda^*.$$

The proof depends on the following identity:

$$(54) \qquad DJ\xi = \frac{\partial}{\partial u} \left(\nu^* X_u - \mu^* X_v \right) - \frac{\partial}{\partial v} \left(\mu^* X_u - \lambda^* X_v \right).$$

We first notice that the Codazzi equations can be written in terms of λ^*, μ^*, and ν^* in the form

$$\lambda_v^* - \mu_u^* + \Gamma_{22}^2\lambda^* - 2\Gamma_{12}^2\mu^* + \Gamma_{11}^2\nu^* = 0,$$

(55)

$$\mu_v^* - \nu_u^* - \Gamma_{22}^1\lambda^* + 2\Gamma_{12}^1\mu^* - \Gamma_{11}^1\nu^* = 0.$$

We next write the equations of Gauss:

$$X_{uu} - \Gamma_{11}^1 X_u - \Gamma_{11}^2 X_v - D\lambda\xi = 0,$$

(56) $$X_{uv} - \Gamma_{12}^1 X_u - \Gamma_{12}^2 X_v - D\mu\xi = 0,$$

$$X_{vv} - \Gamma_{22}^1 X_u - \Gamma_{22}^2 X_v - D\nu\xi = 0.$$

Multiplying these equations by X_v, $-X_u$, ν^*, $-2\mu^*$, and λ^*, respectively, and adding, we establish Equation (54).

We now write

(57) $$p = Xe_3, \qquad y_1 = XX_u, \qquad y_2 = XX_v,$$

where the right-hand sides are the scalar products of the vectors in question, so that $p(u, v)$ is the oriented distance from the origin to the tangent plane at $X(u, v)$. Equation (54) gives, after taking scalar product with X,

(58) $$DJp = -\nu^*E + 2\mu^*F - \lambda^*G$$
$$+ (\nu^*y_1 - \mu^*y_2)_u - (\mu^*y_1 - \lambda^*y_2)_v.$$

Let C be a closed curve on S. It divides S into two domains, D_1 and D_2, both having C as boundary. Moreover, if D_1 and D_2 are coherently oriented, C appears as a boundary in opposite senses. To each of these domains, say D_1, we apply Green's theorem, and get

(59) $$\iint\limits_{D_1} Jp \, dA = \iint\limits_{D_1} (-\nu^*E + 2\mu^*F - \lambda^*G) \, du \, dv$$
$$+ \int_C (+\mu^*y_1 - \lambda^*y_2) \, du + (\nu^*y_1 - \mu^*y_2) \, dv.$$

Adding this equation to a similar one for D_2, the line integrals cancel, and we have

$$\iint\limits_S Jp \, dA = \iint\limits_S (-\nu^*E + 2\mu^*F - \lambda^*G) \, du \, dv.$$

By Equation (52),

(60) $$\iint\limits_S Jp \, dA = -2 \iint\limits_S H^* \, dA.$$

In particular, this formula is valid when S and S^* are identical, and we have

(61) $$\iint\limits_{S} 2Kp \, dA = -2 \iint\limits_{S} H \, dA.$$

Subtracting these two equations, we get

(62) $$\iint\limits_{S} \begin{vmatrix} \lambda^* - \lambda & \mu^* - \mu \\ \mu^* - \mu & \nu^* - \nu \end{vmatrix} p \, dA = 2 \iint\limits_{S} H^* \, dA - 2 \iint\limits_{S} H \, dA.$$

To complete the proof, we need the following elementary lemma.

LEMMA: *Let*

(63) $ax^2 + 2bxy + cy^2$ *and* $a'x^2 + 2b'xy + c'y^2$

be two positive definite quadratic forms, with

(64) $ac - b^2 = a'c' - b'^2.$

Then

(65) $\begin{vmatrix} a' - a & b' - b \\ b' - b & c' - c \end{vmatrix} \leqq 0,$

and the equality sign holds only when the two forms are identical.

As proof, we observe that the statement of the lemma remains unchanged under a linear transformation of the variables. Applying such a linear transformation when necessary, we can assume $b' = b$. Then the left-hand side of Equation (65) becomes

$$(a' - a)(c' - c) = -\frac{c}{a'}(a' - a)^2 \leqq 0,$$

as to be proved. Moreover, the quantity equals 0 only when we also have $a' = a$ and $c' = c$.

We now choose the origin to be inside S, so that $p > 0$. Then the integrand in the left-hand side of Equation (62) is nonpositive, and it follows that

$$\iint\limits_{S} H^* \, dA \leqq \iint\limits_{S} H \, dA.$$

Since the relation between S and S^* is symmetrical, we must also have

$$\iint_S H \, dA \leqq \iint_S H^* \, dA.$$

Hence,

$$\iint_S H \, dA = \iint_S H^* \, dA.$$

It follows that the integral at the left-hand side of Equation (62) is 0, and hence, that

$$\lambda^* = \lambda, \qquad \mu^* = \mu, \qquad \nu^* = \nu,$$

completing the proof of Cohn-Vossen's theorem.

By Hadamard's theorem, we see that the Gauss map $g : S \to \Sigma_0$ (see Section 6) is one-to-one for a closed surface with $K > 0$. A point on S can therefore be regarded as a function of its normal vector ξ, and the same is true with any scalar function on S. Minkowski's theorem expresses the unique determination of S when $K(\xi)$ is known.

THEOREM: *Let S be a closed convex surface with Gaussian curvature $K > 0$. The function $K(\xi)$ determines S up to a translation.*

Proof: We shall give a proof of this theorem modeled after the above—that is, by an integral formula [see S. S. Chern, *American Journal of Mathematics* 79 (1957), pp. 949–50]. Let u and v be isothermal parameters on the unit sphere Σ_0, so that we have

(66) $$\xi_u^2 = \xi_v^2 = A > 0 \text{ (say)}, \qquad \xi_u \xi_v = 0.$$

Through the mapping g^{-1} we regard u and v also as parameters on S. Since ξ_u and ξ_v are orthogonal to ξ and are linearly independent, every vector orthogonal to ξ can be expressed as their linear combination. This fact, taken with the relation $X_u \xi_v = X_v \xi_u$, allows us to write

(67) $$\begin{aligned} -X_u &= a\xi_u + b\xi_v, \\ -X_v &= b\xi_u + c\xi_v. \end{aligned}$$

Forming scalar products of these equations with ξ_u and ξ_v, we have

(68) $$Aa = L, \qquad Ab = M, \qquad Ac = N.$$

Moreover, taking the vector product of the two relations in Equation (67), we find

$$X_u \times X_v = (ac - b^2)(\xi_u \times \xi_v).$$

But

(69) $$X_u \times X_v = D\xi, \qquad \xi_u \times \xi_v = A\xi,$$

so that, combining with Equation (68), we have

$$D = A(ac - b^2) = \frac{KD^2}{A},$$

which gives

(70) $$A = KD, \qquad ac - b^2 = \frac{1}{K}.$$

Since $A \, du \, dv$ and $D \, du \, dv$ are, respectively, the volume elements of Σ_0 and S, the first relation in Equation (70) expresses the well-known fact that K is the ratio of these volume elements.

Suppose S^* is another convex surface with the same function, $K(\xi)$. We set up a homeomorphism between S and S^*, so that they have the same normal vector at corresponding points. Then the parameters u and v can be used for both S and S^*, and corresponding points have the same parameter values. We denote by asterisks the vectors and functions for the surface S^*. Since $K = K^*$, we have from Equation (70), $ac - b^2 = a^*c^* - b^{*2}$ and $D = D^*$.

Let

(71) $$p = X \cdot \xi \quad \text{and} \quad p^* = X^* \cdot \xi$$

be the distances from the origin to the tangent planes of the two surfaces. Our basic relation is the identity

$$(X, X^*, X_u)_v - (X, X^*, X_v)_u$$

$$= A\{2(ac - b^2)p^* + (-ac^* - a^*c + 2bb^*)p\}$$

$$= A\left\{2(ac - b^2)(p^* - p) + \begin{vmatrix} a - a^* & b - b^* \\ b - b^* & c - c^* \end{vmatrix} p\right\}$$

which follows immediately from Equations (67), (69), (70), and (71). From it, we find, by Green's theorem, the integral formula

(72) $\displaystyle\int_{\Sigma_0} \left\{ 2(ac - b^2)(p^* - p) + \begin{vmatrix} a - a^* & b - b^* \\ b - b^* & c - c^* \end{vmatrix} p \right\} A \, du \, dv = 0.$

By translations if necessary, we can suppose the origin to be inside both surfaces, S and S^*, so that $p > 0$ and $p^* > 0$. Since

$$\begin{pmatrix} a & b \\ b & c \end{pmatrix} \quad \text{and} \quad \begin{pmatrix} a^* & b^* \\ b^* & c^* \end{pmatrix}$$

are positive definite matrices, it follows from our algebraic lemma that

$$\begin{vmatrix} a - a^* & b - b^* \\ b - b^* & c - c^* \end{vmatrix} \leqq 0.$$

Hence,

(73) $\displaystyle\int_{\Sigma_0} (ac - b^2)(p^* - p) \, A \, du \, dv \geqq 0.$

But the same relation is true when S and S^* are interchanged. Hence, the integral at the left-hand side of Equation (73) must be identically 0. It follows from Equation (72) that

$$\int_{\Sigma_0} \begin{vmatrix} a - a^* & b - b^* \\ b - b^* & c - c^* \end{vmatrix} p \, A \, du \, dv = 0,$$

possible only when $a = a^*, b = b^*$, and $c = c^*$. The latter implies that

$$X_u^* = X_u \quad \text{and} \quad X_v^* = X_v,$$

which means that S and S^* differ by a translation.

For further reading, see:

1. S. S. Chern, "Integral formulas for hypersurfaces in euclidean space and their applications to uniqueness theorems," *Journal of Mathematics and Mechanics*, 8 (1959), pp. 947–55.

2. T. Otsuki, "Integral formulas for hypersurfaces in a Riemannian manifold and their applications," *Tôhoku Mathematical Journal*, 17 (1965), pp. 335–48.

3. K. Voss, "Differentialgeometrie geschlossener Flächen im euklidischen Raum," *Jahresberichte deutscher Math. Verein.*, 63 (1960–1961), pp. 117–36.

8. BERNSTEIN'S THEOREM
ON MINIMAL SURFACES

A minimal surface is a surface which locally solves the Plateau problem—that is, it is the surface of smallest area bounded by a given closed space curve. Analytically, it is defined by the condition that the mean curvature is identically 0. We suppose the surface to be given by

$$(74) \qquad z = f(x, y),$$

where the function $f(x, y)$ is twice continuously differentiable. Then a minimal surface is characterized by the partial differential equation,

$$(75) \qquad (1 + q^2)r - 2pqs + (1 + p^2)t = 0,$$

where

$$(76) \quad p = \frac{\partial f}{\partial x}, \qquad q = \frac{\partial f}{\partial y}, \qquad r = \frac{\partial^2 f}{\partial x^2}, \qquad s = \frac{\partial^2 f}{\partial x \partial y}, \qquad t = \frac{\partial^2 f}{\partial y^2}.$$

Equation (75), called the minimal surface equation, is a nonlinear "elliptic" differential equation.

Bernstein's theorem is the following "uniqueness theorem."

THEOREM: *If a minimal surface is defined by Equation* (74) *for all values of x and y, it is a plane. In other words, the only solution of Equation* (75) *valid in the whole* (x, y)-*plane is a linear function.*

Proof: We shall derive this theorem as a corollary of the following theorem of Jörgens [*Math Annalen* 127 (1954), pp. 130–34].

THEOREM: *Suppose the function* $z = f(x, y)$ *is a solution of the equation*

$$(77) \qquad rt - s^2 = 1, \quad r > 0,$$

for all values of x and y. Then $f(x, y)$ *is a quadratic polynomial in x and y.*

For fixed (x_0, y_0) and (x_1, y_1), consider the function

$$h(t) = f(x_0 + t(x_1 - x_0), y_0 + t(y_1 - y_0)).$$

We have

$$h'(t) = (x_1 - x_0)p + (y_1 - y_0)q,$$

$$h''(t) = (x_1 - x_0)^2 r + 2(x_1 - x_0)(y_1 - y_0)s + (y_1 - y_0)^2 s \geqq 0,$$

where the arguments in the functions p, q, r, s, t are $x_0 + t(x_1 - x_0)$ and $y_0 + t(y_1 - y_0)$. From the last inequality, it follows that

$$h'(1) \geqq h'(0)$$

or

(78) $$(x_1 - x_0)(p_1 - p_0) + (y_1 - y_0)(q_1 - q_0) \geqq 0.$$

where

(79) $$p_i = p(x_i, y_i) \quad \text{and} \quad q_i = q(x_i, y_i), \quad i = 0, 1.$$

Consider the transformation of Lewy:

(80) $$\xi = \xi(x, y) = x + p(x, y), \qquad \eta = \eta(x, y) = y + q(x, y).$$

Setting

(81) $$\xi_i = \xi(x_i, y_i), \qquad \eta_i = \eta(x_i, y_i), \qquad i = 0, 1,$$

we have, by Equation (78),

(82) $$(\xi_1 - \xi_0)^2 + (\eta_1 - \eta_0)^2 \geqq (x_1 - x_0)^2 + (y_1 - y_0)^2.$$

Hence, the mapping

(83) $$(x, y) \rightarrow (\xi, \eta)$$

is distance-increasing.

Moreover, we have

(84) $$\xi_x = 1 + r, \qquad \xi_y = s$$

$$\eta_x = s, \qquad \eta_y = 1 + t,$$

so that

(85) $$\frac{\partial(\xi, \eta)}{\partial(x, y)} = 2 + r + t \geqq 2,$$

and the mapping in Equation (83) is locally one-to-one. It follows easily that Equation (83) is a diffeomorphism of the (x, y)-plane onto the (ξ, η)-plane.

We can therefore regard the function $f(x, y)$, which is a solution of Equation (77), as a function in ξ and η. Let

(86) $$F(\xi, \eta) = x - iy - (p - iq),$$

(87) $$\zeta = \xi + i\eta.$$

It can be verified by a computation that $F(\xi, \eta)$ satisfies the Cauchy-Riemann equations, so that $F(\zeta) = F(\xi, \eta)$ is a holomorphic function in ζ. Moreover, we have

(88) $$F'(\zeta) = \frac{t - r + 2is}{2 + r + t}.$$

From the last relation, we get

$$1 - |F'(\zeta)|^2 = \frac{4}{2 + r + t} > 0.$$

Thus $F'(\zeta)$ is bounded in the whole ζ-plane. By Liouville's theorem, we have

$$F'(\zeta) = \text{const.}$$

On the other hand, by Equation (88) we have

(89) $$r = \frac{|1 - F'|^2}{1 - |F'|^2}, \qquad s = \frac{i(\overline{F'} - F')}{1 - |F'|^2}, \qquad t = \frac{|1 + F'|^2}{1 - |F'|^2}.$$

It follows that r, s, and t are all constants, and Jörgens's theorem is proved.

Bernstein's theorem is an easy consequence of Jörgens's theorem. In fact, let

(90) $$W = (1 + p^2 + q^2)^{1/2}.$$

Then the minimal surface equation is equivalent to each of the following equations:

(91) $$\frac{\partial}{\partial x} \frac{-pq}{W} + \frac{\partial}{\partial y} \frac{1 + p^2}{W} = 0,$$

$$\frac{\partial}{\partial x} \frac{1 + q^2}{W} + \frac{\partial}{\partial y} \frac{-pq}{W} = 0.$$

It follows that there exists a C^2-function, $\varphi(x, y)$, such that

(92) $$\varphi_{xx} = \frac{1}{W}(1 + p^2), \qquad \varphi_{xy} = \frac{1}{W}pq, \qquad \varphi_{yy} = \frac{1}{W}(1 + q^2).$$

These partial derivatives satisfy the equation

$$\varphi_{xx}\varphi_{yy} - \varphi_{xy}^2 = 1, \quad \varphi_{xx} > 0.$$

By Jörgens's theorem, φ_{xx}, φ_{xy}, and φ_{yy} are constants. Hence, p and q are constants, and $f(x, y)$ is a linear function. [This proof of Bernstein's theorem is that of J. C. C. Nitsche, *Annals of Mathematics*, 66 (1957), pp. 543–44.]

Minimal surfaces have an extensive literature. See the following expository article:

1. J. C. C. Nitsche, "On new results in the theory of minimal surfaces," *Bulletin of the American Mathematical Society*, 71 (1965), pp. 195–270.

ON CONJUGATE AND CUT LOCI

Shoshichi Kobayashi

Let us fix a point p of a complete Riemannian manifold M and a geodesic g starting at p. Then the cut point of p along g is the first point q on g such that, for any point r on g beyond q, there is a shorter geodesic from p to r than g. On the other hand, a point q is a conjugate point of p along g if there is a 1-parameter family of geodesics from p to q neighboring g. (A precise definition is given in Section 2.) For a point r on g which lies beyond a conjugate point q of p, g is no longer a minimizing geodesic.

At the time when this chapter was written for the first edition of this volume, there were few books treating conjugate and cut loci. The purpose of this chapter was to fill this gap by giving a self-contained account of basic properties of cut and conjugate loci. Although several books dealing with global properties of geodesics have appeared in the past twenty years, we hope that this chapter will continue to serve as a quick introduction to the subject.

The earliest paper on cut points was written by Poincaré (17). Although the subject was taken up again in the mid 1930's by

Myers (15, 16), and by J. H. C. Whitehead (28), the recent interest in cut points results from Klingenberg's papers (8, 9) on Riemannian manifolds with positive curvature. Material for the present chapter was taken largely from Schoenberg (21), Synge (23), Preismann (18), Rauch (19), lecture notes by Ambrose and Singer at MIT (see (29)), and Volume 2 of Kobayashi-Nomizu (44).

For better understanding, rigorous proofs are occasionally preceded by sketchy but intuitive proofs. When rigorous proofs require large machineries or extensive knowledge, less satisfactory but more elementary proofs are given instead. In the last section we offer suggestions for further reading.

1. CUT LOCI

If we take two points, p and q, of a (connected) Riemannian manifold M and join them by a (continuous) piecewise differentiable curve, then we can measure the arc length of this curve using the Riemannian metric (often called the element of arc in older books). We consider all possible piecewise differentiable curves joining p and q and define the distance $d(p, q)$ between p and q as the infimum of their arc lengths. Then the distance function d satisfies the usual three axioms of a metric space, and we can talk about Cauchy sequences of points of M and also the completeness of M.

Hereafter, we shall assume that M is a complete Riemannian manifold. Since a compact metric space is complete, a compact Riemannian manifold is always complete. There are a few basic facts about geodesics of a complete Riemannian manifold which we must use in this article. If we say "a geodesic $g(t)$," we understand that it is parametrized by arc length and t is a parameter. A geodesic $g(t)$, which is originally defined only for some interval $a \leqq t \leqq b$, is said to be infinitely extendable if it can be extended to a geodesic $g(t)$ defined for the whole interval, $-\infty < t < \infty$. If we take as an example the $(x\text{-}y)$-plane with the usual metric and delete the origin, we then obtain an incomplete Riemannian manifold and the positive x-axis is not infinitely extendable.

The first basic fact we need is that, on a complete Riemannian manifold, every geodesic is infinitely extendable. This fact is intuitively clear and not difficult to prove. Although its converse is also true, its proof is much harder and we do not need it here. The second basic fact we need is that, on a complete Riemannian manifold, any two points can be joined by a minimizing geodesic. [A geodesic from p to q is said to be minimizing if its arc length gives the distance $d(p, q)$.] Two proofs of this fact are known, neither of which is very easy. The first is the work of Hilbert (6). [See also Hopf and Rinow (7) and Cartan (4).] The second proof is that of de Rham (20). [See also Kobayashi and Nomizu (13).] The first systematic study of completeness of a Riemannian manifold was made by Hopf and Rinow (7). [See also de Rham (20) and Kobayashi and Nomizu (13), which follows closely de Rham.] The statements of these facts are clear and we advise the reader to proceed assuming their validity.

We shall now explain exponential maps. At each point p of a complete Riemannian manifold M, we define a mapping of the tangent space $T_p(M)$ at p onto M in the following manner. If X is a tangent vector at p, we draw a geodesic $g(t)$ starting at p in the direction of X; we parametrize the geodesic in such a way that $g(0) = p$. If X has length a, then we map X into the point $g(a)$ of the geodesic. We denote this mapping by $\exp_p$ or simply exp,

$$\exp_p: T_p(M) \to M$$

and call it the exponential map at p. Thus, $\exp_p$ maps a line in the tangent space $T_p(M)$ through its origin onto the geodesic of M through p in the direction of the line. Since every point q of M can be joined by a geodesic to p, $\exp_p$ maps $T_p(M)$ onto M.

To the reader who might wonder where the name "exponential map" originates, we point out that $\exp_p$ may be considered as a sort of generalization of the usual exponential function in the following way. In the theory of Lie groups, the exponential map of a Lie group G is a mapping of the tangent space $T_e(G)$ of G at the identity element e into G which sends a line through the origin of $T_e(G)$ onto the 1-parameter subgroup in the direction of the line. Since the tangent space $T_e(G)$ may be considered as the Lie algebra of G, the exponential map sends the Lie algebra of G into G. The ordinary exponential function is nothing but the exponential map of the multiplicative group of positive real numbers the Lie algebra of which is just the real line. If G is a compact Lie group, we can construct a Riemannian metric on G so that the 1-parameter subgroups are the geodesics through the identity element e and the exponential map as a Lie group is the exponential map at e as a Riemannian manifold. In passing, we might point out here that since every element of a compact G can be joined to e by a geodesic according to the aforementioned theorem of Hilbert, every element of a compact G lies on a 1-parameter subgroup of G.

We fix a point p of M. We shall now define the cut locus $C(p)$ of p. Take a geodesic $g(t)$, $0 \leqq t < \infty$, starting at p. Then the first point on this geodesic where the geodesic ceases to minimize its arc length is called the *cut point of p along the geodesic $g(t)$*. To be more precise, we let A be the set of positive real numbers s such that the geodesic $g(t)$, $0 \leqq t \leqq s$, is minimizing—that is, $s = d(p, g(s))$. It is easy to see that either $A = (0, \infty)$ or $A = (0, r]$, where r is some positive number. If $A = (0, r]$, then $g(r)$ is the cut point of p along the geodesic $g(t)$. If $A = (0, \infty)$, then we say that p has no cut point along the geodesic $g(t)$. Thus, if q is a point on $g(t)$ which comes after the cut point $p' = g(r)$—that is, $q = g(s)$ with $s > r$—then we can find a geodesic from p to q which is shorter than $g(t)$. On the other hand, if q is a point which comes before the cut point p', then not only do we not find a shorter geodesic from p to q but we cannot have another geodesic from p to q of the same length. In other words, if q comes before p', then $g(t)$ is the *unique* minimizing geodesic joining p and q, for if there is another minimizing geodesic g' from p to q, then, by moving from p to q along g' and continuing from q to p' along g, we obtain a nongeodesic curve c from p to p' with an arc length equal to the distance $d(p, p')$. We choose a point p_1 on g' before q and also a point p_2 on g after q. Taking both p_1 and p_2 sufficiently close to q, we replace the portion of c from p_1 to p_2 by the minimizing geodesic joining p_1 and p_2 and obtain a curve from p to p' with an arc length less than the distance $d(p, p')$, which is absurd. The set of cut points of p is called the *cut locus* of p and is denoted by $C(p)$. Corresponding to the cut point p' of p along $g(t)$, we consider the point $X \in T_p(M)$, that is, the tangent vector X given by

$$g(t) = \exp_p (tX) \quad \text{and} \quad p' = \exp_p (X),$$

and call it the *cut point of p in $T_p(M)$* corresponding to the cut point p'.

If M is a compact Riemannian manifold, then its diameter is finite and no geodesic of length greater than the diameter is minimizing. If M is compact, on each geodesic starting from a point p we find the cut point of p. Correspondingly, in the tangent space $T_p(M)$ we find a cut point in every direction starting from p. It is quite reasonable to expect that the cut locus in $T_p(M)$ (not in M) is homeomorphic to a hypersphere in $T_p(M)$, provided that M is compact. To prove that this is actually the case, we have to show some kind of connectedness of the cut locus of p in $T_p(M)$, which we shall accomplish in the following theorem.

THEOREM 1.1: *Let S be the unit sphere in the tangent space $T_p(M)$ with center at the origin. We define a function f on S by assigning to each unit vector $X \in S$ the distance from p to the cut point in the direction of X. Then f is a continuous function.*

This theorem holds even for a noncompact manifold. However, if M is not compact, for some direction $X \in S$ the cut point may not exist. Then we set $f(X) = \infty$. Thus, we consider f as a mapping of S into $[0, \infty]$, where $[0, \infty]$ is the 1-point-compactification of $[0, \infty)$. To prove the theorem, we need properties of conjugate points as well as those of cut points; therefore, we postpone its proof to Section 4. We shall derive here some consequences from the theorem.

Assume that M is compact. As we have already stated, the cut locus of p in $T_p(M)$, denoted by $C^*(p)$, is homeomorphic to the unit sphere S in $T_p(M)$. Let E^* be the domain in $T_p(M)$ bounded by the cut locus $C^*(p)$. If we denote by E the image of E^* by $\exp_p$, then $\exp_p$ gives a homeomorphism of E^* onto E and we have, obviously, $M = E \cup C(p)$. Thus, M is a disjoint union of an open cell (that is, E) and a closed subset [that is, $C(p)$], which is a continuous image of an $(n-1)$-sphere [that is, $C^*(p)$], where $n = \dim M$. In particular, the cut locus $C(p)$ is connected.

The importance of cut loci lies in the fact that they inherit a number of topological properties of the manifold M. For example, let us consider the homotopy groups $\pi_k(M)$ and $\pi_k(C(p))$. We shall show that the natural mapping (that is, the imbedding) of $C(p)$ into M induces an isomorphism of $\pi_k(C(p))$ onto $\pi_k(M)$ for $k \leqq n-2$ and a homomorphism of $\pi_{n-1}(C(p))$ onto $\pi_{n-1}(M)$. Let S^k be the k-sphere and consider the element of $\pi_k(M)$ given by a smooth map $h: S^k \to M$. If $k \leqq n-1$, then the image $h(S^k)$ cannot cover all of M, and we may assume without loss of generality that $h(S^k)$ misses the point p. Since we can shrink $M - \{p\}$ into $C(p)$ by pushing it along the geodesic rays emanating from p toward $C(p)$, we can deform h to a continuous map $h': S^k \to C(p)$, showing that $\pi_k(C(p))$ is mapped onto $\pi_k(M)$ for $k \leqq n-1$. Let $f: S^k \to C(p)$ and assume that f can be extended to $g: B^{k+1} \to M$, where B^{k+1} is the solid ball with the boundary S^k. If $k + 1 \leqq n-1$, then the image $g(B^{k+1})$ cannot cover M and we may assume that it misses the point p. We can deform q to $g': B^{k+1} \to C(p)$ as before, so that g' is an extension of f, showing that $\pi_k(C(p)) \to \pi_k(M)$ is an isomorphism for $k \leqq n-2$. We have a very natural element in $\pi_{n-1}(C(p))$, which belongs to the kernel $\pi_{n-1}(C(p)) \to \pi_{n-1}(M)$. Since the cut locus $C^*(p)$ in the tangent space $T_p(M)$ is homeomorphic with the $(n-1)$-sphere S^{n-1}, $\exp_p: C^*(p) \to C(p)$ gives an element of $\pi_{n-1}(C(p))$ that obviously belongs to the kernel.

Before we proceed, we should perhaps find out what cut loci look like for some simple Riemannian manifolds.

Examples

1. If M is an n-dimensional unit sphere, then the geodesics are the so-called great circles. If p is the north pole, then the cut locus $C(p)$ reduces to the south pole. The cut locus $C^*(p)$ in the tangent space $T_p(M)$ is the sphere of radius π with a center at the origin.

2. Identifying each point of the unit sphere S^n with its antipodal point, we obtain an n-dimensional real projective space, which we shall denote by M. The Riemannian metric of S^n induces a Riemannian metric on M in a natural manner so that the projection of S^n onto M is a local isometry. If p is the point of M which corresponds to the north and south poles of S^n, then the image of the equator of S^n by the projection $S^n \to M$ is the cut locus $C(p)$. In other words, $C(p)$ is the so-called hyperplane at the infinity which is an $(n-1)$-dimensional real projective space. The cut locus $C^*(p)$ in the tangent space $T_p(M)$ is the sphere of radius $\pi/2$ with center at the origin and is a double covering space of $C(p)$ under $\exp_p$.

3. In the euclidean plane R^2 with coordinate system (x, y), consider the closed unit square $0 \leqq x, y \leqq 1$. By identifying $(x, 0)$ with $(x, 1)$ and also $(0, y)$ with $(1, y)$, we obtain a flat torus, which we denote here by M. We are considering, of course, the Riemannian metric on M induced in a natural way from the euclidean metric of R^2. Let p be the point with coordinates $(\frac{1}{2}, \frac{1}{2})$. Then the cut locus $C^*(p)$ in $T_p(M)$ can be identified with the boundary of the unit square under the natural identification of $T_p(M)$ with R^2. The cut locus $C(p)$ consists of two intersecting circles which generate the first homology group of the torus.

4. In the (x, y)-plane, consider a curve $y = f(x)$, $a \leqq x \leqq b$, such that $f(a) = f(b) = 0, f(x) > 0$, for $a < x < b$, and such that it is tangent to the lines $x = a$ and $x = b$. If we place this curve in the (x, y, z)-space and revolve it about the x-axis, then we obtain a smooth closed surface M which is topologically a sphere. Let p be the point with coordinates $(a, 0, 0)$. Since p is invariant under the revolution, which is a 1-parameter group of isometries, so must be the cut locus $C(p)$. On the other hand, $M - C(p)$ must be connected. These two restrictions on $C(p)$ imply immediately that $C(p)$ reduces to a single point—that is, the point with coordinates $(b, 0, 0)$. This fact applies in particular to the ellipsoid $(x/a)^2 + (y/b)^2 + (z/c)^2 = 1$, with $b = c$. (On the general ellipsoid, the cut locus of any of the six poles is known to be an arc on the longer of two principal ellipses through the given pole and containing the opposite pole as midpoint. Also, on the general ellipsoid, the cut locus of an umbilical point is the diametrically opposite umbilical point.)

The reader unfamiliar with complex or quaternionic projective spaces is advised to skip Example 5.

5. Let M be a complex (resp. quaternionic) projective space of real dimension $2n$ (resp. $4n$). If we normalize the metric so that the maximal sectional curvature is 1 (and hence the minimum sectional curvature is $\frac{1}{4}$), then the cut locus $C^*(p)$ in the tangent space $T_p(M)$ is the sphere of radius π with a center at the origin of $T_p(M)$. The cut locus $C(p)$ is the hyperplane at infinity—that is, the complex (resp. quaternionic) projective space of real dimension $2(n-1)$ (resp. $4(n-1)$) consisting of points at infinity with respect to p. The exponential map $\exp_p : C^*(p) \to C(p)$ defines a fiber bundle with fiber S^1 (resp. S^3) and is nothing but the so-called Hopf-fibering.

The cut locus $C(p)$ is not a submanifold in general even if M is a homogeneous Riemannian manifold. (A Riemannian manifold is said to be homogeneous if the group of isometries acts transitively.) In Example 5, $C(p)$ is a nice submanifold because the group of isometries of M leaving p fixed is transitive on the directions at p—that is, transitive on the unit sphere in the tangent space $T_p(M)$ with a center at the origin. Even for a homogeneous Riemannian manifold it is usually difficult to determine the cut locus $C(p)$ of a point of p. The cut locus $C(p)$ is, in general, not triangulable, see Section 5.

2. CONJUGATE LOCI

We shall begin with a geometric definition of conjugate point. As in Section 1, we denote by M a complete Riemannian manifold. Given a geodesic $g = g(t)$, $a \leqq t \leqq b$, consider a *variation* (or more precisely, a geodesic variation) of g—that is, a 1-parameter family of geodesics $g_s = g_s(t)$, $a \leqq t \leqq b$ and $-\epsilon < s < \epsilon$, such that $g = g_0$. For each fixed s, $g_s(t)$ describes a geodesic when t moves from a to b. Each variation gives rise to an *infinitesimal variation*, that is, a certain vector field defined along g. More explicitly, for each fixed t, consider the vector tangent to the curve $g_s(t)$, $-\epsilon < s < \epsilon$, at $s = 0$; it is a vector at the point $g(t)$ of g. Roughly speaking, the endpoints $g(a)$ and $g(b)$ of the geodesic g are said to be conjugate to each other if there is a variation g_s of g passing through these endpoints—that is, $g_s(a) = g(a)$ and $g_s(b) = g(b)$ for $-\epsilon < s < \epsilon$. But we do not actually require that g_s passes through these endpoints; it will be enough if the variation g_s passes these endpoints to infinitesimal order 1. We may now state the definition of conjugate points more precisely: The points $g(a)$ and $g(b)$ of g are said to be *conjugate* if there is a variation g_s which induces an infinitesimal variation vanishing at $t = a$ and $t = b$.

There is another equally geometric way of defining a conjugate point. Let p be a point of M and $\exp_p : T_p(M) \to M$ the exponential map defined in Section 1. A point X of $T_p(M)$—that is, a tangent vector at p—is called a conjugate point of p in $T_p(M)$ if $\exp_p$ is degenerate at X—that is, the Jacobian matrix of $\exp_p$ is singular at X. Its image, $p' = \exp_p(X)$ by $\exp_p$, is called a *conjugate point* of p along the geodesic $g(t) = \exp_p(tX)$.

It is easy to see that if p' is conjugate to p along g in the second sense, it is also so in the first sense. Since $\exp_p$ is degenerate at X, we can find a curve X_s through X in $T_p(M)$, such that $\exp_p$ annihilates the vector tangent to the curve X_s at $X = X_0$. We may also assume that for each s, the length of X_s is equal to that of X. Then $g_s(t) = \exp_p(tX_s/\|X\|)$ gives us a variation g_s of g with the desired property. The converse is a little harder to prove. Let p be the point $g(0)$ of a geodesic g and $p' = g(b)$ a conjugate point of p along g in the first sense. Let g_s be a variation of g such that the induced infinitesimal variation vanishes at p and p'. If we have exactly $q_s(0) = p$, then we have no difficulty in showing that p' is conjugate to p along g in the second sense. Considering the 1-parameter family of rays in $T_p(M)$ corresponding to the 1-parameter family of geodesics g_s through p, and taking the points with distance b from the origin on the rays, we obtain a curve X_s in $T_p(M)$ the tangent vector of which is annihilated by $\exp_p$ at X_0. This condition shows that X_0 is conjugate to p in $T_p(M)$, and, consequently, p' is conjugate to p in the second sense. It remains to prove that we can find a variation g_s such that not only the induced infinitesimal variation vanishes at p and p' but also $g_s(0) = p$. To achieve this proof, we give the third definition of conjugate point, which is less geometric but more analytic.

If Y is a vector field defined along geodesic g, then we denote by Y' its covariant derivative along g and by Y'' the second covariant derivative along g. For each geodesic g we have a naturally associated vector field along g—that is, the field of tangent vectors to g, which we shall denote by G. Since g is parametrized by its arc length, G is a unit vector field. By definition of a geodesic, we have $G' = 0$. A vector field Y along g is called a *Jacobi field* if it satisfies the following second-order, ordinary, linear, differential equation (called the Jacobi equation),

$$Y'' + R(Y, G)G = 0,$$

where $R(Y, G)$ denotes the curvature transformation determined by Y and G. (For any ordered pair of vectors at a point of M, the curvature determines a linear transformation of the tangent space at that point.) We say that points p and p' on a geodesic g are *conjugate* to each other if there is a nonzero Jacobi field Y along g which vanishes at p and p'. The equivalence of this definition with the first follows from the fact that a vector field Y defined along a geodesic g is a Jacobi field if and only if it is an infinitesimal variation of g.

We prove this fact as follows. Let g_s be a variation of g and set $f(s, t) = g_s(t)$. We denote by $D/\partial t$ and $D/\partial s$ the covariant differentiations with respect to t and s, respectively. Since g_s is a geodesic for each s, we have

$$\frac{D}{\partial t}\frac{\partial f}{\partial t} = 0.$$

Hence,

$$0 = \frac{D}{\partial s}\frac{D}{\partial t}\frac{\partial f}{\partial t} = \frac{D}{\partial t}\frac{D}{\partial s}\frac{\partial f}{\partial t} + R\left(\frac{\partial f}{\partial t}, \frac{\partial f}{\partial s}\right)\frac{\partial f}{\partial t}$$

$$= \frac{D}{\partial t}\frac{D}{\partial t}\frac{\partial f}{\partial s} + R\left(\frac{\partial f}{\partial t}, \frac{\partial f}{\partial s}\right)\frac{\partial f}{\partial t},$$

showing that an infinitesimal variation is a Jacobi field. (The conscientious reader will find the preceding calculation somewhat ambiguous; since f is not a one-to-one mapping into M, neither $\partial f / \partial t$ nor $\partial f / \partial s$ is well defined as a vector field on the image of f. Returning to the definition of covariant differentiation and curvature, we can explain the ambiguities. (The use of differential forms and the so-called structure equations of Cartan is a more satisfactory proof, but the use of such machineries is inappropriate in this article and we have to be satisfied with the preceding proof.)

Conversely, let Y be a Jacobi field along g. We choose two parameter values a and b close to each other so that $g(a)$ and $g(b)$ are in a convex neighborhood (any two points of which can be joined by a unique minimizing geodesic). We choose a curve $g_s(a)$, $-\epsilon < s < \epsilon$, through the point $g(a)$ which is tangent to Y at $g(a)$. Similarly, we let $g_s(b)$, $-\epsilon < s < \epsilon$, be a curve through $g(b)$ tangent to Y at $g(b)$. We can also arrange it so that the distance $d(g_s(a), g_s(b))$ is equal to $d(g(a), g(b))$ for each s. Letting $g_s = g_s(t)$ be the unique minimizing geodesic joining $g_s(a)$ and $g_s(b)$, we find that the infinitesimal variation induced by the variation g_s of g coincides with Y. The construction shows that if Y vanishes at $g(a)$, we can find a variation g_s with the additional property $g_s(a) = g(a)$. We have thus shown simultaneously the equivalence of the first and second definitions.

Perhaps the most important method of obtaining Jacobi fields is by infinitesimal isometries. Let φ_s be a 1-parameter group of isometries of M. Given a geodesic g, we obtain a variation g_s by $g_s = \varphi_s(g)$. The induced infinitesimal variation is nothing but the infinitesimal transformation (restricted to g) which generates φ_s. If p and p' are points on g which are left fixed by φ_s, then p and p' are conjugate to each other, provided that the Jacobi field along g induced by φ_s is nontrivial.

Let M^* be a Riemannian covering manifold of M; it is a covering space of M with the naturally induced Riemannian structure. Let g^* be a geodesic in M^* and g its image on M under the natural projection of M^* onto M. Then two points on g^* are conjugate along g^* if and only if their images on g are conjugate along g. This relation is evident when we consider Jacobi equations.

We shall now consider a few simple examples.

Examples

1. If M is an n-dimensional unit sphere and p is the north pole, then both p itself and the south pole p' are conjugate to p along any geodesic—that is, great circle—through p. We see easily that no other points are conjugate to p.

2. Let M be the real projective space of dimension n as in Example 2, Section 1, and let p be a point of M. Then p itself is conjugate to p along any geodesic through p and no other points are conjugate to p, because a sphere is a Riemannian covering manifold of M.

3. Let M be a flat torus as in Example 3, Section 1. Since the universal covering manifold of M is a euclidean space and there are no conjugate points in a euclidean space, there are no conjugate points on M.

4. Let M be a surface of revolution as in Example 4, Section 1. Since the revolution about the x-axis, which is a 1-parameter group of isometries of M, induces an infinitesimal variation vanishing at $p = (a, 0, 0)$ and at $q = (b, 0, 0)$ (and at no other points), p and q are conjugate to each other. Because the cut locus $C(p)$ of p consists of q only, and because along any geodesic starting at p the first conjugate point of p generally never comes before the cut point of p (as we shall see in Section 4), the locus of first conjugate points of p consists of q only. This condition applies in particular to the ellipsoid $(x/a)^2 + (y/b)^2 + (z/c)^2 = 1$ with $b = c$. [On the general ellipsoid, the situation is more complicated. See the picture in Struik (22), p. 143, and see also Braunmühl (3).]

5. If M is a complex or quaternionic projective space as in Example 5 of Section 1, the first conjugate locus of a point p coincides with the cut locus $C(p)$ of p.

In spite of some of the preceding examples, the cases where the cut locus $C(p)$ coincides with the first conjugate locus of p are exceptional rather than general. See Section 5 for further discussion.

3. CURVATURE AND CONJUGATE POINTS

By looking at the Jacobi equation, we can see it is natural to expect a close relationship between the curvature and the distribution of conjugate points. The Jacobi equation is a tensorial analogue of a differential equation of the type

$$y'' + K(x)y = 0,$$

which has been studied by Sturm. For another differential equation

$$y'' + L(x)y = 0$$

of the same type such that

$$K(x) \leq L(x),$$

the comparison theorem of Sturm says that if $u(x)$ is a solution of the first equation having m zeros in the interval $a \leq x \leq b$, and if $v(x)$ is a solution of the second equation with

$$u(a) = v(a) \quad \text{and} \quad u'(a) = v'(a),$$

then $v(x)$ has at least m zeros in the same interval. If $K(x)$ is a positive constant function K, then a solution of the first equation is of the form $u(x) = C_1 \sin \sqrt{K}x + C_2 \cos \sqrt{K}x$ and, hence, the distance between any two consecutive zeros of a solution $v(x)$ of the second equation is, at most, $\pi/\sqrt{K}$. On the other hand, if $L(x)$ is a positive constant function L, then the distance between two consecutive zeros of a solution $u(x)$ of the first equation is at least $\pi/\sqrt{L}$. Analogously, we have the following theorem.

THEOREM 3.1 (BONNET): *Let g be a geodesic in M and $K(P)$ the sectional curvature of a plane section P tangent to g. If*

$$0 < L \leq K(P) \leq H$$

for all such plane sections P, then the distance d along g of any two consecutive conjugate points satisfies the following inequalities:

$$\frac{\pi}{\sqrt{H}} \leq d \leq \frac{\pi}{\sqrt{L}}.$$

The theorem follows from the comparison theorem of Sturm. The reader will find the proof for surfaces in a number of standard textbooks on elementary differential geometry of surfaces. The proof for higher dimensional manifolds may be reduced to that for surfaces by Synge's lemma which states: If V is a (2-dimensional) surface imbedded in M and g is a geodesic of M lying in V, then the Gaussian curvature of V at a point of g is less than, or equal to, the sectional curvature of M for the tangent plane to V at that point; the equality holds when and only when the field of tangent planes to V is parallel along g in M. This lemma of Synge is an immediate consequence of the equations of Gauss and Coddazi for V imbedded in M, but we shall prove the comparison theorem of Rauch which implies immediately Bonnet's theorem.

THEOREM 3.2 (RAUCH'S COMPARISON THEOREM): *Let M and N be Riemannian manifolds of the same dimension n. Let $g = g(t)$ be a geodesic of M and X a Jacobi field along g. Let $h = h(t)$ be a geodesic of N and Y a Jacobi field along h. Assume that: (1) X is perpendicular to g and vanishes at $g(0)$, Y is perpendicular to h and vanishes at $h(0)$; (2) X' and Y' have the same length at $t = 0$; (3) neither g nor h has conjugate points in the interval $(0, b)$; (4) for each t in the interval $(0, b)$,*

$$K_M(P) \geqq K_N(Q),$$

where $K_M(P)$ (resp. $K_N(Q)$) is the sectional curvature by an arbitrary tangent plane P (resp. Q) at $g(t)$ tangent to g (resp. at $h(t)$ tangent to h). Under these four assumptions, Y at $h(b)$ is longer than, or equal to, X at $g(b)$.

Before we proceed with the proof, we remark that Theorem 3.1 follows from Theorem 3.2 if we take M to be the sphere of radius H and N to be M of Theorem 3.1, and if we then take N to be the sphere of radius L.

Proof: Let $u(t)$ and $v(t)$ be the square of the lengths of X at $g(t)$ and Y at $h(t)$, respectively. We wish to show $u(b) \leqq v(b)$. It is sufficient to prove

(3.1)
$$\lim_{t \to 0} \frac{u(t)}{v(t)} = 1;$$

(3.2)
$$\left(\frac{u(t)}{v(t)}\right)' \leqq 0,$$

that is, $u(t)/v(t)$ is decreasing. Since $u' = 2(X, X')$, $v' = (Y, Y')$, $u'' = 2(X', X') + 2(X, X'')$, and $v'' = 2(Y', Y') + 2(Y, Y'')$, we apply l'Hospital's rule twice to $u(t)/v(t)$ and obtain (3.1).

Proving (3.2) is equivalent to proving $u'(t)/u(t) \leqq v'(t)/v(t)$. We fix a parameter value s, $0 < s < b$. By the third assumption, $u(s) \neq 0$ and $v(s) \neq 0$. We may thus set $U = X/u(s)^{1/2}$ and $V = Y/v(s)^{1/2}$. Since X and Y are Jacobi fields, so are U and V. We have, therefore,

$$\frac{u'(s)}{u(s)} = (U, U)'(s) = \int_0^s (U, U)'' \, dt$$

$$= 2 \int_0^s [(U', U') - K_M(P)(U, U)] \, dt,$$

where $K_M(P)$ denotes the sectional curvature of the planes P spanned by U and the vectors tangent to the geodesic g. Similarly,

$$\frac{v'(s)}{v(s)} = 2 \int_0^s [(V', V') - K_N(Q)(V, V)] \, dt.$$

Let $\bar{U}$ be a vector field defined along g such that

$$(\bar{U}, \bar{U})(t) = (V, V)(t), \qquad (\bar{U}', \bar{U}')(t) = (V', V')(t).$$

(We may establish the existence of such a $\bar{U}$ by choosing an isomorphism of the normal space to g at $g(0)$ onto the normal space to h at $h(0)$, extending it to an isomorphism of the normal space to g at $g(t)$ onto the normal space to h at $h(t)$, and then considering the vector field along g corresponding to V.) We shall show that

$$\frac{u'(s)}{u(s)} \leqq 2 \int_0^s [(\bar{U}', \bar{U}') - K_M(\bar{P})(\bar{U}, \bar{U})] \, dt \leqq \frac{v'(s)}{v(s)}.$$

The second inequality follows from the fourth assumption and from the preceding integral expression for $v'(s)/v(s)$. To prove the first inequality, we need the following lemma.

LEMMA 1: *If X and Y are Jacobi fields along a geodesic g, then*

$$(X, Y') - (X', Y) = const.$$

If, moreover, X and Y vanish at a point simultaneously, then

$$(X, Y') - (X', Y) = 0.$$

A simple direct calculation shows that the derivative of $(X, Y') - (X', Y)$ vanishes identically. The second statement of the lemma follows trivially from the first.

Using Lemma 1, we shall prove the following lemma which will obviously imply the first inequality.

LEMMA 2: *Let $g = g(t)$, $0 \leq t \leq s$, be a geodesic without conjugate point of $g(0)$ in the interval $0 \leq t \leq s$. Let U be a Jacobi field along g perpendicular to g, and $\overline{U}$ a vector field along g perpendicular to g. If $U(0) = \overline{U}(0) = 0$, and if $U(s) = \overline{U}(s)$, then*

$$\int_0^s [(U', U') - K_M(P)(U, U)] \, dt$$

$$\leq \int_0^s [(\overline{U}', \overline{U}') - K_M(\overline{P})(\overline{U}, \overline{U})] \, dt,$$

where $K_M(P)$ and $K_M(\overline{P})$ denote the sectional curvatures of the planes P spanned by U and the vectors tangent to g, and of the planes $\overline{P}$ spanned by $\overline{U}$ and the vectors tangent to g. The equality holds only when $U = \overline{U}$.

Consider the vector space of Jacobi fields Z perpendicular to g and vanishing at $g(0)$. Being a solution of a second order differential equation, Z is uniquely determined by its initial conditions $Z(0)$ and $Z'(0)$. Since $Z(0) = 0$ and $Z'(0)$ must be perpendicular to g, the vector space of these Z's is of dimension $n - 1$. Let $Z_1, \cdots, Z_{n-1}$ be a basis for this vector space. Because there exists no conjugate point of $g(0)$ in the interval $(0, s)$, $Z_1, \cdots, Z_{n-1}$ are linearly independent at each point $g(t)$, $0 < t \leq s$. We may write, therefore,

$$U = a_1 Z_1 + \cdots + a_{n-1} Z_{n-1},$$

$$\overline{U} = f_1 Z_1 + \cdots + f_{n-1} Z_{n-1},$$

where $a_1, \cdots, a_{n-1}$ are constants and $f_1, \cdots, f_{n-1}$ are functions. We have

$$(U', U') = (\Sigma f_i'Z_i, \Sigma f_i'Z_i) + 2(\Sigma f_i'Z_i, \Sigma f_iZ_i')$$
$$+ (\Sigma f_iZ_i', \Sigma f_iZ_i'),$$
$$-(R(\overline{U}, G)G, \overline{U}) = -\Sigma f_i(R(Z_i, G)G, \overline{U}) = \Sigma f_i(Z_i'', \overline{U})$$
$$= (\Sigma f_iZ_i'', \Sigma f_iZ_i),$$

where G denotes the vector field tangent to g, and we also have

$$(\Sigma f_iZ_i, \Sigma f_iZ_i')' = (\Sigma f_iZ_i, \Sigma f_iZ_i') + (\Sigma f_iZ_i', \Sigma f_iZ_i')$$
$$+ (\Sigma f_iZ_i, \Sigma f_i'Z_i') + (\Sigma f_iZ_i, \Sigma f_iZ_i'').$$

Combining these three equalities and using Lemma 1, we obtain

$$\int_0^s [(\overline{U}', \overline{U}') - K_M(\overline{P})(\overline{U}, \overline{U})] \, dt = \int_0^s (\Sigma f_i'Z_i, \Sigma f_i'Z_i) \, dt$$
$$+ (\Sigma f_iZ_i, \Sigma f_iZ_i')_s.$$

In the preceding calculation, we replace $\overline{U}$ by U, $\overline{P}$ by P, and f_i by a_i, and observe that $a_i' = 0$. Then we obtain

$$\int_0^s [(U', U') - K_M(P)(U, U)] \, dt = \int_0^s (\Sigma a_iZ_i, \Sigma a_iZ_i')_s.$$

Lemma 2 follows from the two preceding equalities and from the observation $a_i = f_i(s)$.

4. RELATIONS BETWEEN CUT POINTS AND CONJUGATE POINTS

The first general statement we can make about relations between cut points and conjugate points is that along any geodesic g starting at a point, say p, the cut point of p comes no later than the first conjugate point of p. In other words, we prove the following theorem.

THEOREM 4.1: *A geodesic g starting at a point p does not minimize distance to p beyond the first conjugate point of p.*

Proof: Let $g(c)$ be the first conjugate point of p and let $c < b$. We wish to show that the distance $d(g(0), g(b))$ is less than the arc length b of g from $t = 0$ to $t = b$, which is intuitively obvious. Suppose we have a variation $g_s = g_s(t)$ of the geodesic g such that $g_s(0) = g(0)$ and $g_s(c) = g(c)$. We shall show that there is a shorter way to go from $p = g(0)$ to $g(b)$ than along g. We fix a small e and first consider a broken path from $p = g_e(0)$ to $g_e(c) = g(c)$ via g_e, and then from $g(c)$ to $g(b)$ via g, which has length $c + (b - c) = b$. Smoothing out the corner of this path at $g(c)$, we obtain a shorter path from p to $g(b)$. However, in general we have only an infinitesimal variation or a Jacobi field vanishing at $g(0)$ and $g(c)$ but not necessarily a variation with the fixed points $g(0)$ and $g(c)$. Therefore, the rigorous proof is more delicate. We first prove the following lemma.

LEMMA 1: *Let g, b, and c be defined as previously. Then there exists a vector field X along g such that:* (1) X *is perpendicular to g;* (2) X *vanishes at $t = 0$ and at $t = b$;* (3) $\int_0^b [(X', X') - (R(X, G)G, X)] \, dt < 0$, *where G denotes the vector field tangent to g.*

Proof: Let Y be a nonzero Jacobi field along g vanishing at $t = 0$ and $t = c$. Take a small positive number d such that $g(c - d)$ has no conjugate points along g between $g(c - d)$ and $g(c + d)$ and such that there is a Jacobi field Z along g with $Z(c - d) = Y(c - d)$ and $Z(c + d) = 0$. (It is sufficient to take a positive number d such that $g(c - d)$ and $g(c + d)$ are in a convex neighborhood U of $g(c)$ such that each point of U has a normal coordinate neighborhood containing U.) We define a vector field X along $g = g(t)$, $0 \leqq t \leqq b$, as follows:

$$X = Y \quad \text{for } 0 \leqq t \leqq c - d,$$

$$X = Z \quad \text{for } c - d \leqq t \leqq c + d,$$

$$X = 0 \quad \text{for } c + d \leqq t \leqq b.$$

To simplify the calculation, we introduce the following notation: If W is a vector field along f, we set

$$I_r^s(W) = \int_r^s [(W', W') - (R(W, G)G, W)] \, dt.$$

Integrating

$$0 = (Y'' + R(Y, G)G, Y) = (Y'', Y) + (R(Y, G)G, Y)$$

$$= (Y', Y)' - (Y', Y') + (R(Y, G)G, Y)$$

from $t = 0$ to $t = c$, we obtain

$$I_0^c(Y) = 0.$$

Thus,

$$I_0^b(X) = I_0^b(X) - I_0^c(Y) = I_0^{c-d}(Y) + I_{c-d}^{c+d}(Z)$$
$$-I_0^{c-d}(Y) - I_{c-d}^c(Y)$$
$$= I_{c-d}^{c+d}(Z) - I_{c-d}^c(Y).$$

Let $\overline{Y}$ be the vector field along g from $t = c - d$ to $t = c + d$ defined by

$$\overline{Y} = Y \quad \text{for } c - d \leqq t \leqq c,$$
$$\overline{Y} = 0 \quad \text{for } c \leqq t \leqq c + d.$$

Applying Lemma 2 of the preceding section to $\overline{Y}$ and Z, we have

$$I_{c-d}^{c+d}(Z) < I_{c-d}^{c+d}(\overline{Y}) = I_{c-d}^c(Y).$$

Hence,

$$I_0^b(X) < 0.$$

Finally, we have to show that X is perpendicular to g—that is, Y and Z are perpendicular to g. It is sufficient to prove that, in general, a Jacobi field V along g which is perpendicular to g at two points is perpendicular to g everywhere. Consider the function (V, G) of t, which vanishes for two distinct values of t. Differentiating twice, we have

$$(V, G)'' = (V'', G) = (R(V, G)G, G) = 0.$$

Hence, (V, G) is identically zero, which completes the proof of Lemma 1.

In the intuitive but incomplete proof of Theorem 4.1 given previously, we constructed a path from $p = g(0)$ to $g(b)$ which is shorter than the geodesic segment $g(t)$, $0 \leqq t \leqq b$. The vector field X in Lemma 1 may be considered an infinitesimal analogue of that path.

Let X be the vector field along g given in Lemma 1. Let $g_s = g_s(t)$ be a 1-parameter family of curves from $p = g(0)$ to $g(b)$, such that (1) $g_0 = g$, and (2) g_s induces X—that is, for each fixed t, $X(t)$ is the vector tangent to the curve described by $g_s(t)$, $-\epsilon < s < \epsilon$. We remark that for $s \neq 0$ g_s need not be a geodesic. For each fixed s, let $L(s)$ be the length of the curve $g_s = g_s(t)$, $0 \leq t \leq b$. (For $s \neq 0$, g_s is not necessarily parametrized by arc length and $L(s)$ may be different from b.) For completion of the proof of Theorem 4.1, it is sufficient to show that the function L of s attains a local maximum at $s = 0$.

LEMMA 2: $L'(0) = 0$ and $L''(0) = I_0^b(X) < 0$.

Proof: The proof may be given by calculation in a manner similar to that in Section 2. (Here, again, a more satisfactory proof can be given with the use of differential forms and Cartan's structure equations.) As previously, we set $f(s, t) = g_s(t)$ and denote by $D/\partial t$ and $D/\partial s$ the covariant differentiations with regard to t and s respectively. Then

$$L(s) = \int_0^b \left(\frac{\partial f(s, t)}{\partial t}, \frac{\partial f(s, t)}{\partial t} \right)^{1/2} dt.$$

To differentiate $L(s)$ with respect to s at $s = 0$, we first differentiate the integrand with respect to s. We have

$$\frac{\partial}{\partial s} \left(\frac{\partial f}{\partial t}, \frac{\partial f}{\partial t} \right)^{1/2} = \left(\frac{D}{\partial s} \frac{\partial f}{\partial t}, \frac{\partial f}{\partial t} \right) \left(\frac{\partial f}{\partial t}, \frac{\partial f}{\partial t} \right)^{-1/2}.$$

At $s = 0$, $[(\partial f/\partial t), (\partial f/\partial t)] = 1$, because $g = g_0$ is a geodesic. We can also substitute $(D/\partial t)(\partial f/\partial s)$ for $(D/\partial s)(\partial f/\partial t)$. Hence, integrating by parts, we obtain

$$L'(0) = \int_0^b \left(\frac{D}{\partial t} \frac{\partial f}{\partial s}, \frac{\partial f}{\partial t} \right)_{s=0} dt$$

$$= \left(\frac{\partial f}{\partial s}, \frac{\partial f}{\partial t} \right)_{s=0} \Big|_{t=0}^{t=b} - \int_0^b \left(\frac{\partial f}{\partial s}, \frac{D}{\partial t} \frac{\partial f}{\partial t} \right)_{s=0} dt$$

$$= (X(b), G(b)) - (X(0), G(0)) - \int_0^b (X, G') \, dt.$$

Since $X(b) = 0$ and $X(0) = 0$, and because g is a geodesic—that is, $G' = 0$—we have $L'(0) = 0$. Referring to the preceding calculation and also to the calculation in Section 2, the reader should easily be able to obtain the equality $L''(0) = I_0^b(X)$. By Lemma 1, $I_0^b(X) < 0$, and the proof of Theorem 4.1 is complete.

Another theorem on cut points and conjugate points follows.

THEOREM 4.2: *If $g(b)$ is the cut point of a point $p = g(0)$ along a geodesic $g = g(t)$, $0 \leqq t < \infty$, then one of the following (or possibly both) holds: (1) $g(b)$ is the first conjugate point of $g(0)$ along g; (2) there exist at least two minimizing geodesics from $g(0)$ to $g(b)$.*

Proof: Let $a_1, a_2, \cdots$ be a monotone decreasing sequence of real numbers converging to b. For each k, let $\exp tX_k$, $0 \leqq t \leqq b_k$, be a minimizing geodesic from $g(0)$ to $g(a_k)$, where X_k is a unit tangent vector at $g(0)$ and b_k is the distance $d(g(0), g(a_k))$. Write $g(t) = \exp tX$, where X is the unit vector tangent to g at $g(0)$. Since $g(b)$ is the cut point of $g(0)$ along g and $a_k > b$, we have

$$X \neq X_k \quad \text{and} \quad a_k > b_k.$$

Since $b_k = d(g(0), g(a_k))$, we have

$$b = \lim b_k.$$

Hence, the set of vectors $b_k X_k$ is contained in a compact subset of the tangent space $T_{g(0)}(M)$. By taking a subsequence if necessary, we may assume that $b_1 X_1, b_2 X_2, \cdots$ converges to a vector of length b—say bY—where Y is a unit vector. Then $\exp tY$, $0 \leqq t \leqq b$, is a minimizing geodesic from $g(0)$ to $g(b)$, because

$$\exp bY = \lim \exp b_k X_k = \lim g(a_k) = g(b).$$

If $X \neq Y$, then we have two minimizing geodesics, $\exp tX$ and $\exp tY$ from $g(0)$ to $g(b)$. If $X = Y$, then $g(b)$ is conjugate to $g(0)$ along g. In fact, assume the contrary. Then $\exp : T_{g(0)}(M) \to M$ is a diffeomorphism of a neighborhood U of bX in $T_{g(0)}(M)$ onto a neighborhood of $g(b)$ in M. If k is large enough so that both $a_k X$ and $b_k X_k$ are in U, then $\exp a_k X = g(a_k) = \exp b_k X_k$, and, hence, $a_k X = b_k X_k$, which is a contradiction. By Theorem 4.1, $g(b)$ must be the *first* conjugate point of $g(0)$ along g, which completes the proof.

From Theorem 4.2 it follows rather easily that if $g(b)$ is the cut point of $g(0)$ along g, then $g(0)$ is the cut point of $g(b)$ along g (in the reverse direction).

We are now in position to prove Theorem 1.1 as we promised in Section 1. With the notations used in Theorem 1.1, we wish to prove that the function f defined on the sphere S is continuous. Actually we shall prove a slightly stronger statement. At each point p of M we consider the unit sphere S_p in the tangent space $T_p(M)$ and let $S(M)$ be the unit sphere bundle—that is, $S(M) = \bigcup_p S_p$.

THEOREM 4.3: *Let $f: S(M) \to [0, \infty]$ be the function defined as follows: For each unit vector $X \in S(M)$ at p, $f(X)$ is the distance from p to its cut point along the geodesic issued in the direction of X. Then f is continuous.*

Proof: Theorem 1.1 asserts that the restriction of f to each S_p is continuous. To prove the theorem, we assume that f is not continuous at $X \in S_p$ and let $X_1, X_2, \cdots$ be a sequence of points of $S(M)$ converging to X, such that $f(X) \neq \lim f(X_k)$. By taking a subsequence if necessary, we may assume that $\lim f(X_k)$ exists in $(0, \infty)$.

We first consider the case $f(X) > \lim f(X_k)$. We set

$$a_k = f(X_k), \qquad a = \lim a_k.$$

Let $T(M)$ denote the tangent bundle of M, that is, $T(M) = \bigcup_p T_p(M)$. We define a mapping $E: T(M) \to M \times M$ by

$$E(Z) = (\pi(Z), \exp Z),$$

where π is the projection from $T(M)$ onto M, $\pi(T_p(M)) = p$. Since $f(X) > a$, $\exp aX$ cannot be conjugate to p along the geodesic $\exp tX$. Hence, E maps a neighborhood, say U, of aX diffeomorphically onto a neighborhood of $(\pi(X), \exp aX)$. We may assume, by omitting a finite number of $a_k X_k$ if necessary, that all $a_k X_k$ are in U. Then $\exp a_k X_k$ cannot be conjugate to $\pi(X_k)$ along the geodesic $\exp tX_k$. By Theorem 4.2, there is another minimizing geodesic $\exp tY_k$ from $\pi(X_k)$ to $\exp a_k X_k$, where Y_k is a unit vector at $\cdot \pi(X_k)$ such that $Y_k \neq X_k$ and $\exp a_k Y_k = \exp a_k X_k$. Since $E: U \to M \times M$ is injective, $a_k Y_k$ is not in U. By taking a subsequence if necessary, we may assume that $Y_1, Y_2, \cdots$ converges to a unit vector, say Y, at p. Then aY is the limit of $a_1 Y_1, a_2 Y_2, \cdots$ and does not lie in U. We have

$$\exp aY = \exp (\lim a_k Y_k) = \lim (\exp a_k Y_k) = \lim (\exp a_k X_k)$$
$$= \exp (\lim a_k X_k) = \exp aX.$$

Hence, both $\exp tX$ and $\exp tY$ are minimizing geodesics from p to $\exp aX = \exp aY$. Consequently, if b is any number greater than a, the geodesic $\exp tX$, $0 \leq t \leq b$, is not minimizing, in contradiction to the assumption $f(X) > a$.

Next we consider the case $f(X) < \lim f(X_k)$. As before, we set $a_k = f(X_k)$ and $a = \lim a_k$. Then

$$d(p, \exp aX) = \lim d(\pi(X_k), \exp a_k X_k) = \lim a_k = a,$$

showing that $\exp tX$, $0 \leq t \leq a$, is a minimizing geodesic in contradiction to the assumption $f(X) < a$. The proof of Theorem 4.3 is thus complete.

As an immediate consequence, we have the following corollary.

COROLLARY: *The distance $d(p, C(p))$ between a point p and its cut locus $C(p)$ is a continuous function of p.*

For a special point of $C(p)$, we can sharpen the result of Theorem 4.2 as follows.

THEOREM 4.4: *Let q be a point on the cut locus $C(p)$ of p which is closest to p. Then either q is conjugate to p with respect to a minimizing geodesic joining p and q, or q is the midpoint of a geodesic starting and ending at p.*

Proof: Assuming that q is not conjugate to p, we let g and g' be two minimizing geodesics from p to q and b be the distance $d(p, q)$. We let K be a cone formed by geodesics of length b issuing from p and neighboring g. Similarly, for g' we consider a cone K'. The endpoints of the family of geodesics defining K give a hypersurface through q, with a tangent space at q perpendicular to g. The endpoints of the geodesics defining K' give another hypersurface through q, with a tangent space at q perpendicular to g'. We assume that g and g' meet at q with an angle not equal to π. Then the two tangent hyperplanes at q intersect (they do not coincide), as do the two hypersurfaces. It follows that K and K' intersect. We let r be a point in $K \cap K'$ near q. Then r is joined by two geodesics, one neighboring g and the other neighboring g'

and each being shorter than g and g'. Thus, r is a cut point of p closer to p than is the point q, in contradiction to the choice of q.

Taking a special point p, we shall go one step further.

THEOREM 4.5: *Let p be a point such that $d(p, C(p))$ is the smallest, and let q be a point on $C(p)$ which is closest to p. Then either q is conjugate to p with respect to a minimizing geodesic joining p and q, or q is the midpoint of a closed geodesic starting and ending smoothly at p.*

Proof: We assume that q is not conjugate to p. Applying Theorem 4.4, we see that q is the midpoint of a geodesic starting and ending at p. To see that the geodesic returns to p with angle 0, we reverse the roles of p and q in the proof of Theorem 4.4.

5. CONCLUDING REMARKS

Let M be a compact, simply connected Riemannian manifold the sectional curvature of which, $K(P)$, satisfies the inequalities $0 < K(P) \leqq 1$. Let p be a point of M such that the distance from p to its cut locus $C(p)$ is the smallest. Let q be a point on $C(p)$ which is closest to p. According to Theorem 4.5, two possibilities exist. If q is conjugate to p, then $d(p, C(p)) \geqq \pi$ by Theorem 3.1. If p and q are opposite points of a closed geodesic, then $d(p, C(p)) \geqq \pi$ under the additional assumption that the dimension of M is even [see Klingenberg (8)]. If we assume $\frac{1}{4} \leqq K(P) \leqq 1$, then $d(p, C(p)) \geqq \pi$, even if the dimension of M is odd [see Klingenberg (10 and 11)].

The inequality $d(p, C(p)) \geqq \pi$ is essential in proving the so-called sphere theorem, which states that if M is a compact, simply connected Riemannian manifold with $\frac{1}{4} < K(P) \leqq 1$, then M is homeomorphic with a sphere. The proof may be divided into the following three lemmas.

LEMMA 1: *If $\frac{1}{4} < K(P) \leqq 1$, then $d(p, C(p)) \geqq \pi$ for every $p \in M$.*

LEMMA 2: *Let h and a be positive numbers such that*

$$\frac{1}{4} < h \leqq K(P) \leqq 1 \quad and \quad \frac{\pi}{2\sqrt{h}} < a < \pi.$$

If p and q are points of M such that the distance $d(p, q)$ is equal to the diameter of M, then $d(p, x) < a$ or $d(q, x) < a$ for every $x \in M$.

LEMMA 3: *If there exist two points p and q and a positive number a such that*

(5.1) $$d(p, C(p)) > a \quad and \quad d(q, C(q)) > a;$$

(5.2) $$d(p, x) < a \quad or \quad d(q, x) < a \quad for \ every \ x \in M,$$

then M is homeomorphic with a sphere.

It is evident that Lemmas 1 through 3 imply the sphere theorem. For the proofs of Lemmas 2 and 3, we refer the reader to Tsukamoto (27) and Klingenberg (8), respectively. References for Lemma 1 have been given previously. The second lemma was proved originally by Berger (1) using results of Toponogov (24 and 25). For the proof of the sphere theorem, see Toponogov (26). By analyzing the distribution of conjugate points and by using Morse theory instead of the method we have described, it is possible to prove that, under the same assumption as in the sphere theorem, M is a homotopy sphere [see Klingenberg (12)]. For an improvement of Rauch's comparison theorem and further properties of cut loci, see Berger (2).

We have seen some examples in which the cut locus $C(p)$ of a point p coincides with the first conjugate locus of p. This phenomenon occurs for every simply connected Riemannian symmetric space [see Crittenden (5)]. For more recent results on cut and conjugate loci of compact symmetric spaces, see Naitoh (45), Sakai (48, 49), Takeuchi (53).

For a general Riemannian manifold, the cut locus and the first conjugate locus can be disjoint. In fact, Weinstein (57) has shown that on every compact manifold M not homeomorphic to the 2-sphere there exist a Riemann metric and a point p such that the cut locus $C(p)$ does not intersect the conjugate locus $Q(p)$ of p. It seems to be unknown if on every compact homogeneous Riemannian manifold the cut locus and the first conjugate locus intersect. Sugahara (52) studied Riemannian manifolds with the property that $C(p) \cap Q(p)$ is empty.

While Warner (30) studied the stratification of conjugate loci into regular and singular conjugate points, Bishop (33) considered the decomposition of the cut loci into ordinary and singular cut points. Ozols (47) analyzed the local structure of the cut locus near a nonconjugate cut point. In the dimensions less than or equal to 6, the local structure of the generic cut locus has been completely determined by Buchner (35, 36). However, the cut locus can be very complicated at a conjugate point. Generalizing the results of Myers (15, 16) on cut loci of a real analytic 2-dimensional Riemannian manifold, Buchner (34) has shown that the cut locus $C(p)$ of a real analytic n-dimensional Riemannian manifold is a simplicial complex. On the other hand, if the metric is not real analytic, $C(p)$ may not be triangulable according to Gluck and Singer (38, 51).

For more advanced studies on the topics treated here, see Cheeger and Ebin's monograph (37). In his survey article (50) Sakai summarizes more recent results on comparison theorems (of Berger, Warner, Heintze, Karcher), differentiable sphere theorems (of Shikata, Calabi, Gromoll, Sugimoto, Shiohama, Karcher, Grove, ImHof), finiteness theorems (of Weinstein, Cheeger) and related results of Gromov. His article also contains a very extensive bibliography.

BIBLIOGRAPHY

1. Berger, M., "Les variétés Riemanniennes ($\frac{1}{4}$)-pincées," *Annali della Scuola Normale Sup. di Pisa, Ser. III* 14 (1960), pp. 161–70.

2. Berger, M., "An extension of Rauch's metric comparison theorem and some applications," *Illinois Journal of Mathematics*, 6 (1962), pp. 700–12.

3. Braunmühl, A. v., "Über Enveloppen geodätischer Linien," *Math. Ann.*, 14 (1879), pp. 557–66.

4. Cartan, É., *Leçons sur la géométrie des espaces de Riemann*. Paris: Gauthier-Villars, 1928, 1946.

5. Crittenden, R., "Minimum and conjugate points in symmetric spaces," *Canadian Journal of Mathematics*, 14 (1962), pp. 320–28.

6. Hilbert, D., "Über das Dirichletsche Prinzip," *J. Reine Angew. Math.*, 129 (1905), pp. 63–67.

7. Hopf, H., and W. Rinow, "Über den Begriff der vollständigen differential geometrischen Fläche," *Comm. Math. Helv.*, 3 (1931), 209–25

8. Klingenberg, W., "Contributions to Riemannian geometry in the large," *Ann. of Mathematics*, 69 (1959), 654–66.

9. Klingenberg, W., "Über kompakte Riemannsche Mannigfaltigkeiten," *Math. Ann.*, 137 (1959), 351–61.

10. Klingenberg, W., "Über Riemannsche Mannigfaltigkeiten mit positiver Krümmung," *Comm. Math. Helv.*, 35 (1961), pp. 47–54.

11. Klingenberg, W., "Über Riemannsche Mannigfaltigkeiten mit nach oben beschränkter Krümmung," *Annali di Mat.*, 60 (1963), pp. 49–60.

12. Klingenberg, W., "Manifolds with restricted conjugate locus," *Ann. of Math.*, 78 (1963), pp. 527–47.

13. Kobayashi, S., and K. Nomizu, *Foundations of Differential Geometry*. New York: Wiley-Interscience, 1963.

14. Morse, M., *The Calculus of Variations in the Large*. Providence, R.I.: American Mathematical Society, 1934. (Reprinted 1965.)

15. Myers, S. B., "Connections between differential geometry and topology I," *Duke Mathematics Journal*, 1 (1935), 376–91.

16. Myers, S. B., "Connections between differential geometry and topology II," *Duke Mathematics Journal*, 2 (1936), 95–102.

17. Poincaré, H., "Sur les lignes géodésiques des surfaces convexes," *Transactions of the American Mathematics Society*, 6 (1905), 237–74.

18. Preismann, A., "Quelques propriétés globales des espaces de Riemann," *Comm. Math. Helv.*, 15 (1943), 175–216.

19. Rauch, H. E., "A contribution to differential geometry in the large," *Ann. of Mathematics*, 54 (1951), 38–55.

20. Rham, G. de, "Sur la réductibilité d'un espace de Riemann," *Comm. Math. Helv.*, 26 (1952), 328–44.

21. Schoenberg, J. M., "Some applications of the calculus of variations to Riemannian geometry," *Ann. of Mathematics*, 33 (1932), 485–95.

22. Struik, D. T., *Lectures on Classical Differential Geometry*. Reading, Massachusetts: Addison-Wesley Publishing Co., Inc., 1950.

23. Synge, J. L., "The first and second variations of the length in

Riemannian spaces," *Proceedings of the London Mathematical Society*, 25 (1926), 247–64.

24. Topogonov, V. A., "On the convexity of Riemannian spaces with positive curvature," *Dokl. Nauk*, 115 (1957), 674–76.

25. Topogonov, V. A., "Riemannian spaces which have their curvature bounded from below by a positive number," *Dokl. Nauk*, 120 (1958), 719–21.

26. Topogonov, V. A., "Dependence between curvature and topological structure of Riemannian spaces of even dimensions," *Dokl. Nauk*, 133 (1960), 1031–33.

27. Tsukamoto, Y., "On Riemannian manifolds with positive curvature," *Memoirs of Fac. Sci. Kyushu Univ. Ser. A, Math.*, 15 (1962), 90–96.

28. Whitehead, J. H. C., "On the covering of a complete space by the geodesics through a point," *Ann. of Mathematics*, 36 (1935), 679–704.

Since the preparation of this manuscript, the following articles on the subject have appeared.

29. Bishop, R., and R. Crittenden, *Geometry of Manifolds*. New York: Academic Press Inc., 1964. This book is largely based on the lectures given by Ambrose, mentioned in the introduction.

30. Warner, F. W., "The conjugate locus of a Riemannian manifold," *American Journal of Mathematics*, 87 (1965), 575–604.

31. Warner, F. W., "Extensions of the Rauch comparison theorem to submanifolds," *Transactions of the American Mathematical Society*, 122 (1966), 341–56.

Among the books and papers that have appeared since 1966, we mention the following:

32. Besse, A. L., *Manifolds All of Whose Geodesics Are Closed*, Springer-Verlag, 1978.

33. Bishop, R., Decomposition of cut loci, *Proc. Amer. Math. Soc.*, 65 (1977), 133–136.

34. Buchner, M., Simplicial structure of the real analytic cut locus, *Proc. Amer. Math. Soc.*, 64 (1977), 118–121.

35. Buchner, M., Stability of the cut locus in dimension less than or equal to 6, *Invent. Math.*, 43 (1977), 199–231.

36. Buchner, M., The structure of the cut locus in dimension less than or equal to six, *Compos. Math.*, 37 (1978), 103–119.

37. Cheeger, J. and Ebin, D., *Comparison Theorems in Riemannian Geometry*, North-Holland Math. Library, vol. 9, 1975.

38. Gluck, H. and Singer, D., Scattering of geodesic fields, I, II, *Ann. of Math.*, 108 (1978), 347–372; 110 (1979), 205–225.

39. Gromoll, D., Klingenberg, W., and Meyer, W., *Riemannsche Geometrie im Grossen*, Lecture Notes in Math. 55 (1968), Springer-Verlag.

40. Itoh, J., Some considerations on the cut locus of Riemannian manifolds, *Advanced Studies in Pure Math.*, 3 (1984), Geometry of Geodesics and Related Topics, 29–46.

41. Klingenberg, W., *Lectures on Closed Geodesics*, Springer-Verlag, 1978.

42. Karcher, H., Schnittort und konvexe Mengen in Riemannsche Mannigfaltigkeiten, *Math. Ann.*, 177 (1968), 105–121.

43. Klingenberg, W., *Riemannian Geometry*, Walter de Gruyter, 1982.

44. Kobayashi, S. and Nomizu, K., *Foundations of Differential Geometry*, Vol. 2, John Wiley & Sons, 1969.

45. H. Naitoh, On cut loci and first conjugate loci of the irreducible symmetric *R*-spaces and the irreducible compact Hermitian symmetric spaces, *Hokkaido Math. J.*, 6 (1977), 230–242.

46. Nakagawa, H. and Shiohama, K., On Riemannian manifolds with certain cut loci I, II, *Tohoku Math. J.*, 22 (1970), 14–23; 357–361.

47. Ozols, V., Cut loci in Riemannian manifolds, *Tohoku Math. J.*, 26 (1974), 219–227.

48. Sakai, T., On cut loci of compact symmetric spaces, *Hokkaido Math. J.*, 6 (1977), 136–161.

49. Sakai, T., On the structure of cut loci in compact Riemannian symmetric spaces, *Math. Ann.*, 235 (1978), 129–148.

50. Sakai, T., Comparison and finiteness theorems in Riemannian geometry, *Advanced Studies in Pure Math.*, 3 (1984) (Geometry of Geodesics and Related Topics), 125–181.

51. Singer, D. and Gluck, H., The existence of nontriangulable cut loci, *Bull. Amer. Math. Soc.*, 82 (1976), 599–602.

52. Sugahara, K., On the cut locus and the topology of Riemannian manifolds, *J. Math. Kyoto Univ.*, 14 (1974), 391–411.

53. Takeuchi, M., On conjugate and cut loci of compact symmetric spaces I, II, *Tsukuba Math. J.*, 2 (1978), 35–68; 3 (1979), 1–29.

54. Thom, R., Sur le cut-locus d'une variété plongée, *J. Diff. Geom.*, 6 (1972), 577–586.

55. Warner, F., Conjugate loci of constant order, *Ann. of Math.*, 86 (1967), 192–212.

56. Weinstein, A. D., The cut locus and conjugate locus of a Riemannian manifold, *Ann. of Math.*, 87 (1968), 29–41.

57. Weinstein, A. D., The generic conjugate locus, *Global Analysis*, Proc. Symp. Pure Math., Amer. Math. Soc., 15 (1970), 299–301.

58. Wolter, F.-E., Distance function and cut loci on a complete Riemannian manifold, *Arch. Math.*, 32 (1979), 92–96.

RIEMANNIAN COMPARISON CONSTRUCTIONS

Hermann Karcher

When S. Kobayashi wrote his chapter for the first edition of this book he reported on amazing results which had just been obtained. Those results triggered a rapid development of comparison theory: in 1975 the book by Cheeger and Ebin [CE] dealing mostly with curvature > 0 appeared. In 1978 Gromov's theorem classifying almost flat manifolds appeared, see [BK]. In 1985 the case of curvature < 0 was treated in a book by Ballman, Gromov, and Schroeder [BGS]. In survey articles, by Burago and Zalgaller [BZ] and M. Berger [B], the field was explained to a wider audience. The article, "Comparison and Finiteness Theorems in Riemannian Geometry," by T. Sakai [S] contains a very complete bibliography.

I have been asked to present, "with complete proofs," part of this development. I selected material which the reader will, hopefully, see as a direct continuation of the research portrayed by Kobayashi. It therefore seemed natural to assume his chapter as background; I refer to its sections as (K.1), (K.2), etc. In particular I will use the exponential map, minimizing geodesics, conjugate

170

and cut points, Jacobi fields, and the second variation formula without further motivation and almost without separate introduction. The only other prerequisite I am aware of is: not to be scared by the covariant derivative. I am using notation which resembles as closely as possible the use of Euclidean directional derivatives. The following three examples should explain the notation.

0.1 The fact that the Riemannian scalar product $g(\ ,\ )$ is *parallel* for the covariant derivative D but usually not constant for the local derivative ∂ of a chart is expressed by the following formula (U and V are vector fields):

$$\partial_X(g(U,V)) = \begin{cases} g(D_XU,V) + g(U, D_XV) \\ (\partial_Xg)(U,V) + g(\partial_XU,V) + g(U, \partial_XV). \end{cases}$$

0.2 The so-called *symmetry* of the covariant derivative is expressed as

$$D_XY - D_YX = [X, Y] \quad (= \partial_XY - \partial_YX),$$

or with the help of a map $c: I \times I \to M$ as $(D/\partial s)(\partial/\partial t)c = (D/\partial t)(\partial/\partial s)c$. Both identities follow from the local expression of D in terms of the chart derivative and the symmetric Christoffel symbols:

$$D_XY = \partial_XY + \Gamma(X,Y), \qquad \Gamma(X,Y) = \Gamma(Y,X).$$

Usually this symmetry is (axiomatically) assumed; then one finds with (0.1):

$$2g(\Gamma(X,Y), Z) = -(\partial_Zg)(X,Y) + (\partial_Xg)(Y,Z)$$

$$+(\partial_Yg)(Z,X).$$

0.3 The gradient of a function f is the vector field defined by

$$\partial_Xf = g(\operatorname{grad} f, X).$$

Further covariant differentiation gives the Hessian

$$D^2_{X,Y}f := \partial_X(\partial_Y f) - \partial_{D_X Y}f = g(D_X \operatorname{grad} f, Y)$$

$$= \partial^2_{X,Y}f - \partial_{\Gamma(X,Y)}f \quad \text{(local chart expression)}.$$

The local formula shows the symmetry of D^2f; $D \operatorname{grad} f$ is a symmetric endomorphism field. A function f with a positive definite Hessian is convex along any geodesic $c(D_{c'}c' = 0)$: $(f \circ c)'' = \partial_{c'}(g(\operatorname{grad} f, c')) = g(D_{c'}\operatorname{grad} f, c') > 0$.

CONTENTS

Guideline: observe how global properties are concluded from infinitesimal (= curvature) assumptions.

1. The setup to get curvature control started.

Natural functions, Jacobi equation, Riccati equation, constant curvature case, reduction of the n-dimensional case to a 1-dimensional discussion, the Riccati inequality, principal curvature—and Hessian bounds, generalized Rauch bounds, Bishop-Gromov volume bounds.

2. Immediate applications.

The Hadamard-Cartan theorem, fixed points of isometries, growth of the fundamental group, the Ricci-diameter bound with equality discussion.

3. Busemann functions.

$K \geqslant 0$ and compact totally convex exhaustion, Ricci $\geqslant 0$ and the splitting of a line as factor.

4. Triangle comparison theorems.

Angle and secant comparisons with upper or lower curvature bounds. A new proof of Toponogov's theorem.

5. Applications of the triangle theorems.

Bound for the number of generators of the fundamental group; critical points of the distance function, cut locus estimates (Klingenberg, Cheeger, Toponogov), sphere theorems (Rauch, Berger, Klingenberg, Shikata, Grove-Shiohama).

6. Complex projective space and its distance spheres.
Description from scratch: metric, embedding, equivariant isometries, curvature tensor of $\mathbb{C}P^n$, curvature tensor and short closed geodesics of the distance spheres.

1. THE SETUP TO GET CURVATURE CONTROL STARTED

1.1 *Natural maps and functions.* If one tries to generalize the arclength parametrization of curves to get good coordinates for a Riemannian manifold M, then a natural map *from* a Euclidean space $\mathbb{R}^n$ (e.g. T_pM) *into* M is the exponential map (K.1). It is defined via an initial value problem for geodesics from a point p:

$$\exp_p: T_pM \to M$$

$$\exp_p(v) = c(1), \quad \text{where } c \text{ is the geodesic with} \quad (1.1.1)$$

$$c(0) = p, \qquad c'(p) = v.$$

Particularly natural functions from (parts of) a Riemannian manifold M into $\mathbb{R}$ are distance functions (from a point or a submanifold), i.e. functions which satisfy

$$|\operatorname{grad} f| = 1 \qquad \text{(distance function)}. \qquad (1.1.2)$$

Integral curves of distance functions are geodesics; namely, let u be an arbitrary tangent vector then

$$0 = \partial_u g(\operatorname{grad} f, \operatorname{grad} f) \underset{(0.1)}{=} 2g(D_u \operatorname{grad} f, \operatorname{grad} f)$$

$$\underset{(0.3)}{=} 2g(D_{\operatorname{grad} f} \operatorname{grad} f, u) \qquad \text{(symmetry of } D \operatorname{grad} f\text{)},$$

so indeed

$$D_{\operatorname{grad} f} \operatorname{grad} f = 0. \qquad (1.1.3)$$

The level surfaces are therefore called a family of "parallel" surfaces.

1.2 *Connections with Jacobi fields* (K.2). The differential equation for Jacobi fields J along a geodesic c,

$$J'' - R(J, c')c' = 0 \qquad (J'' = D_{c'}(D_{c'}J)), \qquad (1.2.1)$$

is the linearization of the geodesic equation along c. It is thus by definition that estimates on Jacobi fields as in Rauch's theorem (K.3.2) contain estimates of the differential of the exponential map or the Hessian of distance functions. We only have to relate the *initial conditions* for the Jacobi equation to the relevant families of geodesics.

First, for the exponential map we get

$$\partial \exp_p\Big|_{s \cdot v}(sw) := \frac{\partial}{\partial t}\exp_p(s \cdot (v + tw))\Big|_{t=0} =: J(s), \qquad (1.2.2)$$

$$J(0) = 0, \qquad J'(0) = w$$

since

$$J'(0) = \frac{D}{\partial s}\frac{\partial}{\partial t}\exp_p(s(v + tw))\Big|_{0,0}$$

$$\underset{(0.2)}{=} \frac{D}{\partial t}\left(\frac{\partial}{\partial s}\exp_p(s(v + tw))\Big|_0\right)\Big|_0$$

$$= \frac{D}{\partial t}(v + tw) = w.$$

Rauch type estimates $|J'(0)| \cdot h(s) \leqslant |J(s)| \leqslant |J'(0)| \cdot H(s)$ therefore translate into $|w| \cdot h(s) \leqslant |\partial \exp_p|_{sv}(sw)| \leqslant |w| \cdot H(s)$.

Second, for distance functions one has a natural unit normal field N along the level surfaces, $N := \operatorname{grad} f$, and the shape operator S with respect to that normal is the Hessian of f:

$$S \cdot u := D_u N = D_u \operatorname{grad} f \qquad (u \text{ tangential to a level surface}).$$

$$(1.2.3)$$

The eigenvalues of $S = D \operatorname{grad} f$ are called principal curvatures (of the level surface).

The sign convention in (1.2.3) is such that the principal curvatures of the sphere are positive for the outer normal. This is the better choice when dealing with level surfaces.

The family of normal geodesics defines natural diffeomorphisms from one level surface $f = \text{const.}$ to nearby ones:

$$E_s(p) := \exp_p s \cdot N(p), \qquad p \in f^{-1}(\text{const.}). \qquad (1.2.4)$$

As in (1.2.2) one can describe the differentials of the E_s by Jacobi fields and the specifics of the construction determine the initial conditions: let $p(t)$ be a curve in a level surface then

$$\partial E_s\Big|_p \cdot \dot{p}(0) := \frac{\partial}{\partial t} E_s(p(t))\Big|_{t=0} =: J(s), \qquad (1.2.5)$$

$$J(0) = \dot{p}(0), \qquad J'(0) = S \cdot \dot{p}(0)$$

since

$$J'(0) \underset{(0.2)}{=} \frac{D}{\partial t} \frac{\partial}{\partial s} \exp_p(s \cdot N(p(t)))\Big|_{s=0} = \frac{D}{\partial t} N(p(t)) \underset{(1.2.3)}{=} S \cdot \dot{p}(0).$$

Since $E_s(p(t))$ is a curve in another level surface $f^{-1}(s + \text{const.})$, (1.2.5) in fact describes all the shape operators S_s of the level surfaces (along each normal geodesic):

$$S_s \cdot J(s) = J'(s). \qquad (1.2.6)$$

Following definition (1.2.3), interpretations of the shape operator usually emphasize the turning speed of the normal along the level surface. But the shape operator also controls the change of length of the geodesic projections E_s between level surfaces:

$$\frac{d}{ds} g\left(\frac{\partial}{\partial t} E_s(p(t)), \ \frac{\partial}{\partial t} E_s(p(t)) \right)$$

$$\underset{(1.2.5)}{=} 2g\left(\frac{\partial}{\partial t} E_s(p(t)), S_s \cdot \frac{\partial}{\partial t} E_s(p(t)) \right). \qquad (1.2.7)$$

1.3   *The Riccati equation for the shape operators of the level surfaces of a distance function.*   The claim of this heading, namely that the shape operators of the level surfaces of a distance function satisfy the *first*-order equation (1.3.1) along the normal geodesics is a key observation. Together with (1.2.7) it allows us to split Rauch's estimates for a second-order equation into two first-order steps—each with a geometric interpretation. Because of its importance I give three derivatives of (1.3.1) which connect three different points of view of the basic comparison construction.

First, differentiate (1.2.6) and use the Jacobi equation:

$$J''(s) = \frac{D}{\partial s}\left(S_s \cdot J(s)\right) = S_s' \cdot J(s) + S_s \cdot J'(s)$$

$$= -R(J, N)N(s) = \left(S_s' + S_s^2\right) \cdot J(s).$$

We abbreviate $R(J, N)N =: R_N \cdot J$; also, for fixed s we can consider $J(s)$ as an arbitrary tangent vector to a level surface. Then the last equation is a *Riccati equation* for S_s:

$$S_s' = -R_N - S_s^2. \qquad (1.3.1)$$

Second, insert a parallel vector field $U(s) \perp N$ into (1.2.3) and differentiate in the direction $\partial/\partial s = N = \operatorname{grad} f$:

$$\frac{D}{\partial s}\left(S_s \cdot U(s)\right) = S_s' \cdot U(s)$$

$$= \frac{D}{\partial s}\left(D_{U(s)}\operatorname{grad} f\right) = D_{N,U}^2 \operatorname{grad} f$$

$$\left(\text{note } \frac{D}{\partial s}U = 0\right).$$

Insert in this equality the definition of the curvature tensor $D_{N,U}^2 \operatorname{grad} f = D_{U,N}^2 \operatorname{grad} f - R(U, N)\operatorname{grad} f$ and the definition of the second differential of a vector field $D_{U,N}^2 \operatorname{grad} f \underset{\text{(Def.)}}{=}$

$D_U(D_N \operatorname{grad} f) - D_{D_U N}\operatorname{grad} f \underset{(1.1.3,\, 1.2.3)}{=} -S^2 \cdot U$ to obtain (1.3.1) again.

Third, consider a 2-dimensional Riemannian metric in Gaussian form:

$$ds^2 = du^2 + G^2(u, v) \, dv^2, \qquad \text{with curvature } K = -\frac{G_{uu}}{G}.$$

The u-lines are geodesics and their orthogonal trajectories are parallel curves. The rate of change of their lengths gives their geodesic curvature:

$$\frac{d}{du} \int G \, dv = \int \frac{G_u}{G} G \, dv = \int \kappa_g \, ds \quad \text{or} \quad \kappa_g = \frac{G_u}{G}.$$

One further differentiation gives the Riccati equation,

$$\frac{d}{du} \kappa_g = -K - \kappa_g^2. \tag{1.3.2}$$

The 2-dimensional formula has been known a long time; we also shall see that the n-dimensional case is close to (1.3.2).

1.3.3.   It turns out to be rather easy to reduce the *comparison* discussion of (1.3.1) both for upper or lower curvature bounds to 1-dimensional Riccati inequalities—as if we had upper or lower curvature bounds in (1.3.2) (see (1.5) below). So the original Rauch line is replaced by:

Step 1.   Prove inequalities for the principal curvatures of level surfaces of distance functions via 1-dimensional Riccati inequalities.

Step 2.   Use Step 1 to integrate (1.2.6) or (1.2.7).

1.4   *The constant curvature case.*   The Rauch comparison theorem (K.3.2) is formulated in such a way that upper or lower bounds for the curvature seem to play a completely symmetric role. Most applications so far have gone via constant curvature models. Integration of the Rauch estimates to distance or volume control always requires in the case of *upper* curvature bounds some size restriction, e.g. stay away from some cut locus. That such restrictions are *not* needed in comparisons with *smaller* curvature models

(Toponogov, Bishop-Gromov) made these results prime tools in the development of comparison theory.

I summarize the relevant constant curvature data.

1.4.1. We denote the hyperbolic space of curvature $\kappa < 0$, Euclidean space ($\kappa = 0$), or the sphere of curvature $\kappa > 0$ either jointly by M_κ, or, if the sign of κ is specified, by H_κ^n, $\mathbb{R}^n$, S_κ^n.

1.4.2. Similarly, we use a common notation for the functions which control the trigonometry of those spaces. Denote the solutions of the differential equation

$$f'' + \kappa \cdot f = 0$$

which have the same initial conditions as sin, respectively, cos by

$$s_\kappa \text{ resp. } c_\kappa; \qquad s_\kappa' = c_\kappa, \qquad c_\kappa' = -\kappa s_\kappa.$$

1.4.3. Distance spheres of radius r have

$$\text{principal curvatures} = \frac{s_\kappa'}{s_\kappa}(r) =: ct_\kappa(r); \quad ct_\kappa' = -\kappa - ct_\kappa^2;$$

$$\text{the length of their great circles} = 2\pi s_\kappa(r).$$

Parallel surfaces at distance r from totally geodesic hyperplanes have

$$\text{principal curvatures} = \frac{c_\kappa'}{c_\kappa}(r).$$

The Hessian of the distance function has in the radial direction the eigenvalue 0. For the proof of Toponogov's theorem I need to rescale the distance function so that all eigenvalues of the Hessian become equal. This "modified distance" function is:

$$md_\kappa(r) := \int_0^r s_\kappa = \begin{cases} \frac{1}{2}r^2 & \text{if } \kappa = 0 \\ 1 - \cos r & \text{if } \kappa = 1 \\ \cosh r - 1 & \text{if } \kappa = -1 \end{cases} .$$

This function also avoids case distinctions in the *cosine formula* for triangles (of edge lengths a, b, c, in M_κ):

$$md_\kappa(c) = md_\kappa(a - b) + s_\kappa(a) \cdot s_\kappa(b) \cdot (1 - \cos \gamma).$$

1.4.4. If $\kappa \leqslant \Delta$ then we have in appropriate intervals

$$s_\Delta(r) \leqslant s_\kappa(r), \qquad c_\Delta(r) \leqslant c_\kappa(r), \qquad ct_\Delta(r) \leqslant ct_\kappa(r),$$

which expresses comparisons of lengths, areas and principal curvatures by explicit formulas (compare 1.4.3).

1.4.5. Rauch's estimates as well as the ones we shall prove are formulated for Jacobi fields perpendicular (or "normal") to their geodesics. This is all one needs since the tangential part of a Jacobi field is always a *linear* field and hence explicitly known from the initial data (independently from the curvature tensor):

$$J_{\text{tan}} := g(J, c') \cdot c' \qquad (\text{normalization } |c'| = 1).$$

Indeed, the skew symmetries of R imply $J''_{\text{tan}} = 0$ and $R(J_{\text{tan}}, c')c' = 0$. This and the Pythagorean theorem extend estimates for normal Jacobi fields to arbitrary ones.

1.5 *Reduction of the discussion of* (1.3.1) *to a 1-dimensional inequality.* All estimates in this section are pointwise. One can therefore allow that the lower and upper curvature bounds vary from point to point. This generalization will not be pursued.

1.5.1. Assume a lower bound $\delta \leqslant K$.
To discuss $S' = -R_N - S^2$ let u be a parallel unit field along a geodesic normal to the level surfaces. Then we obtain a first-order Riccati inequality as follows:

$$g(Su, u)' = -g(R(u, N)N, u) - g(S^2 u, u),$$

$$g(Su, u)' \leqslant -K(u \wedge N) - g(Su, u)^2 \leqslant -\delta - g(Su, u)^2.$$

1.5.2.   Assume an upper curvature bound $K \leqslant \Delta$.
To discuss $S' = -R_N - S^2$ let $p(t)$ be a curve in one level surface of the distance function. Consider the 2-dimensional "ruled" surface

$$F(s,t) := \exp_{p(t)} s \cdot N(p(t)).$$

Since the s-lines are geodesics we have a Gaussian parametrization and the geodesic curvatures of the parallel t-curves satisfy (by 1.3.2)

$$\kappa'_g = -K^F - \kappa^2_g,$$

$$\kappa'_g \geqslant -\Delta - \kappa^2_g \quad \text{from (1.5.3).}$$

Bounds derived from $K \leqslant \Delta$ therefore have a 2-dimensional geometric interpretation!

1.5.3.   For the ruled surface of (1.5.2),

$$K^F \leqslant K^M\left(\frac{\partial}{\partial s} F \wedge \frac{\partial}{\partial t} F\right).$$

Proof.   The s-lines are geodesics in M and hence in F. The t-derivative of this family, therefore, gives Jacobi fields $J(s)$ for both spaces. The covariant derivative D^F in F is the *orthogonal* projection of the covariant derivative D^M in M. The computation

$$\frac{\partial^2}{(\partial s)^2} g(J(s), J(s))$$

$$= \begin{cases} 2g\left(\dfrac{D^M}{ds}J, \dfrac{D^M}{ds}J\right) - 2g\left(R^M(J, F')F', J\right) \\[4mm] 2g\left(\dfrac{D^F}{ds}J, \dfrac{D^F}{ds}J\right) - 2g\left(R^F(J, F')F', J\right) \end{cases}$$

therefore proves

$$g\left(R^F(J, F')F', J\right) \leqslant g\left(R^M(J, F')F', J\right).$$

Volume estimates also fit in this discussion. For any family of invertible linear maps, $L(s)$, between Euclidean spaces we have (Flanders 7.10)

$$\frac{d}{ds}\det(L(s)) = \text{trace}(L' \cdot L^{-1}) \cdot \det(L(s)).$$

We apply this to the differentials ∂E_s of (1.2.4) and with the abbreviation

$$a(s) := \det(\partial E_s)$$

we get with (1.2.5, 1.2.6)

$$\frac{d}{ds}a(s) = (\text{trace } S_s) \cdot a(s). \tag{1.5.4}$$

This says that trace S_s is the growth rate of the hypersurface volume $a(s)$—tangent space-wise.

On the other hand, taking the trace of the Riccati equation (1.3.1) for S gives

$$\frac{d}{ds}\text{trace } S = -\text{trace } R_N - \text{trace } S^2. \tag{1.5.5}$$

1.5.6. Recall: $H := \dfrac{1}{n-1} \cdot \text{trace } S$ is the mean curvature of the level hypersurface.

ricci$(N, N) := \text{trace}(Y \to R(Y, N)N)$ is the Ricci curvature of M^n in the direction N.

Note that (1.5.4) and (1.5.5) are again controls in two first-order steps: (1.5.4) controls the volume growth of the level surfaces in terms of their mean curvature, and (1.5.5) controls the change of the mean curvature in terms of the Ricci curvature, except that trace S^2 and trace S are only related through

Schwarz' inequality for endomorphisms

$$(n-1) \cdot \text{trace } S^2 \geqslant (\text{trace } S)^2, \quad \text{`` = '' iff } S = \lambda \cdot \text{id.} \tag{1.5.7}$$

This and (1.5.5) give:

1.5.8.   Assume a lower bound $(n - 1) \cdot \rho \leqslant \mathrm{ric}(N, N)$. Then we have, for the mean curvature $H = 1/(n - 1)\mathrm{trace}\, S$ of the level surfaces of a distance function (1.1.2) along the geodesics normal to the level surfaces, the Riccati inequality

$$H' \leqslant -\frac{1}{n - 1}\mathrm{ric}(N, N) - H^2 \leqslant -\rho - H^2.$$

Now (1.5.4) and (1.5.8) are a perfect control. Because of the use of (1.5.7) one does not have a corresponding result assuming upper bounds on the Ricci curvature—in fact almost no consequences of such upper bounds are known.

For equality discussions see (2.4.2, 3.6).

1.6.   *The Riccati comparison argument.*   Note the simplicity of the following arguments. The assumptions for (1.6.1) came from (1.5.1, 1.5.2, 1.5.8).

Consider two functions f, F which satisfy on some interval

$$f' \leqslant -\rho - f^2, \qquad F' \geqslant -\rho - F^2. \tag{1.6.1}$$

Then

$$\left((f - F) \cdot e^{\int (f + F)}\right)' \leqslant 0.$$

1.6.2.   **COROLLARIES.**   *If*   a) $f(r_0) \geqslant F(r_0)$,   *respectively,*
b) $f(r_0) \leqslant F(r_0)$ *then*
a) $f(r) \geqslant F(r)$ *for* $r \leqslant r_0$ *(as long as* $f < \infty$, $F > -\infty$*),*
b) $f(r) \leqslant F(r)$ *for* $r \geqslant r_0$ *(as long as* $f > -\infty$, $F < \infty$*).*
1.6.3.   *Assume in addition to (1.6.1)*
a) $\lim_{r \to 0} f(r) = +\infty$,   *respectively*
b) $\lim_{r \to 0} F(r) = +\infty$,   *then*
a) $f(r) \leqslant ct_\rho(r)$ *as long as* $f = \infty$,
b) $F(r) \geqslant ct_\rho(r)$ *as long as* $ct_\rho = \infty$. *(Definition of* ct_ρ *in (1.4.3).)*

Proof.   a) Let f be defined and finite on $(0, R)$ and assume for some $r_0 \in (0, R)$ that $f(r_0) > ct_\rho(r_0)$, i.e. $f(r_0) \geqslant ct_\rho(r_0 - \epsilon)$ for some $\epsilon > 0$. Because of (1.6.2) a) we have on $(0, r_0)$ $f(r) \geqslant ct_\rho(r - \epsilon)$, a contradiction since $\lim_{r \downarrow \epsilon} ct_\rho(r - \epsilon) = + \infty$. The proof of b) is essentially the same.

REMARK. We will apply these estimates to (1.5.1, 1.5.2, 1.5.8) to get estimates for solutions of the Riccati equation (1.3.1). There are other ways to deal with (1.3.1). I hope the reader finds it helpful for an intuitive understanding that the estimates are explained essentially in a 2-dimensional picture.

1.7–1.9 *Basic geometric comparison results.* Depending on the choice of the distance function, we obtain from the just proved estimates various geometric comparison statements. In particular, distance functions from a point, from a closed geodesic (5.3.1), and from a hypersurface have been used. For the purpose of this exposition it will be sufficient to treat here only the *distance function from a point*. Executing step 1 of (1.3.3) I first derive principal curvature estimates; in step 2 these are integrated to length, respectively, volume comparison results. All the explicit bounds are *sharp* for the constant curvature models. Conditions involving $\Delta^{-1/2}$ etc. are to be ignored if $\Delta \leqslant 0$.

1.7. *Bounds for the principal curvatures of distance spheres.* Let $c(r)$, $0 \leqslant r \leqslant R$, be a geodesic arc which does not meet a conjugate point. Within a sufficiently narrow neighborhood U the arc c is length minimizing and we can define on U a "local" distance function f from $p = c(0)$. Sections 1.1 to 1.6 apply to such local distance functions. Estimates for the principal curvatures $\kappa_i(r)$ of the level surfaces or for the *Hessian D grad f* are reformulations of (1.6.3). I find the bounds for the Hessian more useful when they are rewritten for the modified distance function $md_\kappa \circ f$ (1.4.3). This can be done using

$$\text{grad}(h \circ f) = (h' \circ f) \cdot \text{grad } f$$

$$D \text{ grad}(h \circ f) = (h' \circ f) \cdot D \text{ grad } f + (h'' \circ f) \cdot df \otimes \text{grad } f.$$

1.7.1.   If $\delta \leqslant K$, then (1.5.1) and (1.6.3) imply up to the first conjugate point

$$\kappa_i(r) \leqslant ct_\delta(r),$$

or equivalently

$$D \operatorname{grad}(md_\delta \circ f) \leqslant (c_\delta \circ f) \cdot id.$$

If $\delta > 0$ then conjugate points are not farther than $\pi \delta^{-1/2}$ away.

1.7.2.   If $(n-1) \cdot \rho \leqslant \operatorname{ric}$, then (1.5.8) and (1.6.3) imply up to the first conjugate point the mean curvature estimate

$$h(r) := \frac{1}{n-1} \sum \kappa_i(r) \leqslant ct_\rho(r) =: h_\rho(r), \qquad \text{or}$$

$$-\Delta f := \operatorname{trace}(D \operatorname{grad} f) \leqslant (n-1) ct_\rho(r).$$

If $\rho > 0$ then conjugate points are not farther than $\pi \rho^{-1/2}$ away.

1.7.3.   If $K \leqslant \Delta$, then (1.5.2) and (1.6.3) imply for $0 < r < \pi \cdot \Delta^{-1/2}$

$$\kappa_i(r) \geqslant ct_\Delta(r),$$

or equivalently

$$D \operatorname{grad}(md_\Delta \circ f) \geqslant (c_\Delta \circ f) \cdot id.$$

There are no conjugate points in $(0, \pi \cdot \Delta^{-1/2})$.

REMARK.   I repeat that the estimates are sharp in the constant curvature models. Also note, that (0.3) and (1.7.3) give convexity statements for the distance functions from a point, if $r < \pi / 2 \cdot \Delta^{-1/2}$.

1.8.   *Generalized Rauch estimates.*   We execute step 2 of (1.3.3), namely integrate (1.2.6) using the principal curvature estimates of (1.7). If a Jacobi field J is $\neq 0$ on some interval, then $|J|' =$

$\langle J, J' \rangle / |J|$, hence

$$\left(\frac{|J|}{s_\kappa} \right)' = \frac{|J|}{s_\kappa} \cdot \left(\frac{\langle J, J' \rangle}{\langle J, J \rangle} - \frac{s'_\kappa}{s_\kappa} \right).$$

1.8.1.   Assume $\delta \leqslant K$, $J(0) = 0$ and let r_{conj} be the distance to the first conjugate point along c. Then (1.7.1) and (1.2.6) imply

$$\frac{|J(r)|}{s_\delta(r)} \quad \text{is nonincreasing in} \quad 0 < r < r_{\text{conj}},$$

in particular $|J(r)| \leqslant |J'(0)| \cdot s_\delta(r)$ (Rauch).

1.8.2.   Assume $K \leqslant \Delta$, $J(0) = 0$ and $r < \pi \cdot \Delta^{-1/2}$. Then (1.7.3) and (1.2.6) imply

$$\frac{|J(r)|}{s_\Delta(r)} \quad \text{is nondecreasing in} \quad 0 < r < \pi \cdot \Delta^{-1/2},$$

in particular $|J(r)| \geqslant |J'(0)| \cdot s_\Delta(r)$ (Rauch).

REMARKS.   (i) We have already seen in (1.2.2) how these estimates control the change of arclength under the exponential map (still sharp in the constant curvature models). This arclength control will be improved to distance control in Section 4.

(ii) In a situation where the curvatures approach 0 as the distance r from some distinguished point $*$ grows, one wants the curvature bounds $\delta(r)$, $\Delta(r)$ to depend on r. The generalizations of (1.8.1, 1.8.2) are easy. The tricky part (which rarely works) is the improvement to distance control.

1.9.   *Bounds for the volume of distance spheres and balls.*   We just saw that Rauch's original comparison results were later improved to monotonicity statements. The same is true for volumes: first one had Bishop's comparison results, later Gromov pointed out that the corresponding global monotonicity statements are true even beyond conjugate points; they are more powerful and easier to use (see 2.4.2, 3.6 for applications).

We saw in (1.5.4) that the trace of the shape operator is the growth rate of the tangent space-wise hypersurface volume $a(s)$. This growth rate is controlled in (1.7.2, 1.7.3). One further radial integration,

$$v(r) := \int_0^r a(s)\, ds, \qquad (1.9.1)$$

gives the ray-wise contribution of the geodesic in question to the volume of the ball. Finally, an integration over all directions at the center of the ball gives the volumes of spheres and balls. I first state the ray-wise inequalities.

1.9.2. *Assume $(n-1)\rho \leqslant$ ric. Then (1.5.4) and (1.7.2) imply up to the first conjugate point*

$$\frac{a}{a_\rho}(r) \quad \textit{is nonincreasing},$$

where $a_\rho(r) = s_\rho(r)^{n-1}$ is the integral of (1.5.4) in the constant curvature model M_ρ. Applying this to (1.9.1) gives

$$\frac{v}{v_\rho}(r) \quad \textit{is nonincreasing}.$$

Proof.

$$\left(\frac{a}{a_\rho}\right)'(r) = \frac{a}{a_\rho} \cdot (h - h_\rho) \leqslant 0.$$

$$\frac{v}{v_\rho}(r) = \frac{\displaystyle\int_0^r \frac{a}{a_\rho}(s) \cdot a_\rho(s)\, ds}{\displaystyle\int_0^r 1 \cdot a_\rho(s)\, ds}$$

is monotone since $(a/a_\rho)(s)$ is monotone.

1.9.3. *Assume $K \leqslant \Delta$ and $r < \pi \cdot \Delta^{1/2}$. Then (1.5.4), (1.7.3) and (1.9.1) imply (with the same argument as in 1.9.2)*

$$\frac{a}{a_\Delta}(r) \qquad and \qquad \frac{V}{V_\Delta} \qquad \textit{are nondecreasing}.$$

At the final step (the integration over all directions at the center of the ball) the cut locus interferes: Beyond the cut point (of the center) a geodesic ray does not contribute to the volume of distance spheres and balls. This cannot be rescued without further assumptions in the case of lower volume bounds; on the other hand, this actually helps in integrating (1.9.2): If a geodesic hits a cut point in M we can set $a(r) = 0$ for larger r and can integrate up to the cut point distance in M_ρ even if (1.9.2) is not true that far! Therefore we have Gromov's global monotonicity result for distance balls B_r.

1.9.4. *Assume* $(n-1) \cdot \rho \leqslant$ ric. *Then*

$$\frac{\text{vol}_{n-1}(\partial B_r \subset M)}{\text{vol}_{n-1}(\partial B_r \subset M_\rho)}$$

and

$$\frac{\text{vol}_n(B_r \subset M)}{\text{vol}_n(B_r \subset M_\rho)}$$

are nonincreasing functions of r with $\lim_{r \to 0} \text{vol-ratio}(r) = 1$.

1.9.5. *Assume* $K \leqslant \Delta$ *and that* B_r *does not meet a cut point of its center. Then*

$$\frac{\text{vol}_{n-1}(\partial B_r \subset M)}{\text{vol}_{n-1}(\partial B_r \subset M_\Delta)}$$

and

$$\frac{\text{vol}_n(B_r \subset M)}{\text{vol}_n(B_r \subset M_\Delta)}$$

are nondecreasing functions of r with $\lim_{r \to 0} \text{vol-ratio}(r) = 1$.

2. IMMEDIATE APPLICATIONS OF THE CURVATURE CONTROLLED BOUNDS

2.1. NONPOSITIVELY CURVED MANIFOLDS (Hadamard, Cartan).
Let M^n *be complete and assume curvature bounds* $K \leqslant \Delta \leqslant 0$.

2.1.1.   *If M is simply connected, then (for each p):* $\exp_p \colon T_p M \to M$ *is an expanding ($\geqslant 1$) diffeomorphism.*

2.1.2.   *If M is not simply connected, then in each homotopy class of paths from p to q there is exactly one geodesic.*

2.1.3.   *If M is simply connected and we take* $f(x) := d(p, x)$ *then* $D \operatorname{grad}(\tfrac{1}{2} f^2) \geqslant id$; *thus* f^2 *is a strictly convex function and all distance balls are strictly convex.*

Proof. From (1.2.2, 1.8.2) and $(1/r)s_\Delta(r) \geqslant 1$ we have $|\partial \exp_p \cdot w| \geqslant |w|$ everywhere. Therefore $\exp_p$ is an expanding local diffeomorphism. If $\exp_p$ were not injective we would have two geodesic arcs γ_1, γ_2 from p to some $q \in M$. By simple connectivity γ_1 and γ_2 are homotopic. The homotopy can be lifted to $T_p M$ since $\exp_p$ is an expanding local diffeomorphism; note that local inverse images of Cauchy sequences in M are Cauchy in $T_p M$. But the lifts of different geodesics give different radial segments and can therefore not have a common endpoint. The contradiction proves injectivity, hence (2.1.1).

It is useful to consider $T_p M$ not only as an Euclidean space, but also with the Riemannian metric $\tilde{g}$ pulled back by the (local) diffeomorphism $\exp_p$. This makes $\exp_p \colon (T_p M, \tilde{g}) \to M$ a locally isometric map and suggests taking $(T_p M, \tilde{g})$ as a metric realization of the universal covering $\tilde{M}$. Different geodesic arcs in the same homotopy class would then give different geodesic connections in $(T_p M, \tilde{g})$ for which (2.1.1) holds, proving (2.1.2).

(2.1.3) combines (0.3, 1.7.3) in the case $\Delta = 0$, i.e. $md_\Delta \circ f = (1/2)f^2$ (which is differentiable because of (2.1.1)).

2.2   Fixed Points of Isometries (Cartan 1928).

Let M be simply connected and assume $K \leqslant 0$. *Then every bounded set is contained in a unique smallest convex ball. In particular, every isometry group of M which has a bounded orbit has a fixed point.*

Proof (Eberlein).   The intersection of two closed balls B_1, B_2 of radius r and midpoints m_1, m_2 is contained in a smaller ball. Let $p \in B_1 \cap B_2$ have maximal distance from the midpoint m between m_1, m_2; the strict convexity (2.1.3) of $x \to d(p, x)^2$ together with $d(p, m_1) \leqslant r, \ d(p, m_2) \leqslant r$ implies $d(p, m) < r$. A ball of smallest radius which contains a given bounded set is therefore unique. If the bounded set is the orbit of an isometry group, then all isometries of the group map a smallest containing ball to a smallest containing ball; the center of the unique such ball is a fixed point. Note that the arguments are, verbatim, the same as for the Euclidean case.

2.3. *Growth of the fundamental group* (Svarc 1955, Milnor 1968).

2.3.1 Let $\{\gamma_1, \ldots, \gamma_N\}$ be a finite set of generators for a group π. Define the growth function

growth(k) := Number of elements in π which can be written as a product of at most k factors in the generators.

This is justified since if the growth function for one set of generators grows exponentially (respectively, polynomially of degree d) then it does so for all other finite sets of generators (not difficult).

2.3.2. *Let M be compact and assume $K \leqslant \Delta < 0$. Then the fundamental group $\pi_1(M, p)$ grows exponentially.*

2.3.3. *Let M^n be complete and assume $0 \leqslant$ ric. Then every subgroup $G \subset \pi_1(M, p)$ which has a finite set Γ of generators has at most polynomial growth of degree n.*

REMARK. For bounds on the number of generators see (5.1).

Proof. The fundamental group can be considered as a group π_1 which acts isometrically on the universal cover $\tilde{M}$. This is easy for (2.3.2) where we have a nice metric model for $\tilde{M}$, and from a sufficiently abstract point of view (2.3.3) is the same. π_1 is called the group of deck transformations.

For (2.3.2) we have to define a suitable set of generators. Pick some $\tilde{p} \in \tilde{M}$ and define the Dirichlet fundamental domain

$$F := \left\{ q \in \tilde{M};\ d(q, \tilde{p}) \leqslant d(q, \gamma\tilde{p}) \text{ for all } id \neq \gamma \in \pi_1 \right\}.$$

Let D be the diameter of the compact manifold M; clearly F is contained in the ball of radius D around $\tilde{p}$. Now define a finite subset $\Gamma \subset \pi_1$ (which will be shown to generate π_1)

$$\Gamma := \{ \gamma \in \pi_1; \; d(F, \gamma F) \leqslant 1 \},$$

with $d(A, B) := \min\{ d(p, q); \; p \in A, \; q \in B \}$. The open sets $\gamma \cdot \mathring{F}$ are disjoint and by the triangle inequality $\Gamma \cdot F \subset B_{3D+1}(\tilde{p})$. Therefore Γ is finite, namely

$$|\Gamma| \leqslant \mathrm{vol}(B_{3D+1}(\tilde{p})) \cdot \mathrm{vol}(F)^{-1}.$$

Also $\Gamma \cdot F \supset \{ q \in \tilde{M}; \; d(F, q) \leqslant 1 \}$, hence $\Gamma^k \cdot F \supset B_k(\tilde{p})$; therefore $\bigcup_k (\Gamma^k \cdot F) = \tilde{M}$ and Γ generates. Moreover, growth$(k) = |\Gamma^k| \geqslant \mathrm{vol}(B_k(\tilde{p})) \cdot \mathrm{vol}(F)^{-1}$. Finally, (1.9.5) says that $\mathrm{vol}(B_k,(\tilde{p}))$ is at least as large as the volume of a ball of radius k in the hyperbolic space H_Δ^n. Hence

$$\mathrm{vol}(B_k(\tilde{p})) \geqslant \mathrm{vol}(B_k^\Delta) = \mathrm{vol}(S^{n-1}) \cdot \int_0^k s_\Delta^{n-1}(r) \, dr,$$

which grows exponentially, proving (2.3.2).

The proof of (2.3.3) needs only minor modifications. Since the Dirichlet fundamental domain for G may not have finite diameter put $F_1 = \mathring{F} \cap B_1(\tilde{p})$.

From the generating set Γ we need its maximal displacement $L := \max\{ d(\tilde{p}, \gamma\tilde{p}); \; \gamma \in \Gamma \}$. The sets $\gamma \cdot F_1$ are disjoint (for $\gamma \in G$) and $\Gamma^k \cdot F_1 \subset B_{k \cdot L+1}(\tilde{p})$ (triangle inequality). Finally we get from the upper volume bound (1.9.4)

$$|\Gamma^k| \cdot \mathrm{vol}(F_1) \leqslant \mathrm{vol}(B_{k \cdot L+1}(\tilde{p})) \leqslant \mathrm{vol}(B_{k \cdot L+1} \subset \mathbb{R}^n) \leqslant \mathrm{const} \cdot k^n.$$

2.4   RICCI DIAMETER BOUND.   *Let M^n be complete and assume a positive lower bound $0 < (n-1)\rho \leqslant$ ric. Then:*

    2.4.1.   (Myers 1935) $\mathrm{diam}(M) \leqslant \pi \cdot \rho^{-1/2}$ *and M is compact.*

    2.4.2.   (Cheng 1975) *If* $\mathrm{diam}(M^n) = \pi \cdot \rho^{-1/2}$ *then M^n is isometric to S_ρ^n.*

Proof. (2.4.1) follows from (1.7.2) and (K.4.1) (no geodesic minimizes beyond its first conjugate point.)

Proof of (2.4.2) (Itokawa, Shiohama 1983). The argument shows the power of Gromov's global monotonicity extension (1.9.4) of the volume bound. We may normalize to $\rho = 1$ (to have $\pi \cdot \rho^{-1/2} = \pi$). From (1.9.4) we have: The volume ratio $f(r) := \text{vol}(B_r) \cdot \text{vol}(B_r^\rho)^{-1}$ is nonincreasing, in particular $1 = f(0) \geqslant f(r) \geqslant f(\pi)$ for $0 \leqslant r \leqslant \pi$. In the sphere S_ρ^n we have for $\mu(r) := \text{vol}(B_r^\rho) \cdot \text{vol}(S_\rho^n)^{-1}$ obviously $\mu(r) + \mu(\pi - r) = 1$. Let $p, q \in M$ be such that $d(p, q) = \text{diam}(M) = \pi$. The open balls $B_r(p)$, $B_{\pi-r}(q)$ are then *disjoint*, hence

$$\text{vol}(M) \geqslant \text{vol}(B_r(p)) + \text{vol}(B_{\pi-r}(q))$$

$$= (f(r) \cdot \mu(r) + f(\pi - r) \cdot \mu(\pi - r)) \cdot \text{vol}(S_\rho^n)$$

$$\geqslant f(\pi) \cdot \text{vol}(S_\rho^n) = \text{vol}(B_\pi(p)) = \text{vol}(M).$$

Thus we have equality in all estimates involved! First this gives $f(\pi) = 1$, hence $f(r) = 1$ $(0 \leqslant r \leqslant \pi)$. And next $\overline{B}_r(p) \cup \overline{B}_{\pi-r}(q) = M^n$, which says: For each $x \in M^n$ (put $r := d(x, p)$) we have $d(p, x) + d(x, q) = d(p, q)$, so that together the shortest geodesics from p to x and from x to q are segments from p to q. All geodesics starting at p therefore reach q precisely at distance π.

2.4.3. Along all of these segments we must have equality in the estimates leading to (1.9.4). First, the mean curvature of the distance spheres along each segment is $h(r) = h^\rho(r) = ct_\rho(r)$ (1.7.2, 1.9.2). Equality in (1.5.8) requires equality in (1.5.7), i.e. $S_r = h(r) \cdot id$. This gives (1.3.1) $R(Y, N)N = \rho \cdot Y$ (i.e. all sectional curvatures of 2-planes containing tangents to segments from p to q are $= \rho$). For such R_N the Jacobi equation (1.2.1) can be solved explicitly $J(r) = J'(0) \cdot s_\rho(r)$ (up to parallel translation), and (1.2.2) implies that M^n can be mapped isometrically to S_ρ^n by sending segments from p to q in M^n isometrically to meridians from pole to pole in S_ρ^n.

3. BUSEMANN FUNCTIONS

3.1. DEFINITION. The Busemann function b of a ray c is defined as an increasing limit of (shifted) distance functions:

$$b(x) := \lim_{t \to \infty} (t - d(x, c(t))) \leqslant d(x, c(0))$$

Note $d(c(t_2), c(t_1)) = t_2 - t_1$ if $0 \leqslant t_1 \leqslant t_2$, hence $t_2 - d(x, c(t_2)) \geqslant (t_1 - d(x, c(t_1)))$ by the triangle inequality. We have the Lipschitz bound $|b(x) - b(y)| \leqslant d(x, y)$. The sets $\{x \in M;\ b(x) > a\}$ are called horoballs.

3.2. LOCAL SUPPORT FUNCTIONS.

CLAIM: *For every $y \in M$ there is a unit vector $Y \in T_y M$ such that $c_Y(r) := \exp_y r \cdot Y$ is a ray and*

$$b(x) \geqslant b(y) + r - d(x, c_Y(r)) =: b_{Y,r}(x), \qquad (3.2.1)$$

in particular

$$b(c_Y(r)) = b(y) + r.$$

NOTE. By construction $c_Y(r)$ is not in the cut locus $C(y)$ and hence vice versa (K.4.2); the distance function $x \to d(x, c_Y(r))$ is therefore *differentiable* at y and—because of (3.2.1)—its level sphere through y stays inside the horoball. The ray c_Y need not be unique, so b is not differentiable in general, but if so then $\operatorname{grad} b(y) = Y$.

Proof of 3.2. To construct the ray $c_Y(r)$ let $\gamma_t(r)$ be a minimizing geodesic from y to $c(t)$ (c the given ray). For a subsequence ($t_n \to \infty$) we have convergence of the initial unit directions $\gamma'_{t_n}(0)$ to some $Y \in T_y M$, and $c_Y(r) := \exp_y r \cdot Y$ is a ray (since each subarc is a limit of segments). To prove the inequality, we have by definition

$$r = d(y, \gamma_t(r)) = d(y, c(t)) - d(\gamma_t(r), c(t))$$

$$\text{if}\quad 0 \leqslant r \leqslant d(y, c(t))$$

and

$$b(x) - b(y) = \lim_{t \to \infty} \left(t - d(x, c(t)) - t + d(y, c(t)) \right).$$

The triangle estimate $d(x, c(t)) \leqslant d(x, c_Y(r)) + d(c_Y(r), c(t))$ and the first relation are inserted in the second:

$$b(x) - b(y) \geqslant \lim_{t_n \to \infty} \left(-d(x, c_Y(r)) - d(c_Y(r), c(t_n)) \right.$$

$$\left. + r + d(\gamma_{t_n}(r), c(t_n)) \right)$$

$$\geqslant -d(x, c_Y(r)) + r + \lim_{t_n \to \infty} -d(\gamma_{t_n}(r), c_Y(r))$$

$$= -d(x, c_Y(r)) + r \quad \text{(by definition of } c_Y(r)\text{)}.$$

3.3.   *Hessian estimates.*   For the local support functions $b_{Y,r}$ we translate (1.7.1, 1.7.2) into:

$$\text{If ric} \geqslant 0 \text{ on } M \text{ then trace } D \operatorname{grad} b_{Y,r}(y) \geqslant -\frac{n-1}{r}, \quad (3.3.1)$$

$$\text{if } K \geqslant 0 \text{ on } M \text{ then } D \operatorname{grad} b_{Y,r}(y) \geqslant -\frac{1}{r} \cdot id. \quad (3.3.2)$$

Inequality (3.3.1) allows us to use the Calabi-Hopf maximum principle, see (3.7) below. With (3.3.2) we are only one more argument away from the next result (3.5).

3.4.   DEFINITION.   A subset $A \subset M$ is called totally convex if for arbitrary $p, q \in M$ *all* geodesic connections (not just the minimizing ones) are in A.

3.5.   COMPACT   TOTALLY   CONVEX   EXHAUSTION   (Cheeger-Gromoll-Meyer 1969, 1972).   Let M be complete, noncompact, $K \geqslant 0$. Let c_α denote the set of all rays from some point $p \in M$, b_α

the corresponding Busemann functions and $bm := \max_\alpha b_\alpha$. Then:
a) The sublevels of Busemann functions $\{x \in M; \; b(x) \leqslant a\}$ are totally convex.
b) The sublevels of bm are a continuous exhaustion of M by compact totally convex sets.

3.5.1. *Classification* (Gromoll-Meyer 1969). Assume in addition to (3.5) $K > 0$ then:
a) No level surface of a Busemann function or of bm contains a geodesic arc, i.e., the minimal level surface of bm is a point.
b) M^n is diffeomorphic to $\mathbb{R}^n$.

3.5.2 *Classification* (Cheeger-Gromoll 1972). Let M^n be complete, noncompact, and $K \geqslant 0$. Then M^n contains a compact, totally geodesic, totally convex submanifold S ("soul") and M^n is diffeomorphic to the normal bundle of S in M^n.

Proof of (3.5). Let $\gamma: [0, L] \to M$ be a not necessarily minimizing geodesic arc of length L. Assuming $b(\gamma(0)) = a$, $b(\gamma(L)) = a_1 \leqslant a$ we have to show $b \circ \gamma(t) \leqslant a$. By possibly shortening γ we may assume $a_1 = a$. If the continuous function $\tilde{h}(t) := b(\gamma(t)) - a$ has a positive maximum 2μ then

$$h(t) := b(\gamma(t)) - a - \mu \cdot L^{-2} \cdot t(L - t)$$

also has an (interior!) maximum $\geqslant \mu$ at some t_0. Consider the local support functions $b_{Y,r}$ of b at $y = \gamma(t_0)$. The local, smooth functions

$$h_{Y,r}(t) := b_{Y,r}(\gamma(t)) - a - \mu L^{-2} \cdot t(L - t)$$

also have a local maximum at t_0, but from (3.3.2)

$$h''_{Y,r}(t_0) \geqslant -\frac{1}{r} + 2\mu L^{-2} > 0 \qquad \text{for large } r.$$

The contradiction proves (3.5a).

For b) it is clear that the intersection of totally convex sets is totally convex. We have to show that the sublevels of bm are compact. If not we would find a divergent sequence q_n with $bm(q_n) \leqslant a$ (assume $a \geqslant 0 = bm(p)$, p from (3.5)) and segments γ_n from p to q_n; by convexity $bm \circ \gamma_n \leqslant a$. Any limit of $\{\gamma_n\}$ is a *ray* in the sublevel $\{bm \leqslant a\}$, contradicting the definition of bm.

Proof of (3.5.1a). Let b be a Busemann function. We improve (3.3.2) to

3.5.3.   *For every $y \in M$ there exists $\epsilon > 0$ and $R > 0$ such that the local support functions $b_{Y,r}$ (3.2) satisfy*

$$D \text{ grad } b_{Y,r}(y) \geqslant \epsilon \cdot id, \qquad if \quad r \geqslant R.$$

Namely, put $3k := \min\{$sectional curvatures of M in the unit ball around $y\}$, $\epsilon = k$, $R = 2 + (1/k)$. The Riccati equation and inequality (1.5.1) control the principal curvatures of the level surfaces of the local support functions $b_{Y,r}$ (of 3.2). Any solution $\kappa(r)$ with $\kappa(0) \leqslant \epsilon$ of $\kappa' \leqslant -K(u \wedge N) - \kappa^2 \leqslant -3k$ (1.5.1) drops on $[0,1]$ at least to $\epsilon - 3k < -\epsilon$ and, because of (1.6), then stays below $F(r) = (r - 1 - (1/\epsilon))^{-1}$, which solves $F' = -F^2$, $F(1) = -\epsilon$. For $r \geqslant R$ the $b_{Y,r}$ are defined beyond the pole of F so that, by (1.6.3), their level surfaces cannot have principal curvatures $\leqslant \epsilon$ at y. This proves (3.5.3), which in turn implies that no Busemann function can be constant on any geodesic arc (no weak interior maximum) —proving (3.5.1a). (To deal with bm extended 3.2.1 to $\sup b_\alpha(x) \geqslant \sup(b_\alpha(y) + r - b_{Y_\alpha,r}(x))$.)

b) The proof of the diffeomorphism statement is very similar to (5.4.3) and I omit it.

The proof of (3.5.2) still requires a lot of work since the minimal level of bm is not yet the soul. One needs that lower dimensional totally convex sets are top dimensional in some totally geodesic submanifold, see [CE].

3.6.   RICCI SPLITTING THEOREM (Cheeger-Gromoll 1971).   *Let M^n be complete and assume* ric $\geqslant 0$. *If M^n contains a line then this line splits off as a Riemannian factor, $M^n = N^{n-1} \times \mathbb{R}$.*

3.7.   CALABI-HOPF MAXIMUM PRINCIPLE (1957).   *Let (M, g) be a connected Riemannian manifold and f a continuous function on M.*

Assume, that for any $x \in M$ and any $\epsilon > 0$ there exists a C^∞-support function $f_{x,\epsilon} \leqslant f$, $f_{x,\epsilon}(x) = f(x)$ (see 3.2) with

$$-\Delta f_{x,\epsilon} = \text{trace } D \text{ grad } f_{x,\epsilon} \geqslant -\epsilon.$$

Then f attains no maximum unless it is constant.

Proof of 3.6 (Eschenburg-Heintze 1984).   Let γ be a line in M. Consider the two Busemann functions $b_\pm$ for the rays $c_\pm(t) := \gamma(\pm t)$. They satisfy

$$b_+(x) + b_-(x) = \lim_{t \to \infty} \left(t - d(x, \gamma(t)) + t - d(x, \gamma(-t)) \right) \leqslant 0$$

and

$$(b_+ + b_-)(\gamma(t)) = 0 \qquad \text{(maximum!)}.$$

The sums of local support functions for b_+, b_- (3.2) satisfy the assumptions of the maximum principle (3.7) because of (3.3.1), which implies $b_+ + b_- = 0$! The level surfaces of $b_+ = -b_-$ can now be touched by large spheres from both sides:

$$b_{Y,r}(x) \leqslant b_+(x) = -b_-(x) \leqslant -b_{-Y,r}(x), \qquad \text{equality at} \quad x = y.$$

In particular $b_+ = -b_-$ is differentiable, $|\text{grad } b_+| = 1$, the radial rays c_Y (3.2) are the integral curves of grad b_+, and all the rays extend to lines. The proof can be finished with elementary arguments, but another application of the maximum principle works more elegantly. Every Busemann function b is subharmonic (ric $\geqslant 0$!) (indeed, if b agrees with some harmonic function h on the boundary of some geodesic ball, then (3.7) implies that $b - h$ cannot have a positive maximum, i.e. $b \leqslant h$). Therefore $b_+ = -b_-$ is sub- and superharmonic, hence harmonic, hence C^∞ and $\Delta b_+ = 0$. Finally we have another equality discussion as in (2.4.3): b_+ is a distance function (1.1.2) whose level surfaces have constant mean curvature 0. (1.5.8) implies $0 \leqslant -\text{ric} \circ \gamma$ ($\leqslant 0$). Therefore we must have equality in Schwarz' inequality (1.5.7): $S = 0 \cdot id$. From this and (1.3.1) we have $R(\ , \text{grad } b_+)\text{grad } b_+ = 0$. The Jacobi equation

(1.2.1) along the rays has explicit (parallel!) solutions and $F: b_+^{-1}(0) \times \mathbb{R} \to M^n$, $f(x, t) := \exp_x t \cdot \operatorname{grad} b_+(x)$ is an isometry.

4. THE ALEKSANDROW-TOPONOGOV ANGLE COMPARISON THEOREMS

Results closely related to those in this chapter were used by E. Cartan in the twenties and by Preissmann (1943) under curvature assumptions $K \leqslant 0$. Aleksandrow used triangle comparison theorems in the forties and fifties as a substantial tool, in particular in his theory of convex surfaces. Some years after Rauch, Toponogov proved in the Riemannian context the n-dimensional angle comparison for lower curvature bounds $\delta \leqslant K$ (1959). Remarkably, this theorem is true without any size restrictions. The proof, originally long and technical, has been considerably simplified.

4.0. DEFINITION. A triangle T in a Riemannian manifold is given by its three geodesic edges (which I assume minimizing although generalizations can be handled with the same proof). Assume lower curvature bounds $\delta \leqslant K$ or upper bounds $K \leqslant \Delta$. A triangle with the *same edgelengths* as T in the plane of constant curvature M_δ, respectively, M_Δ is called an "Aleksandrow triangle" T_δ, respectively, T_Δ. The two segments and the angle between them is called a hinge; a "Rauch hinge" in M_δ, respectively, M_Δ has the same edgelength, angle, edgelength as occur at one vertex of T.

4.1. TRIANGLE COMPARISON THEOREMS ASSUMING $K \leqslant \Delta$. *Size restrictions on T are necessary, namely, T does not meet the cut locus of its vertices, and, the circumference satisfies $l(T) < 2\pi\Delta^{-1/2}$ (ignore this, if $\Delta \leqslant 0$). Then an Aleksandrow triangle T_Δ exists and the angles of T and T_Δ satisfy*

$$\alpha \leqslant \alpha_\Delta, \qquad \beta \leqslant \beta_\Delta, \qquad \gamma \leqslant \gamma_\Delta. \tag{4.1.1}$$

The third edge c_Δ^* closing a Rauch hinge in M_Δ satisfies

$$|c| \geqslant |c_\Delta^*|. \tag{4.1.2}$$

With the obvious definitions of corresponding points on the edges of T and T_Δ and secants σ, σ_Δ between them one has

$$|\sigma| \leqslant |\sigma_\Delta|. \tag{4.1.3}$$

4.2. Triangle Comparison Theorems Assuming $\delta \leqslant K$ (Toponogov).

4.2.0. *An Aleksandrow triangle T_δ always exists, more precisely: a circumference $l(T) > 2\pi\delta^{-1/2}$ (if $\delta > 0$) does not occur and $l(T) = 2\pi\delta^{-1/2}$ occurs only on S_δ^n; if $l(T) < 2\pi\delta^{-1/2}$ then the three triangle inequalities in T are sufficient for the existence of T_δ.*

The angles of T and T_δ satisfy (take $l(T) < 2\pi\delta^{-1/2}$, 4.2.0)

$$\alpha_\delta \leqslant \alpha, \qquad \beta_\delta \leqslant \beta, \qquad \gamma_\delta \leqslant \gamma. \tag{4.2.1}$$

The third edge c_δ^* closing a Rauch hinge in M_δ satisfies

$$|c| \leqslant |c_\delta^*|. \tag{4.2.2}$$

Secants σ, σ_δ between corresponding points on T, respectively, T_δ satisfy

$$|\sigma| \geqslant |\sigma_\delta|. \tag{4.2.3}$$

4.3. Remarks. Because of the cosine law in M_δ, respectively, M_Δ (e.g. on S^n: $\cos c = \cos a \cos b + \sin a \sin b \cos \gamma$) it is trivial that the opposite edgelength of a hinge varies monotonely increasing with the hinge angle. (4.1.1) and (4.1.2) are therefore immediately equivalent, and so are (4.2.1) and (4.2.2). If one considers the limit of short secants across a vertex, then (4.1.3) implies (4.1.1) and (4.2.3) implies (4.2.1), again immediately. The converse (4.1.1) $\Rightarrow$ (4.1.3) and (4.2.1) $\Rightarrow$ (4.2.3) is also true but needs more trigonometry into which I do not want to go. (4.1.3) and (4.2.3) extend immediately to infinite triangles if $\Delta < 0$ or $\delta < 0$. They also extend to Gromov's limits of Riemannian spaces in which minimizing curves exist but angles cannot be defined. I shall prove (4.1) using estimates on $\partial \exp$ (1.2.2, 1.8.2) and Toponogov's theorem (4.2) with the Hessian bounds (1.7.1). It is enough to prove (4.2.3) for the secants to the opposite vertex, see (4.5.4).

4.4. *Proof* of (4.1.2). Because of the size restrictions we can parametrize the edge c of T in exponential coordinates from the opposite vertex, i.e.

$$c(t) = \exp_p X(t), \qquad X: [0,1] \to T_p M \qquad (|X| < \pi\Delta^{-1/2}).$$

We identify $T_p M$ and $T_{p_\Delta} M_\Delta$ isometrically (e.g. with $\mathbb{R}^n$) and define

$$\tilde{c}(t) := \exp_{p_\Delta} X(t), \qquad X: [0,1] \to T_{p_\Delta} M_\Delta \approx T_p M.$$

Now follow the procedure outlined in (1.2.2):

$$\dot{c}(t) = \partial \exp_p\Big|_{X(t)} \cdot \dot{X}(t) = \frac{\partial}{\partial t}\left(\exp_p s \cdot X(t)\right)\Big|_{s=1} = J_t(1),$$

$$\dot{\tilde{c}}(t) = \partial \exp_{p_\Delta}\Big|_{X(t)} \cdot \dot{X}(t) = \tilde{J}_t(1);$$

and the Jacobi fields have initial conditions

$$J_t(0) = 0 = \tilde{J}_t(0), \qquad \frac{D}{\partial s} J_t(0) = \dot{X}(t) = \frac{D}{ds} \tilde{J}_t(0)$$

so that (1.8.2), combined with (1.4.5), proves

$$|\dot{c}(t)| \geq |\dot{\tilde{c}}(t)|, \qquad \text{hence length } (c) \geq \text{length } (\tilde{c}).$$

Finally length $(\tilde{c}) \geq |c_\Delta^*|$ since $\tilde{c}$ is some connection of its endpoints and c_Δ^* is the shortest one.

REMARK. The corresponding proof of (4.2) gets considerably more complicated, mainly because one does not want the size restrictions. Also, the triangle inequality at the very end has to be used in M rather than in M_δ, i.e. the proof has to start by parametrizing c_δ^*. This causes the corresponding definition of $\tilde{c}$ in M to run into problems with conjugate points.

4.5.   *Proof* of (4.2.3).

4.5.1.   Because of (2.4.2) (and its simple proof) we can start with the alternative: either $\mathrm{diam}(M) = \pi \cdot \delta^{-1/2}$, then M^n is isometric to S_δ^n or else $\mathrm{diam}(M) \leqslant \pi \cdot \delta^{-1/2} - 2\epsilon$. Of course we know the sphere, therefore we have during the following proof all segments $\leqslant \pi \delta^{-1/2} - 2\epsilon$ which eliminates many separate considerations. It helps the exposition without lengthening the full argument very much if I treat the simpler cases separately.

4.5.2.   First assume that no edge of T meets the cut locus of a vertex. Let r, respectively, r_δ denote the (differentiable) distance function from the vertex $p \in T$ to get the modified distance functions $md_\delta \circ r$, respectively, $md_\delta \circ r_\delta$. Their restrictions to the opposite edge c, respectively, c_δ are—because we stay away from the cut locus—differentiable functions

$$h := md_\delta \circ r \circ c, \qquad H := md_\delta \circ r_\delta \circ c_\delta. \qquad (4.5.3)$$

They satisfy

$$h'' + \delta h \leqslant 1, \qquad H'' + \delta H = 1 \quad \text{with}$$
$$h(0) = H(0), \qquad h(|c|) = H(|c|)$$

because of the Hessian estimate (1.7.1).

4.5.4.   *Claim.* The difference $\lambda := h - H$ satisfies

$$\lambda'' + \delta \cdot \lambda \leqslant 0, \qquad \lambda(0) = 0, \qquad \lambda(|c|) = 0,$$

which implies

$$\lambda \geqslant 0 \qquad \text{on} \quad [0, |c|].$$

(In fact, $\lambda(0) \geqslant 0$, $\lambda(|c|) \geqslant 0$ is enough, and handles the case of general secants, (4.3).)

Proof.   a) If $\delta < 0$ then $\lambda'' \leqslant (-\delta) \cdot \lambda$ shows that λ cannot have a negative minimum, i.e. $h \geqslant H$ (4.2.3).

b) If $\delta = 0$ then λ cannot have a negative minimum $= -2\mu$ since

$$\tilde{\lambda} := \lambda + \mu \cdot |c|^{-2} \cdot t(|c| - t)$$

would also have an (interior) minimum $\leqslant -\mu$, contradicting $\tilde{\lambda}'' \leqslant -2\mu|c|^{-2} < 0$.

c) If $\delta > 0$ we have from (4.5.1) $|c| \leqslant \pi\delta^{-1/2} - 2\epsilon$, hence

$$\sigma_\epsilon(t) := s_\delta(t + \epsilon) - s_\delta\left(\frac{\epsilon}{2}\right) > 0 \qquad \text{for } 0 \leqslant t \leqslant |c|.$$

Again, suppose λ has some negative minimum. Then $\tilde{\lambda} := \lambda/\sigma_\epsilon$ also has a negative minimum $-\mu$ at, say, $t_0 \in (0, |c|)$. But this contradicts

$$\tilde{\lambda}'(t_0) = (\lambda'\sigma_\epsilon - \lambda\sigma_\epsilon') \cdot \sigma_\epsilon^{-2}(t_0) = 0,$$

$$\tilde{\lambda}''(t_0) = (\lambda''\sigma_\epsilon - \lambda\sigma_\epsilon'') \cdot \sigma_\epsilon^{-2}(t_0)$$

$$= \left((\lambda'' + \delta\lambda) \cdot \sigma_\epsilon + \delta \cdot \lambda \cdot s_\delta\left(\frac{\epsilon}{2}\right)\right) \cdot \sigma_\epsilon^{-2}(t_0)$$

$$\leqslant -\delta \cdot \mu \cdot s_\delta\left(\frac{\epsilon}{2}\right) \cdot \sigma_\epsilon^{-1}(t_0) < 0.$$

4.5.5.   In a second step we allow the edge c to meet the cut locus of the opposite vertex p, but *not* in a conjugate point. Then each segment from p to $c(t_0)$ is *locally* minimizing beyond $c(t_0)$. So we get differentiable local distance functions

$$r_{\text{loc}} \geqslant r, \qquad \text{defined in particular near } c(t_0).$$

The principal curvature estimates (1.7.1) were stated for these local distance functions. In addition to (4.5.3) we now also have for each $t_0 \in (0, |c|)$

$$h_{\text{loc}} := md_\delta \circ r_{\text{loc}} \circ c|_{[t_0-\tau, t_0+\tau]},$$

$$h_{\text{loc}}'' + \delta \cdot h_{\text{loc}} \leqslant 1, \qquad h_{\text{loc}} \geqslant h, \qquad h_{\text{loc}}(t_0) = h(t_0). \tag{4.5.6}$$

This is enough to make the proof of (4.5.4) work. At that t_0 at which the negative minimum of $\tilde{\lambda}$ occurs we replace h by h_{loc}, and get also $\lambda_{\text{loc}} \geqslant \lambda$ and $\tilde{\lambda}_{\text{loc}} \geqslant \tilde{\lambda}$, $\tilde{\lambda}_{\text{loc}}(t_0) = \tilde{\lambda}(t_0)$. The smooth function $\tilde{\lambda}_{\text{loc}}$ therefore also has a local interior minimum at t_0, but $\tilde{\lambda}''_{\text{loc}}(t_0) < 0$ as before.

4.5.7. In the last step we deal with conjugate endpoints of segments from p to $c(t_0)$. We cannot get local superfunctions which satisfy the same differential inequality as before, but they almost do.

Let $\gamma(s)$ be a segment from p to $c(t_0)$ and define for small $\eta > 0$ local superdistance functions, differentiable near $c(t_0)$, by

$$r_{\text{loc}, \eta}(x) := d(x, \gamma(\eta)) + \eta \geqslant r(x) = d(x, \gamma(0)).$$

The Hessian of $md_\delta \circ r_{\text{loc}, \eta}$ has (as in the computation for (1.7.1)) the radial eigenvalue $c_\delta(r_{\text{loc}, \eta})$ and spherical eigenvalues

$$\kappa_{sp} \leqslant ct_\delta(r_{\text{loc}, \eta} - \eta) \cdot s_\delta(r_{\text{loc}, \eta}) = c_\delta(r_{\text{loc}, \eta}) + \frac{s_\delta(\eta)}{s_\delta(r_{\text{loc}} - \eta)}.$$

$$(4.5.8)$$

We can assume $r_{\text{loc}}(c(t_0)) > 0$ (since we do not need to prove anything if the vertex p lies on the opposite edge, i.e. $|a| + |b| = |c|$) and replace the denominator in (4.5.8) by a constant; this also uses (4.5.1) once more! With this we get instead of (4.5.6)

$$h_{\text{loc}, \eta} := md_\delta \circ r_{\text{loc}, \eta} \circ c\big|_{[t_0 - \tau,\, t_0 + \tau]},$$

$$h_{\text{loc}, \eta} \geqslant h, \qquad h_{\text{loc}, \eta}(t_0) = h(t_0), \qquad (4.5.9)$$

$$h''_{\text{loc}, \eta} + \delta \cdot h_{\text{loc}, \eta} \leqslant 1 + \text{const} \cdot s_\delta(\eta), \qquad \text{const. independent of } \eta.$$

This is *still* good enough to make the proof of (4.5.4) work. At t_0 where the continuous function $\tilde{\lambda}$ has its negative minimum $\leqslant -\mu$, the smooth functions $\tilde{\lambda}_{\text{loc}, \eta}$ all have a minimum. As before

$$\text{if} \quad \delta \leqslant 0 \quad \text{then } \tilde{\lambda}''_{\text{loc}, \eta}(t_0) \leqslant -2\mu |c|^{-2} + \text{const} \cdot s_\delta(\eta),$$

$$\text{if} \quad \delta > 0 \quad \text{then } \tilde{\lambda}''_{\text{loc}, \eta}(t_o) \leqslant -\mu \cdot \delta \cdot s_\delta\left(\frac{\epsilon}{2}\right) \cdot \sigma_\epsilon(t_0)^{-1}$$

$$+ \text{const} \cdot s_\delta(\eta),$$

which gives the same contradiction ($\tilde{\lambda}''_{\text{loc}}(t_0) < 0$) for sufficiently small η.

4.6. *Triangles T with circumference $= 2\pi\delta^{-1/2}$ occur only on S_δ^n and triangles with circumference $l(T) > 2\pi\delta^{-1/2}$ do not occur.*

Proof. (i) Again with (4.5.1) we may assume $\text{diam}(M) \leqslant 2\pi\delta^{-1/2} - 2\epsilon$. The proof given in (4.5) formally includes the case $l(T) = 2\pi\delta^{-1/2}$, but this does not occur: T_δ would have to be a great circle so that each edge meets the antipodal point of the opposite vertex. Now (4.2.3) requires a segment of length $\pi \cdot \delta^{-1/2}$ in M. (ii) A triangle with $l(T) > 2\pi\delta^{-1/2}$ by continuity gives a triangle with $l(T') = 2\pi\delta^{-1/2}$, so $M = S_\delta^n$—contradicting $l > 2\pi\delta^{-1/2}$.

5. APPLICATIONS OF THE TRIANGLE THEOREMS

5.1. *Number of generators for $\pi_1(M, p)$* (Gromov 1978).

5.1.1. *Let M^n be complete and assume $K \geqslant 0$. Then the fundamental group $\pi_1(M, p)$ can be generated by $N \leqslant 2 \cdot 5^{(1/2)n}$ elements (compare 2.3.3).*

5.1.2. *Let M^n be compact. Assume curvature and diameter bounds $-\Lambda^2 \leqslant K$, $\text{diam}(M) < \frac{1}{2}D$. Then the fundamental group $\pi_1(M, p)$ can be generated by $N \leqslant 2 \cdot (3 + 2\cosh \Lambda D)^{(1/2)n}$ elements. (If $n = 2$ then this bound is $\geqslant 5 \cdot \text{genus}^4$, hence never sharp.)*

Proof. Define for each $\alpha \in \pi_1(M, p)$ the "length" $|\alpha|$ as the length of a shortest geodesic loop in the homotopy class α. Now define a "short" set of generators $\{\alpha_1, \ldots, \alpha_N, \ldots\}$:
  (i) α_1 is a shortest element in $\pi_1(M, p) \setminus \{id\}$.
  (ii) If $\alpha_1, \ldots, \alpha_k$ are already chosen, denote by $\langle \alpha_1, \ldots, \alpha_k \rangle$ the subgroup of $\pi_1(M, p)$ which they generate. Then α_{k+1} is a shortest element in $\pi_1(M, p) \setminus \langle \alpha_1, \ldots, \alpha_k \rangle$.

5.1.3. CLAIM. *No short set of generators contains more elements than the bound in (5.1.1) respectively, (5.1.2).*

Proof of the claim.   By definition $|\alpha_1| \leqslant |\alpha_2| \leqslant \cdots$ and also

$$\left| \alpha_i \cdot \alpha_j^{-1} \right| \geqslant \max \{ |\alpha_i|, |\alpha_j| \} \qquad (5.1.4)$$

otherwise α_i or α_j was not chosen minimally.

Now apply Toponogov's theorem to the triangle T obtained in the universal cover $\tilde{M}$ by lifting the shortest geodesic loops which represent the classes $\alpha_i, \alpha_j, \alpha_i\alpha_j^{-1}$; the edgelengths of T are then $|\alpha_i|, |\alpha_j|, |\alpha_i\alpha_j^{-1}|$. (4.2.1) gives a lower bound for the angle Φ opposite the longest edge:

$$(K \geqslant 0) \quad \left| \alpha_i \cdot \alpha_j^{-1} \right| \leqslant |\alpha_i|^2 + |\alpha_j|^2 - 2|\alpha_i| \, |\alpha_j| \cos \Phi, \quad (5.1.5)$$

i.e. $\Phi \geqslant 60°$;

$$(K \geqslant -\Lambda^2) \quad \cos \Phi \leqslant \frac{\cosh \Lambda|\alpha_i| \cdot \cosh \Lambda|\alpha_j| - \cosh \Lambda|\alpha_i\alpha_j^{-1}|}{\sinh \Lambda|\alpha_i| \cdot \sinh \Lambda|\alpha_j|},$$

$$(5.1.6)$$

i.e.

$$\cos \Phi \leqslant \frac{\cosh \Lambda D}{1 + \cosh \Lambda D} \quad \text{or} \quad \sin^{-2}\frac{1}{2}\Phi \leqslant 2 + 2\cosh \Lambda D.$$

(The last inequality uses $|\alpha_j| \leqslant D$, i.e. generators can always be chosen $\leqslant 2 \cdot \text{diam}(M) + \epsilon$, by dividing any loop from p into segments shorter than ϵ and joining the dividing points with p, back and forth.)

Finally, consider the initial unit vectors of the short loops representing $\alpha_1, \alpha_2, \ldots$. They are points on the unit sphere in T_pM with mutual distance $\geqslant \Phi$. The open balls of radius $\sin \frac{1}{2}\Phi$ are therefore disjoint and the inner halves of these balls are all contained in the ball of radius $(1 + \sin^2\frac{1}{2}\Phi)^{1/2}$. Therefore their number N is bounded by the volume ratio, i.e.

$$N \leqslant 2 \cdot \left(1 + \sin^2\frac{1}{2}\Phi \right)^{n/2} \cdot \left(\sin\frac{1}{2}\Phi \right)^{-n} = 2 \left(1 + \sin^{-2}\frac{1}{2}\Phi \right)^{+n/2}.$$

$$(5.1.7)$$

Insertion of (5.1.5) respectively, (5.1.6) into (5.1.7) proves the theorem.

5.2. *Critical points of the distance function.* The following definition has turned out to be very useful and suggestive. The arguments in this section have been developed in connection with the sphere theorems.

5.2.1. DEFINITION. A point $q \in M$, $q \neq p$, is called *not critical* for the distance function from p, if the initial tangents of *all* segments qp lie in an open halfspace in $T_q M$. Otherwise q is called critical.

The following fact justifies the name.

5.2.2. *Existence of a gradient like vector field.* Let A be a compact set in M with no critical points for the distance function from p. Then one has a vector field X on an open neighborhood of A such that the distance from p is strictly decreasing along the integral curves of X.

Proof. For every $x \in A$ we have by definition a vector $Y_x \in T_x M$ and $\epsilon_x > 0$ such that the angle between Y_x and any segment xp is $\leqslant (\pi/2) - 2\epsilon_x$. By radial parallel translation we extend Y_x to a smooth local vector field X_x on some ball $B(x)$. We can choose $B(x)$ so small that for all $y \in B(x)$ the angle between $X_x(y)$ and any segment yp is $\leqslant (\pi/2) - \epsilon_x$. Denote by $B'(x)$ the concentric balls of half the radius as $B(x)$ and choose (by compactness of A) finitely many $B'(x_i)$ which cover A. With nonnegative C^∞-functions φ_i which are 1 on $B'(x_i)$ and 0 outside $B(x_i)$ we define a smooth vector field

$$X := \frac{1}{\Sigma \varphi_i} \cdot \Sigma \varphi_i X_{x_i}.$$

The following now holds for each $x \in \cup B'(x_i)$: Each segment xp has an angle $< \pi/2$ with all $\varphi_i X_{x_i}(x)$ which are $\neq 0$; therefore $X(x) \neq 0$ and the angle between xp and $X(x)$ is $< \pi/2$. So X is the desired vector field on $\cup B'(x_i) \supset A$.

Of course, (5.2.2) alone cannot be of too much help. The discovery was that the existence of "large" distances in M can be combined with *Toponogov's theorem* (4.2.2) to draw rather strong conclusions about these critical points.

5.2.3. ASSUMPTION. Let M be complete and $0 < \delta \leqslant K$. We shall consider critical points for the distance function from p. Let q be a *farthest* point from p and assume $D := d(p,q) > (\pi/2)\delta^{-1/2}$. (Note: In general p is not farthest from q.)

CONCLUSIONS.
5.2.4. q is a unique farthest point and critical for the distance function from p (compare 6.1.2).
5.2.5. A point $x \neq p, q$ is *not* critical for the distance from p if

a) D is maximal, i.e. $D = \text{diam}(M)$

or

b) if x is far from p, i.e. $d(x, p) \geqslant (\pi/2)\delta^{-1/2}$.

Any fixed segment xq can be used to define the open halfspace (in 5.2.1) since all segments xp have angles $\varphi > \pi/2$ (see 5.2.7) with the chosen segment xq.
5.2.6. Every $x \in M$ which is far from q is close to p

$$d(x,q) > \frac{\pi}{2}\delta^{-1/2} \Rightarrow d(p,x) < \frac{\pi}{2}\delta^{-1/2}.$$

In the middle the angle estimate is more precise than in (5.2.5),

$$d(p,x) = d(x,q) \Rightarrow \sphericalangle(p,x,q) \geqslant D \cdot \delta^{1/2}.$$

Proofs. For uniqueness in (5.2.4) we show that the midpoint m on a segment between two (assumed) farthest points $q, \bar{q}$ (from p) has to be at a greater distance. Let $p_\delta q_\delta \bar{q}_\delta$ be the Aleksandrow triangle of $pq\bar{q}$ (4.0) and m_δ the midpoint of $q_\delta \bar{q}_\delta$. (4.2.3) implies $d(p, m) \geqslant d(p_\delta, m_\delta)$ and from $d(p_\delta, q_\delta) = d(p_\delta, \bar{q}_\delta) > (\pi/2)\delta^{-1/2}$ we have $d(m_\delta, p_\delta) > d(p_\delta, q_\delta) = d(p,q)$, as claimed.

Proof of (5.2.5). For each of the triangles pxq with $\varphi :=$ $\measuredangle(pxq)$ we have from (4.2.1),

$$\cos \varphi \leqslant \frac{\cos \delta^{1/2}D - \cos(\delta^{1/2}|qx|) \cdot \cos(\delta^{1/2}|px|)}{\sin(\delta^{1/2}|qx|) \cdot \sin(\delta^{1/2}|px|)} < 0, \quad (5.2.7)$$

where the last strict inequality follows since uniqueness in (5.2.4) supplies the following strict inequalities:

a) If $D = \text{diam}(M)$ then $|px|, |qx| < D$.

b) If $d(x, p) \geqslant (\pi/2) \cdot \delta^{-1/2}$ then also $|px| < D$.

Now (5.2.5b) forces q to be critical in (5.2.4): For any $Y \in T_q M$ and all $0 < \epsilon < D - (\pi/2)\delta^{-1/2}$ we can take $x := \exp_q \epsilon \cdot Y$ in (5.2.5b). Any limit of segments xp (as $\epsilon \to 0$) is a segment qp. For this limit segment we conclude from $\measuredangle(xq, xp) > (\pi/2)$ (see 5.2.7) that $\measuredangle(-Y, qp) \geqslant (\pi/2)$—in other words: In every closed half-space of $T_q M$ some segment to p starts.

Proof of (5.2.6). Fix a segment qx and choose with (5.2.4) a segment qp such that $\alpha := \measuredangle(qp, qx) \leqslant (\pi/2)$. For the Rauch hinge in M_δ corresponding to xqp in M we have (4.2.2)

$$\cos \delta^{1/2}|px|$$

$$\underset{(4.2.2)}{\geqslant} \cos(\delta^{1/2}D) \cdot \cos(\delta^{1/2}|qx|) + \sin(\delta^{1/2}D) \cdot \sin(\delta^{1/2}|qx|)\cos \alpha$$

$$> 0.$$

Finally, if $d(p, x) = d(q, x)$ then (5.2.7) improves to

$$\cos \varphi \leqslant \frac{\cos(\delta^{1/2}D) - \cos^2(\delta^{1/2}|qx|)}{1 - \cos^2(\delta^{1/2}|qx|)} \leqslant \cos(\delta^{1/2}D). \quad (5.2.8)$$

5.3. *Cut locus estimates* (see K.5).

5.3.1. (Klingenberg 1960). *Let M^n be compact, orientable, even dimensional. Assume curvature bounds $0 < K \leqslant 1$. Then*

The cut locus distance satisfies $d(p, C(p)) \geqslant \pi$.

This estimate is sharp *not* only for round spheres, see (6.1.2).

5.3.2. (Klingenberg 1961). *Let M^n be compact, simply connected (and odd dimensional). Assume curvature bounds $\frac{1}{4} \leqslant K \leqslant 1$. Then*

$$d(p, C(p)) \geqslant \pi.$$

The (current) best counterexamples to an extension of (5.3.2) are given in (6.1.4).

5.3.3. (Easy version of 5.3.2). *Let M^n be compact. Assume curvature bounds $\frac{1}{4} < \delta \leqslant K \leqslant 1$ and $\mathrm{diam}(M) > (\pi/2)\delta^{1/2}$. Then*

$$d(p, C(p)) \geqslant \pi.$$

5.3.4. (Cheeger 1970). *Let M^n be compact. Assume curvature bounds $\delta \leqslant K \leqslant 1$. Then*
a) $\delta > 0$: $d(p, C(p)) \geqslant \min(\pi, \pi \cdot \delta^{-1/2} \cdot \mathrm{vol}(M) \cdot \mathrm{vol}(S_\delta^n)^{-1})$, *this is sharp for round spheres.*
b) $\delta \leqslant 0$: $d(p, C(p)) \geqslant$

$$\min\left(\pi, \mathrm{vol}(M) \cdot \mathrm{vol}(S^{n-2})^{-1} \cdot (n-1) \cdot s_\delta^{1-n}(\mathrm{diam}(M))\right),$$

this is never sharp.

5.3.5. (Toponogov). *Let M^n be complete and noncompact. Assume curvature bounds $0 < K \leqslant 1$ (i.e. $\inf K = 0$ by 3.2.1). Then*

$$d(p, C(p)) \geqslant \pi.$$

5.3.6. (Compact version of (5.3.5)). *Let M be compact. Assume curvature bounds $0 < \delta \leqslant K \leqslant 1$, and that for each $p \in M$ there is a $q \in M$ with $d(p, q) > (\pi/2)\delta^{-1/2}$. Then*

$$d(p, C(p)) \geqslant \pi.$$

REMARK. (5.3.3) is included since its proof is a nice application of critical points and the proof of (5.3.2) is too long to be given here. (5.3.6) is included since it explains why in the noncompact case (5.3.5) the parity of the dimension plays no role. (5.3.6) also says that the counterexamples to an extension of (5.3.2) are "small"—compared to what the curvature bounds would allow.

Proofs. 5.3.7. All proofs start with Klingenberg's result that a closest nonconjugate cut point q of p is the midpoint of a geodesic loop (K.4.4). If the cut locus distance happens to have a minimum $m < \pi$ at p, then (K.4.5) implies (by reversing the role of p and q) that we have a closed geodesic γ through p and q of length $2m = 2\pi - 2\eta$, $\eta > 0$. All proofs then show that such a closed geodesic cannot exist.

The proof of (5.3.1) uses the famous

5.3.8. SYNGE LEMMA (1926).   *Let M^n be compact, orientable, even dimensional. Assume $K > 0$. Then any closed geodesic γ has shorter parallel curves.*

REMARK.   Synge used his lemma to conclude that if M were *not* simply connected there would exist a *shortest closed* geodesic in a nontrivial homotopy class—contradicting (5.3.8).

Proof of (5.3.8).   Parallel translation around γ is an orientation preserving isometry of the odd dimensional subspace $(\gamma'(0))^{\perp} \subset T_{\gamma(0)}M$ and therefore has a fixed vector—in other words, we have a closed parallel unit vector field $v(t) \perp \gamma'(t)$ along γ. The strip $c(\epsilon, t) := \exp_{\gamma(t)} \epsilon \cdot v(t)$ is similar to a band around the equator of S^2, i.e. the closed geodesic γ has *shorter* parallel curves $t \to c(\epsilon, t)$, $\epsilon > 0$ small. This follows from the second variation formula (K.4 Lemmata 1, 2):

$$\frac{d^2}{d\epsilon^2} L(\epsilon) \bigg|_{\epsilon=0} = -\int_0^{2m} g(R(v, \gamma')\gamma', v)\,dt < 0,$$

since we assumed $K > 0$.

We finish the proof of (5.3.1) with (4.1.1): Choose three equidistant points $\gamma(t_i)$ on γ (from (5.3.7)) and consider for small $\epsilon > 0$ the triangle T with vertices $c(\epsilon, t_i)$ $(i = 1, 2, 3)$ and unique edges of length $< \frac{2}{3}m$ (using the second variation again). The circumference of T is short enough so that no edge meets the cut locus of the opposite vertex and (4.1.1) applies. As $\epsilon \to 0$ the edges of T converge to the arcs of γ between the $\gamma(t_i)$ (there are no other segments between the $\gamma(t_i)$) and thus all angles of T converge to π. By (4.1.1) the same is true for the angles of the Aleksandrow triangle T_Δ; its edgelengths converge (as $\epsilon \to 0$) to $\frac{2}{3}m = \frac{2}{3}(\pi - \eta)$. This contradicts spherical trigonometry: $\lim a(\epsilon) = \lim b(\epsilon) = \lim c(\epsilon) = \frac{2}{3}(\pi - \eta)$ and the spherical cosine law imply

$$\lim_{\epsilon \to 0} \cos \gamma(\epsilon) = \frac{\cos \dfrac{2}{3}(\pi - \eta)}{1 + \cos \dfrac{2}{3}(\pi - \eta)} > -1.$$

Proof of (5.3.4) (Heintze-Karcher (1978). Either $d(p, C(p)) \geqslant \pi$ and we are done, or there is the closed geodesic γ of (5.3.7) and we proceed by estimating the volume of M using the level surfaces of the distance function from γ (compare (1.9)). For every $p \in M$ there is a segment from p to γ which (necessarily) meets γ perpendicularly. The set Z of points p for which these segments are not unique has volume zero and $M \setminus Z$ is covered by unique minimizing geodesics to γ. Along these we exploit (1.5.1, 1.6). Choose orthonormal parallel vector fields $u_1, \ldots, u_{n-1}$, $u_1(0) = \gamma'$ along each segment $c \perp \gamma$. As in (1.7.1) we get for the shape operator S of the distance tubes around γ

$$g(Su_1, u_1)(r) \leqslant -\delta \frac{s_\delta}{c_\delta}, \quad g(Su_j, u_j) \leqslant \frac{c_\delta}{s_\delta}(r) \qquad (j = 2, \ldots, n-1),$$

$$\text{(5.3.9)}$$

$$\text{trace } S_r \leqslant -\delta \frac{s_\delta}{c_\delta}(r) + (n-2)\frac{c_\delta}{s_\delta}(r).$$

As in (1.9.2) this implies (along each normal segment c)

$$a(r) \leqslant c_\delta \cdot s_\delta^{n-2}(r), \qquad 0 \leqslant r \leqslant l(c),$$

where $l(c) \leqslant \text{diam}(M)$ denotes the distance up to which c is a minimizing segment to γ. If $\delta > 0$ we do not need separate diameter bounds (because of (1.7.1) no segment c of length $> (\pi/2) \cdot \delta^{-1/2}$ can be minimizing to γ). Since nonminimizing geodesics no longer contribute to the volume of the (outer) tubes we only worsen the volume estimate if we integrate the bound for $a(r)$ for $r \leqslant (\pi/2) \cdot \delta^{-1/2}$ respectively, $r \leqslant \mathrm{diam}(M)$ first at each $\gamma(t)$ over all normal directions and then along γ:

$$\mathrm{vol}_{n-1} \text{ (level surface at distance } r \text{ from } \gamma)$$

$$\leqslant l(c) \cdot \mathrm{vol}(S^{n-2}) \cdot c_\delta(r) \cdot s_\delta(r)^{n-2},$$

and a final radial integration proves (5.3.4):

  b) $\mathrm{vol}(M) \leqslant l(c) \cdot \mathrm{vol}(S^{n-2}) \cdot (n-1)^{-1} s_\delta(\mathrm{diam}(M))^{n-1}$,

  a) $\mathrm{vol}(M) \leqslant l(c) \cdot \mathrm{vol}(S^{n-2}) \cdot (n-1)^{-1} s_\delta((\pi/2) \cdot \delta^{-1/2})^{n-1}$

$$\leqslant \frac{1}{2\pi} \cdot \delta^{1/2} \cdot l(c) \cdot \mathrm{vol}(S^n_\delta).$$

The work to prove (5.3.5) was already done in (3.5): If for some $p \in M$ we had $d(p, C(p)) < \pi$ then we could find a compact totally convex set B with $p \in B$ (3.5b). Changing names we assume that the cut locus distance assumes its nonconjugate minimum on B at p. Again we have the loop γ from p through the closest cut point q. By total convexity this loop is in B, hence $d(q, C(q)) \geqslant d(p, C(p))$ and γ cannot have an angle $< \pi$ at q either (we repeated 5.3.7). By total convexity of the sublevels of the function bm (3.5) γ has to lie on a level of bm which contradicts (3.5.1).

To prove (5.3.6) we have the short closed geodesic γ from (5.3.7). Choose some $p \in \gamma$ and let $\bar{q}$ be a point farthest from p, i.e. $d(p, \bar{q}) > (\pi/2) \cdot \delta^{-1/2}$. Since the case $\delta > \frac{1}{4}$ is covered by (5.3.3) we assume $\delta \leqslant \frac{1}{4}$, i.e. $d(p, \bar{q}) > (\pi/2) \cdot \delta^{-1/2} \geqslant \pi > \frac{1}{2} \mathrm{length}\ (\gamma)$. Next, let $q \in \gamma$ be a *closest* point to $\bar{q}$; note $\gamma \perp q\bar{q}$. We apply (4.2.2) to the 90°-hinge $\bar{q}qp$:

$$0 \underset{\text{(Choice of } \bar{q})}{>} \cos \delta^{1/2} \cdot d(p, \bar{q}) \underset{(4.2.2)}{\geqslant} \cos \delta^{1/2} d(p, q) \cdot \cos \delta^{1/2} d(q, \bar{q})$$

to conclude $d(q, \bar{q}) > \frac{1}{2}\pi \cdot \delta^{-1/2}$ (note $d(p, q) \leqslant \frac{1}{2}\ \mathrm{length}\ (\gamma) < \frac{1}{2}\pi \cdot \delta^{-1/2}$). Now the contradiction arises for the same reason as in the noncompact case (3.5.3): $\bar{q}q$ is too long to even locally minimize the distance from $\bar{q}$ to the closed geodesic γ since the principal curvatures of the distance sphere around $\bar{q}$ through q are negative (1.7.1), i.e. convex to the outside.

To prove (5.3.3) consider the set $G := \{x \in M, d(x, C(x)) \geqslant \pi\}$ which is closed because of the continuity of the cut locus distance (K.4.3 corollary). We show $G \neq \varnothing$ and open, hence (by the assumed connectedness of M) $G = M$. To prove $G \neq \varnothing$ choose $y, \bar{q} \in M$ such that $d(y, \bar{q}) = \text{diam}(M)$; to show G open choose $y \in \{x \in M; d(x, C(x)) > \frac{1}{2}\pi \cdot \delta^{-1/2}\}$, i.e. from an open neighborhood of G. In each case assume by contradiction $d(y, C(y)) < \pi$, so that we have the geodesic loop γ (5.3.7) from y through the nearest cut point q. Of course q is critical for the distance function from y (two segments in opposite directions). But q is so far from y that because of (5.2.5, 5.2.4) there is only one critical point around: the farthest point $\bar{q}$ from y, i.e. $q = \bar{q}$. By definition of q every geodesic from y is minimizing at least up to distance $d(y, q) = d(y, \bar{q})$, so that *all* endpoints are farthest points and by (5.2.4) equal to $\bar{q}$. This makes $q = \bar{q}$ conjugate to y, contradicting $d(y, q) < \pi$.

5.4. *Sphere Theorems.*

5.4.1. (Rauch, Berger, Klingenberg 1951–1961). *Let M^n be complete, simply connected and assume*

$$\tfrac{1}{4} < \delta \leqslant K \leqslant 1 \qquad (\text{``}\tfrac{1}{4}\text{-pinching''}).$$

Then M^n is homeomorphic to S^n. The constant $\frac{1}{4}$ cannot be improved, see (6.1).

5.4.2. (Shikata 1967). *A bi-Lipschitz bound for the homeomorphism goes to 1 as $\delta \to 1$. The homeomorphism can therefore be smoothed to a diffeomorphism if δ is close enough to 1.*

5.4.3. (Grove-Shiohama 1977). *Let M^n be complete. Assume $0 < \delta \leqslant K$ and $\text{diam}(M) > \frac{\pi}{2} \cdot \delta^{-1/2}$. Then M^n is homeomorphic to S^n.*

5.4.4. REMARKS. The $\frac{1}{4}$-pinching theorem (5.4.1) is probably the most widely known comparison result. Rauch's version did not get the sharp constant $\frac{1}{4}$, but he did not use the cut locus estimates (5.3) either—instead he got such estimates as corollaries. Simple proofs of (5.4.1) depend on cut locus estimates. Given those, (5.4.3) is a nice generalization of (5.4.1): upper curvature bounds are replaced by an (implied) lower diameter bound. Note, that the diameter assumption also implies cut locus estimates (5.3.3). Exotic spheres had been discovered a few years earlier, so it was natural to try and replace "homeomorphic" by "diffeomorphic". Gromoll and Calabi proved the first diffeomorphism theorems in 1966; but, given what I have developed in this chapter, Shikata's result is most easily explained. It also points to a difference in the conclusions in (5.4.1) and (5.4.3) which the formulation "homeomorphic" deemphasizes: no metric control on the homeomorphism is obtained in (5.4.3), while (5.4.1) naturally leads to (5.4.2). Some of the arguments which were developed for the proofs of (5.4) I have put, as more general tools, into (5.2) and used them in (5.3).

Proof of (5.4.3). Choose $p, q \in M$ such that $d(p, q) =$ diam(M). Because of (5.2.4, 5.2.5) we have p and q as the *only* *critical points* for the distance function from p. We modify the construction (5.2.2) to obtain a vector field X on M with only one radial source at q and one radial sink at p. First choose two balls $B(p), B(q)$ which do not meet the cut locus $C(p)$ respectively, $C(q)$ and define on these the local vector fields not quite as in (5.2.2): for each $y \in B(p)$ respectively, $y \in B(q)$ let $X_p(y)$ respectively, $X_q(y)$ be the unit tangent vector of the unique segment yp respectively, qy. Now cover M with the concentric balls $B'(p), B'(q)$ and the $B'(x_i)$ of (5.2.2) (for $A = M \setminus (B'(p) \cup B'(q))$). The vector field $X = (1/\Sigma\varphi_i) \cdot \Sigma\varphi_i X_i$ then agrees with X_p near p, X_q near q and has no singularities except p, q. All integral curves run from q to p and their finite arc length depends differentiably on the initial direction. Now identify T_qM isometrically with a tangent space T_NS^n and map the integral curves of X proportional to arc length onto the corresponding meridians of S^n from N. This defines a continuous bijective map $M^n \to S^n$, hence a homeomorphism. The map is of maximal rank differentiable in $M^n \setminus \{q, p\}$;

the problem at q is harmless, but at p no information can be obtained as to how the angle between integral curves in M is related to the angle between the corresponding meridians in S^n.

Proof of (5.4.1). Choose $p, q \in M$ such that $d(p, q) = $ diam(M). Assuming the cut locus estimate $d(y, C(y)) \geqslant \pi$ (5.3) we conclude from (5.2.6) that *every* geodesic which starts from p respectively, q (since it is minimizing up to a distance $\geqslant \pi > (\pi/2)$ $\cdot \delta^{-1/2}$) *reaches* the "equator" set

$$E := \left\{ x \in M; d(p, x) = d(x, q) \quad \left(\leqslant \frac{\pi}{2} \cdot \delta^{-1/2}, 5.2.6 \right) \right\}$$

before it reaches the cut locus $C(p)$ respectively, $C(q)$. (5.2.6) also shows that the gradients of $d(q, x)$ and $-d(p, x)$ make along E an angle $\alpha \leqslant \pi - \delta^{1/2} \cdot \text{diam}(M) \leqslant \pi - \delta^{1/2}\pi < (\pi/2)$ (and $\alpha \to 0$ as $\delta \to 1$). The function $f(x) := d(q, x) - d(p, x)$ is therefore differentiable near E and grad $f \neq 0$, i.e. E is a differentiable submanifold. In particular, the distance from p to E (or q to E) depends differentiably on the initial direction of the segment from p (or q) to E, i.e. the radial map from the unit sphere in $T_p M$ (or $T_q M$) to E is a diffeomorphism. So we call E the equator of M and the segments from p or q to E half meridians. The homeomorphism $M^n \to S^n$ is now clear: Identify $T_q M$ with a tangent space $T_N S^n$ and map the "meridians" of M proportional to arc length onto the corresponding meridians of S^n.

About the proof of (5.4.2): The homeomorphism from M^n to S^n which we just obtained is of maximal rank differentiable on $M \setminus \{E \cup \{p, q\}\}$. The nonsmoothness along E is not serious; more simply than in the proof of (5.4.3) one can combine grad d_q and $-$ grad d_p to a smooth vector field on $M \setminus \{p, q\}$ such that the integral curves are essentially the broken meridians above, only slightly changed near E to smooth the corners. Crucial for the application of Shikata's smoothing result is the control of the "antipodal map" $T_q^1(M) \to T_p^1 M$ which maps the initial directions ($\in T_q M$) of the broken meridians to the final directions ($\in T_p M$). This map is defined by geodesics, its derivative can therefore be described by Jacobi fields along the broken meridians; the Jacobi fields vanish at p, q, they match at E and describe the derivative as

$J'|_q \to J'|_p$. Since Rauch's estimates (1.8) control the maximum and minimum growth of Jacobi fields with $J(0) = 0$, we get the desired bi-Lipschitz bound from the ratio of the upper and lower bound in (1.8)

$$L(\delta) \leqslant \frac{s_\delta}{s_\Delta}\left(\frac{\pi}{2} \cdot \delta^{-1/2}\right) = \sin^{-1}\left(\frac{\pi}{2} \cdot \delta^{-1}\right) \xrightarrow[\delta \to 1]{} 1.$$

Shikata's smoothing construction has been simplified since. It also depends on Hessian bounds for the distance function. But it is high time to get to the examples, so I omit the smoothing arguments.

6.   COMPLEX PROJECTIVE SPACE $\mathbb{C}P^n$ AND ITS DISTANCE SPHERES

The following properties make these examples relevant to the preceding comparison theorems.

6.1.1.   *Complex projective space $\mathbb{C}P^n$ has a natural Riemannian metric (Fubini-Study) which has curvature bounds $1 \leqslant K \leqslant 4$.*

6.1.2.   *Diameter = cut locus distance = $\pi \cdot \Delta^{-1/2}$ (compare 2.4.2, 5.2.4, 5.3.1).*

6.1.3.   *$\mathbb{C}P^n$ is not homeomorphic to S^{2n} (compare 5.3).*

6.1.4.   *The odd dimensional distance spheres of radius r have curvature bounds*

$$0 < \delta(r) := \frac{\cos^2 r}{\sin^2 r} \leqslant K \leqslant 4 + \frac{\cos^2 r}{\sin^2 r} =: \Delta(r) \xrightarrow[r \to (\pi/2)]{} 4$$

and have closed geodesics of length

$$l(r) = 2\pi \cos r \sin r \xrightarrow[r \to (\pi/2)]{} 0.$$

6.1.5.   *If $\delta(r)$: $\Delta(r) < \frac{1}{9}$ then $l(r) < 2\pi \cdot \Delta^{-1/2}$ (compare 5.3.2) and diam(r) $\leqslant \frac{1}{2}\pi \cdot \delta(r)^{-1/2}$ (compare 5.3.6).*

6.2.1.   *Definition as a metric space.* $\mathbb{C}P^n$ is defined as the set of complex lines in $\mathbb{C}^{n+1}$, or, what is the same, as the set of Hopf circles $\{e^{i\varphi} \cdot p; \varphi \in [0, 2\pi]\}_{p \in S^{2n+1} \subset \mathbb{C}^{n+1}}$. Any two circles have, as disjoint compact sets in S^{2n+1}, a natural distance, namely the length of a shortest great circle arc which joins them; this distance is $\leqslant \pi/2$ and $= \pi/2$ if the two Hopf circles lie in totally orthogonal subspaces of $\mathbb{R}^{2(n+1)} = \mathbb{C}^{n+1}$.

6.2.2. *Natural embedding* into a Euclidean sphere. Denote by Sym the vector space of complex linear hermitian symmetric endomorphisms of the Euclidean vector space $\mathbb{C}^{n+1}$ with the scalar product

$$\langle\langle A, B \rangle\rangle = \mathrm{Re\,trace}(\bar{A}^t \cdot B) = \mathrm{Re}\left(\sum_{i,j} \bar{a}_{ij} b_{ij}\right), \qquad A, B \in \mathrm{Sym}.$$

Define the quadratic map

$$V\colon S^{2n+1} \to \mathrm{Sym}, \qquad V(p)(z) := \frac{1}{2}(\langle z, p \rangle \cdot p + \langle z, ip \rangle \cdot ip),$$

that is, $V(p)$ is (up to the factor $\frac{1}{2}$ which is more convenient later) the orthogonal projection onto the complex line $\mathbb{C} \cdot p$, in particular $V(e^{i\varphi} \cdot p) = V(p)$. V gives, therefore, an injective map from the set of Hopf circles into Sym. Since $V(p)$ has two eigenvalues $\frac{1}{2}$ and all others 0 we have $\langle\langle V(p), V(p)\rangle\rangle = \frac{1}{2}$, so the image lies in a sphere.

6.2.3. *Submanifold metric.* The map V has constant rank $2n$ on S^{2n+1}, therefore the image is a $2n$-dimensional submanifold in Sym. We show more. Curves $p(t)$ on S^{2n+1} which are perpendicular to Hopf circles are mapped by V onto curves of the same length in Sym. With its induced submanifold metric the image is therefore isometric to $\mathbb{C}P^n$ (as defined in 6.2.1).

Proof. The assumptions are $\dot{p} \perp p, ip$, hence also $i\dot{p} \perp p, ip$. The tangent vector of $V(p(t))$ is given by

$$\frac{d}{dt}V(p(t))(z) = \frac{1}{2}(\langle z, \dot{p}\rangle p + \langle z, p\rangle\dot{p} + \langle z, i\dot{p}\rangle ip + \langle z, ip\rangle i\dot{p}),$$

hence—with $\{e_i\}$ an orthonormal basis— $\langle\langle \partial V \cdot \dot{p}, \partial V \cdot \dot{p} \rangle\rangle =$

$$\sum_{i=1}^{2n} \left\langle \frac{d}{dt}V(p(t))(e_i), \frac{d}{dt}V(p(t))(e_i) \right\rangle = \frac{1}{4} \cdot 4\langle \dot{p}, \dot{p}\rangle.$$

6.3.1. *Isometries.* Unitary maps $U: \mathbb{C}^{n+1} \to \mathbb{C}^{n+1}$ are isometries of S^{2n+1} which map Hopf circles to Hopf circles, therefore they give isometries of $\mathbb{C}P^n$. These extend to isometries of Sym (conjugations)

$$V(U \cdot p) = U \cdot V(p) \cdot U^{-1},$$

and $\langle\langle U \cdot A, U \cdot B \rangle\rangle = \langle\langle A, B \rangle\rangle = \langle\langle A \cdot U, B \cdot U \rangle\rangle$.

For any fixed orthonormal complex basis $\{v_0, \ldots, v_n\}$ of $\mathbb{C}^{n+1}$ we have a complex conjugation $\overline{\Sigma z_i \cdot v_i} := \Sigma \bar{z}_i \cdot v_i$. These are also isometries which preserve Hopf circles: $\overline{e^{i\varphi} \cdot p} = e^{-i\varphi} \cdot \bar{p}$, hence give isometries of $\mathbb{C}P^n$, and as before extend to isometries of Sym (conjugation of the matrices).

6.3.2. *Totally geodesic submanifolds.* Connected fixed point sets of isometries of Riemannian manifolds are totally geodesic.

a) The fixed point sets in S^{2n+1} of conjugations are equatorial spheres cut out by at most $(n+1)$-dim$_{\mathbb{R}}$ subspaces. These n-spheres are orthogonal to the Hopf circles, so V maps them, preserving length, into the image—but the map is not injective: antipodal points are on the same Hopf circle. The totally geodesic fixed point sets in $\mathbb{C}P^n$ obtained from conjugations in $\mathbb{C}^{n+1}$ are therefore real projective spaces $\mathbb{R}P^n$ of constant curvature 1.

b) To each subspace $\mathbb{C}^{k+1} \subset \mathbb{C}^{n+1}$ we define a complex reflection

$$U: \mathbb{C}^{n+1} \to \mathbb{C}^{n+1}, \qquad U|_{\mathbb{C}^{k+1}} := id, \quad U|_{(\mathbb{C}^{k+1})^\perp} := -id.$$

This map leaves the Hopf circles in $\mathbb{C}^{k+1}$ and $(\mathbb{C}^{k+1})^\perp$ fixed, and no others. As an isometry of $\mathbb{C}P^n$ (6.3.1) it has lower dimensional complex projective subspaces $\mathbb{C}P^k$, $\mathbb{C}P^{n-k-1}$ as totally geodesic fixed point sets.

6.3.3. *The Riemann spheres $\mathbb{C}P^1$ have curvature 4 in $\mathbb{C}P^n$.* Since the metric on any $\mathbb{C}P^1 \subset \mathbb{C}P^n$ can be computed by intersecting S^{2n+1} with the appropriate $\mathbb{C}^2$, it is enough to consider the case $n = 1$ (real dimension 2). The unitary maps of $\mathbb{C}^2$ are transitive on (the Hopf circles of) S^3; this and (6.2.1) give: the metric on $\mathbb{C}P^1$

has constant curvature and diameter $\pi/2$. This leaves two possibilities: S^2 with curvature 4 and $\mathbb{R}P^2$ with curvature 1; but multiplication by i gives $\mathbb{C}P^1$ an orientation—excluding $\mathbb{R}P^2$.

6.3.4.   REMARK.   The parametrization for $S^3 \subset \mathbb{R}^4 = \mathbb{C}^2$

$$F(\alpha, t, t_0) = \begin{pmatrix} \cos\alpha \cdot \begin{pmatrix} \cos(t+t_0) \\ \sin(t+t_0) \end{pmatrix} \\ \sin\alpha \cdot \begin{pmatrix} \cos(t-t_0) \\ \sin(t-t_0) \end{pmatrix} \end{pmatrix}, \qquad i = \left(\begin{array}{cc|cc} 0 & -1 & & \\ 1 & 0 & & 0 \\ \hline & & 0 & -1 \\ & 0 & 1 & 0 \end{array} \right)$$

has the t-lines as Hopf circles since $(\partial/\partial t)F = iF$. The distance between any two Hopf circles can be obtained from the metric which measures the length of curves perpendicular to Hopf circles,

$$ds^2 = d\alpha^2 + \left(\frac{1}{2}\sin 2\alpha \right)^2 dt_0^2.$$

It is the metric of a sphere of curvature 4 in polar coordinates.

6.3.5.   *Symmetric space structure.*   For every $p \in S^{2n+1}$ consider

$$U: \mathbb{C}^{n+1} \to \mathbb{C}^{n+1}, \qquad U\big|_{\mathrm{span}(p,ip)} = id, \qquad U\big|_{\{p,ip\}^\perp} = -id.$$

The differential of this isometry (use the submanifold picture) on the tangent space of the fixed point $V(p)$ is $(-id)$. Riemannian manifolds M which have to every $p \in M$ an isometry $\sigma_p: M \to M$ such that $\sigma_p(p) = p$, $\partial\sigma_p|_p = -id$ are called *symmetric spaces*.

An immediate consequence is that the curvature tensor is parallel. Namely, the differential of an isometry at a fixed point preserves any isometry-invariant tensor; in particular, using $\partial\sigma_p|_p = -id$ we get

$$-(D_X R)(U, V)W = (D_{-X}R)(-U, -V) - W$$

$$= (D_X R)(U, V)W.$$

Similarly, multiplication by i on each tangent space defines a tensor field J $(J(X) := i \cdot X))$ which is compatible with all the unitary isometries, hence it is parallel:

$$-(D_X J)(Y) = (D_{-X} J) \cdot (-Y) = (D_X J)(Y).$$

6.3.6. $\mathbb{C}P^n$ is very different from S^{2n}. Define the ("Kähler"-) 2-form

$$\omega(X, Y) := g(J \cdot X, Y).$$

Because of (6.3.5) it is parallel: $D\omega(X, Y) = g(DJ \cdot X, Y) = 0$, in particular, the exterior derivative vanishes: $d\omega = 0$. Also, ω is obviously the area form when restricted to any $\mathbb{C}P^1$ (if $X \in T_p\mathbb{C}P^1$, $|X| = 1$, then X and $J(X)$ are an orthonormal basis of $T_p\mathbb{C}P^1$ and $\omega(X, J(X)) = 1$), hence $\int_{\mathbb{C}P^1}\omega = (1/4) \cdot 4\pi$.

Observe that in S^{2n} $(n > 1)$ we can differentiably contract any S^2 to a point and then evaluate $\int_{S^2}\omega = \int_{\text{Cone}(S^2)} d\omega$ (Stokes) for any 2-form ω. We showed that this cannot be done in $\mathbb{C}P^n$. In the appropriate language this is expressed as: The second cohomology of $\mathbb{C}P^n$ is not zero.

6.4. The curvature tensor of $\mathbb{C}P^n$. If X, Y, Z are vector fields tangential to a totally geodesic submanifold then $D_Y Z$ is also tangential and hence $R(X, Y)Z$ is tangential. Let Y be any tangent vector of $\mathbb{C}P^n$, then Y, iY span the tangent space of a totally geodesic $\mathbb{C}P^1$ of curvature 4, hence

$$R^{\mathbb{C}P^n}(iY, Y)Y = 4 \cdot iY.$$

If $X \perp Y, iY$, then X and Y span the tangent space of a totally geodesic $\mathbb{R}P^2$ of curvature 1, hence

$$R^{\mathbb{C}P^n}(X, Y)Y = 1 \cdot X \quad \text{if } X \perp Y, iY.$$

Combining,

$$R^{\mathbb{C}P^n}(X, Y)Y = \langle Y, Y \rangle \cdot X - \langle X, Y \rangle \cdot Y + 3\langle X, iY \rangle \cdot iY. \quad (6.4.1)$$

Again for any curvature tensor, the first Bianchi identity gives

$$6R((X,Y)Z) = R(X, Y+Z)Y + Z - R(X, Y-Z)Y - Z$$

$$+ R(X+Z, Y)X + Z - R(X-Z, Y)X - Z.$$

This and (6.4.1) give the full curvature tensor.

$$R^{\mathbb{C}P^n}(X,Y)Z = \langle Y, Z \rangle \cdot X - \langle X, Z \rangle \cdot Y \qquad (6.4.2)$$
$$2 \cdot \langle X, iY \rangle \cdot iZ + \langle iY, Z \rangle \cdot iX - \langle iX, Z \rangle \cdot iY,$$

but already (6.4.1) shows that the curvature range is [1, 4].

6.5.   *Metric and principal curvatures of the distance spheres in* $\mathbb{C}P^n$. Every family of concentric distance spheres $S(r)$ cuts each totally geodesic $\mathbb{C}P^1$ through the midpoint of the $S(r)$ into concentric distance circles of geodesic curvature $2\operatorname{ctg}2\,r$. We call those circles Hopf circles on the distance spheres since their tangent field is obtained by multiplying the radial vector field N by i. Hence

6.5.1.   The Hopf circles on a distance sphere of radius r in $\mathbb{C}P^n$ have length $2\pi \cdot \cos r \cdot \sin r$. They are geodesics and principal curvature lines with principal curvature $\kappa_1(r) = 2\operatorname{ctg}2r$.

The radial direction N and a tangent vector $\perp N, iN$ span a tangent space of a totally geodesic $\mathbb{R}P^2$ of curvature 1. Hence

6.5.2.   The $\mathbb{R}P^2$'s through the midpoints of distance spheres in $\mathbb{C}P^n$ intersect those spheres perpendicular to their Hopf circles in closed geodesics of length $2\pi \cdot \sin r$ (namely distance circles in $\mathbb{R}P^2$). These are also principal curvature lines of curvature $\kappa_2(r) = \operatorname{ctg} r$.

REMARK.   The Hopf circles on the distance spheres in $\mathbb{C}P^n$ shrink by a factor $\cos r$ faster than great circles of distance spheres in S^{2n}. As $r \to \pi/2$ the length goes to zero and the distance spheres in $\mathbb{C}P^n$ get collapsed to $\mathbb{C}P^{n-1}$.

6.6. *The curvature tensor of the distance spheres.* Let S_r denote the shape operator and R_r the curvature tensor of a distance sphere of radius r. The Gauss equations are

$$\langle R^{\mathbb{C}P^n}(X, Y)Y, X\rangle = \langle R_r(X, Y)Y, X\rangle - \langle S_rX, X\rangle\langle S_rY, Y\rangle$$
$$+ \langle S_rX, Y\rangle^2.$$

This and (6.5.1, 6.5.2) give

$$K_r(iN \wedge Y) = 1 + \frac{2\cos 2r}{\sin 2r} \cdot \frac{\cos r}{\sin r} = \left(\frac{\cos r}{\sin r}\right)^2, \quad (6.6.1)$$

$$K_r(X \wedge iX) = 4 + \left(\frac{\cos r}{\sin r}\right)^2 \quad (X \perp iN), \qquad (6.6.2)$$

$$K_r(X \wedge Y) = 1 + \left(\frac{\cos r}{\sin r}\right)^2 \quad (X, Y \perp iN). \qquad (6.6.3)$$

So the curvatures have at least the range claimed in (6.1.4). But since we saw that the eigenspaces of S_r are compatible with the eigenspaces of $R^{\mathbb{C}P^n}(\ , Y)Y$ one checks easily that indeed (6.6.1) gives min K and (6.6.2) max K.

Finally, we get

$$\text{length of Hopf circles } < 2\pi(\max K)^{-1/2}$$

$$\text{iff } \cos^2 r < (1 + 3\sin^2 r)^{-1} \text{ (or } 2 < 3\sin^2 r)$$

$$\text{iff min } K: \max K < \tfrac{1}{9}.$$

REMARK. Berger (1960) discovered the curvature and cut locus properties of these metrics on S^3. Weinstein (1973) observed that they occur as distance spheres in $\mathbb{C}P^n$. They also appear on the quotient $\mathbb{R}P^3 = SO(3)$ as the kinematic metric of a rotating solid body with two equal moments of inertia and one smaller moment.

Acknowledgements. I wish to thank many colleagues, in particular Karsten Grove and Wolfgang Ziller, for many helpful discussions and suggestions. Most of the manuscript was written while I enjoyed the hospitality and ideal working conditions of the I.H.E.S. in Bures.

REFERENCES

[B]   Berger, M.: La Géométrie Métrique des Variétés Riemanniennes, Soc. Math. France, Astérisque, hors série (1985), 9–66.

[BGS]   Ballmann, W., Gromov, M., Schroeder V.: Manifolds of Nonpositive Curvature, Birkhäuser 1985.

[BK]   Buser P., Karcher, H.: Gromov's Almost Flat Manifolds, Soc. Math. France, Astérisque 81 (1981).

[BZ]   Burago, Y., Zalgaller, D.: Convex Sets in Riemannian Spaces of Nonnegative Curvature, Russian Math. Surveys, t. 32-3 (1977), 1–57.

[CE]   Cheeger, J., Ebin, D. G.: Comparison Theorems in Riemannian Geometry, American Elsevier, New York 1975.

[G]   Gromov, M.: Almost Flat Manifolds, J. Diff. Geom. 13 (1978), 231–241.

[S]   Sakai, T.: Comparison and Finiteness Theorems in Riemannian Geometry, Advanced Studies in Pure Mathematics 3 (1984), 125–181.

MANIFOLDS OF NONPOSITIVE CURVATURE

*Patrick Eberlein**

INTRODUCTION

In the past 20 years much progress has been made in understanding the structure of complete Riemannian manifolds of nonpositive sectional curvature, especially those that have finite Riemannian volume. The foundation for much of this research was provided by the paper [**BO**] of R. Bishop and B. O'Neill, which introduced in a systematic way the use of convex functions and convex sets as a tool for studying manifolds of nonpositive sectional curvature. In the special case of symmetric spaces of noncompact type this use of convexity appears earlier in the work of Karpelevic [**K**]. See also [**Mo**]. In this chapter we give a survey of some of the main developments in the study of manifolds of nonpositive sectional curvature, and we illustrate this discussion with examples.

It would be unreasonable to attempt to make a complete survey, and we have chosen to focus on results concerning geodesic flows

*Supported in part by NSF Grant DMS-8601367

and the characterization of symmetric spaces of noncompact type and rank at least 2 (Sections 4 and 6). We describe in some detail the geometry of the symmetric space $\tilde{M}_n = \mathrm{SL}(n, \mathbb{R})/\mathrm{SO}(n, \mathbb{R})$ (Section 5). Not only can one describe the geometry of this space nicely in terms of elementary linear algebra but the intuition one gains by understanding this space is very useful for the understanding of much recent research. Other treatments of symmetric spaces may be found in [**H**], [**K**], [**Mi**], [**Mo**], [**W**] and the appendix of [**E5**].

We now describe very briefly a method that has proved useful in studying manifolds of nonpositive sectional curvature and which developed from the convexity framework introduced in [**BO**] and [**K**]. To every complete, simply-connected Riemannian manifold $\tilde{M}$ of nonpositive sectional curvature one can associate a boundary space $\tilde{M}(\infty)$ that may be identified in a natural way with the $(n-1)$-sphere of unit vectors tangent to $\tilde{M}$ at a fixed point p (Section 2 and [**EO**]). Isometries of $\tilde{M}$ extend to homeomorphisms of $\tilde{M}(\infty)$. Each point p of $\tilde{M}$ also determines a geodesic symmetry diffeomorphism s_p of $\tilde{M}$ that extends to a homeomorphism of $\tilde{M}(\infty)$. The main theme of this chapter is that the geometry of $\tilde{M}$ and its quotient manifolds are reflected by the topological properties of the action on $\tilde{M}(\infty)$ of the isometry group $I(\tilde{M})$ and of the group G^* generated by the geodesic symmetries s_p, $p \in \tilde{M}$. In particular, if $\tilde{M}$ admits a quotient manifold M of the same dimension with finite volume, then the action on $\tilde{M}(\infty)$ of the fundamental group of M, regarded as a discrete subgroup of $I(\tilde{M})$, reflects the action of the geodesic flow on the unit tangent bundle of M (Section 6). Similarly the action on $\tilde{M}(\infty)$ of the symmetry group G^* reflects the action of the holonomy group Φ_p on the unit tangent vectors at an arbitrary point p of $\tilde{M}$ and leads to a characterization of symmetric spaces of noncompact type and rank at least 2 (Section 7).

We describe briefly the organization of this chapter. In Section 1 we recall some definitions and facts of Riemannian geometry. In Sections 2 and 3 we discuss some basic properties of manifolds of nonpositive curvature. The emphasis is on simply connected manifolds and the action and structure of their isometry groups. Our point of view is to study a nonsimply-connected manifold M of nonpositive curvature by studying the action of isometries of the fundamental group of M on the simply-connected Riemannian cover $\tilde{M}$. In Sections 4 and 5 we discuss Riemannian symmetric spaces of nonpositive curvature, the most important class of examples. In Sections 6 and 7 we describe the action of isometries and geodesic symmetries on $\tilde{M}(\infty)$, where $\tilde{M}$ is a simply-connected manifold of nonpositive curvature, and we obtain the applications mentioned above. In the last section we discuss some recent research about manifolds of nonpositive curvature and rank at least 2, and we relate these results to the rest of the chapter.

1. RIEMANNIAN MANIFOLDS

A Riemannian manifold is a differentiable manifold N equipped with an inner product $\langle\ ,\ \rangle p$ on each tangent space T_pN, p a point of N. For convenience we shall assume that all manifolds N are C^∞ and that the corresponding inner products $\langle\ ,\ \rangle$ are C^∞; that is, if X, Y are C^∞ vector fields on N then $p \to \langle X(p), Y(p)\rangle_p$ is a C^∞ function on N.

Given points p, q in a Riemannian manifold N we define $d(p, q)$ to be the infimum of $L(\gamma)$, where $\gamma\colon [a, b] \to N$ is a differentiable curve from p to q and $L(\gamma) = \int_a^b \langle \gamma'(t), \gamma'(t)\rangle^{1/2}dt$ is the length of γ. A Riemannian manifold N is said to be *complete* if it is complete as a metric space relative to the Riemannian metric $d(\ ,\ )$. A theorem of Hopf-Rinow [**CE**, p. 11] says that if N is a complete Riemannian manifold, then the geodesics of N are defined on all of $\mathbb{R}$. We shall not give a precise definition of geodesic except to say that for each tangent vector v at a point p of N there exists a unique geodesic γ_v defined on $(-\epsilon, \epsilon)$ for some $\epsilon > 0$ such that $\gamma_v(0) = p$ and $v = \gamma_v'(0)$, the initial velocity of γ_v. The theorem of Hopf and Rinow also says that if p and q are distinct points in a complete Riemannian manifold N, then there exists at least one geodesic γ that joins p to q and has length equal to $d(p, q)$. In this chapter we assume that all Riemannian manifolds considered are complete.

Exponential Map. Let N be a complete Riemannian manifold. For each point p of N one defines an exponential map

$$\exp_p\colon T_pN \to N$$

by

$$\exp_p(v) = \gamma_v(1)$$

where γ_v denotes the unique geodesic of N with initial velocity v. By the completeness assumption, $\exp_p$ is defined on all of T_pN and is a surjective map. One can also show that $\gamma_v(t) = \exp_p(tv)$ for all vectors v in T_pN and all t in $\mathbb{R}$.

Curvature Tensor. For every pair of tangent vectors u, v in T_pN, N a Riemannian manifold, one defines a skew symmetric curvature transformation

$$R(u, v): T_pN \to T_pN.$$

Briefly, using u, v one defines an appropriate family of geodesic polygons $\{\Delta_t(u, v)\}$ with p as one vertex which shrinks to the point p as $t \to 0$. If $P_t: T_pN \to T_pN$ denotes the orthogonal transformation obtained from parallel translation of vectors around $\Delta_t(u, v)$, then $R(u, v) = d/dt(P_t)|_{t=0}$. See [BC, p. 97] for precise details of the construction of $\Delta_t(u, v)$ and P_t.

Sectional Curvature. Given a 2-dimensional subspace $\pi \subset T_pN$, where N is a Riemannian manifold, one defines the *sectional curvature* $K(\pi) \in \mathbb{R}$ as follows: let $u, v \in T_pN$ be a basis for π and define

$$K(\pi) = \frac{\langle R(u, v)v, u \rangle}{\|u \wedge v\|^2}$$

where $\|u \wedge v\|^2 = \langle u, u \rangle \langle v, v \rangle - \langle u, v \rangle^2$ and $R(u, v)$ is the curvature transformation defined above. It is easy to verify that $K(\pi)$ does not depend on the choice of (u, v). One may also describe $K(\pi)$ as follows: Let Σ_π denote the surface in N which is the union of all geodesics through p that are tangent to π. Then $K(\pi)$ is the Gaussian curvature at p of Σ_π.

Holonomy. Given a point p in a Riemannian manifold N of dimension n one considers all finitely broken differentiable paths γ

that begin and end at p. The collection of all parallel translation operators P_γ around such paths γ forms a subgroup of the orthogonal group in $T_p N$, regarded as a Euclidean space of dimension n. This subgroup, denoted Φ_p, is called the *holonomy group at* p. The Lie algebra of Φ_p contains all curvature transformations $R(u, v)$, $u, v \in T_p N$ as well as the covariant derivatives of all orders of these curvature transformations. One may verify this statement directly from the definition above of $R(u, v)$.

2. SIMPLY-CONNECTED MANIFOLDS OF NONPOSITIVE SECTIONAL CURVATURE

We shall study Riemannian manifolds M of nonpositive sectional curvature ($K \leqslant 0$) by studying the action of the fundamental group $\pi_1(M)$ on the simply-connected Riemannian covering manifold $\tilde{M}$, where $\pi_1(M)$ is regarded as a discrete group of isometries of $\tilde{M}$. See Section 3 below. We begin with a discussion of the geometric properties of a simply-connected manifold of nonpositive sectional curvature. General references are [**BO**] and [**EO**].

In this chapter $\tilde{M}$ will always denote a simply-connected, complete Riemannian manifold of nonpositive sectional curvature. All geodesics of $\tilde{M}$ will be assumed to have unit speed. The unit tangent bundle of $\tilde{M}$ will be denoted by $S\tilde{M}$.

2.1. *Topological structure.* The most basic fact about a complete, simply-connected manifold $\tilde{M}$ of nonpositive sectional curvature is that the exponential map $\exp_p: T_p\tilde{M} \to \tilde{M}$ is a diffeomorphism for each point p of $\tilde{M}$. In particular $\tilde{M}$ is diffeomorphic to a Euclidean space of the same dimension ($T_p\tilde{M}$). Moreover, since the exponential map is one-one there exists a unique unit speed geodesic from p to q.

Notation. Let p, q, r be points in $\tilde{M}$ such that p is distinct from q and r. Let γ_{pq} denote the unique unit speed geodesic such that $\gamma_{pq}(0) = p$ and $\gamma_{pq}(a) = q$, where $a = d(p, q)$. Let $V(p, q) = \gamma'_{pq}(0)$, the initial velocity of γ_{pq}. Let $\sphericalangle_p(q, r) = \sphericalangle(V(p, q), V(p, r))$, the angle subtended at p by q and r.

2.2. *Basic geometric properties.*   In addition to having the same topology and differentiable structure as Euclidean spaces, simply-connected manifolds $\tilde{M}$ with $K \leqslant 0$ have many of the geometric properties of Euclidean spaces. The underlying reason for this is provided by the following special case of a theorem of Toponogov. See [**CE**, Chapter 2] for a proof.

THEOREM 2.2A.   *Let c be a constant that is negative or zero, and let M^* be a complete, simply-connected manifold with constant sectional curvature c. Let $\tilde{M}$ be a complete, simply-connected manifold with all sectional curvatures $\leqslant c$. Let γ_1, γ_2 be unit speed geodesics emanating from a point p in $\tilde{M}$, and let γ_1^*, γ_2^* be unit speed geodesics emanating from a point p^* in M^* such that the angles subtended at p by γ_1, γ_2 and at p^* by γ_1^*, γ_2^* are equal. Then $d(\gamma_1 s, \gamma_2 t) \geqslant d(\gamma_1^* s, \gamma_2^* t)$ for all positive numbers s, t.*

If c is negative then M^* is a hyperbolic space (see Section 2.4 below) while if $c = 0$, then M^* is a Euclidean space with the usual inner product on tangent spaces arising from the dot product. Briefly put, the result above says that if $\tilde{M}$ has sectional curvature $K \leqslant c \leqslant 0$, then the geodesics emanating from a point in $\tilde{M}$ spread apart at least as fast as they do in a simply-connected comparison space M^* with sectional curvature $K \equiv c$.

From the usual law of cosines in a Euclidean space and the result stated above we obtain

2.2B. LAW OF COSINES.   *Let Δ be a triangle in a space $\tilde{M}$ with $K \leqslant 0$ whose sides are geodesics of lengths a, b, c and whose angle opposite c is denoted by θ. Then*

$$c^2 \geqslant a^2 + b^2 - 2ab \cos \theta.$$

From the law of cosines we now obtain two further useful results.

2.2C. ANGLE SUM LAW.   *Let Δ be a geodesic triangle in a space $\tilde{M}$ with $K \leqslant 0$. Then the sum of the interior angles of Δ is at most π.*

PROPOSITION 2.2D.   *Let p be a point of $\tilde{M}$ and let $\{r_n\}$, $\{q_n\}$ be sequences in $\tilde{M}$ such that $d(r_n, p) \to \infty$ and $d(r_n, q_n) \leqslant c$ for all n and some positive constant c. Then $\sphericalangle_p(q_n, r_n) \to 0$ as $n \to \infty$.*

For proofs of these results see for example [**H**, p. 73]. Another important consequence of the law of cosines is a fixed-point theorem due to E. Cartan. The proof given here is somewhat simpler than existing proofs in the literature.

THEOREM 2.2E.   *Let Γ be a group of isometries of a space $\tilde{M}$ such that the orbit $\Gamma(p)$ is bounded in $\tilde{M}$ for some point p of $\tilde{M}$. Then Γ has a fixed point in $\tilde{M}$.*

Proof.   Let $A = \overline{\Gamma(p)}$, the closure of $\Gamma(p)$ in $\tilde{M}$. The completeness theorem of Hopf and Rinow implies that A is compact. Let $r: \tilde{M} \to \mathbb{R}$ be defined by $r(q) = \max\{d(q, a): a \in A\}$, and let q^* be a point in $\tilde{M}$ where r assumes its minimum value r_0. The function r is constant on Γ-orbits in $\tilde{M}$ since A is invariant under Γ. We show that r assumes its minimum at a unique point q^*, which will show that Γ fixes q^*. Suppose r assumes its minimum value r_0 at two points q_1^*, q_2^*, and let q_3^* be the midpoint of the geodesic segment from q_1^* to q_2^*. Given a point a in A we consider the two geodesic triangles with vertices a, q_1^*, q_3^* and a, q_2^*, q_3^*. One of these triangles, say the former, has an angle $\geqslant \pi/2$ at q_3^*. It follows from the law of cosines (2.2B) applied to the triangle with vertices a, q_1^*, q_3^* that $d(a, q_1^*) > d(a, q_3^*)$. The argument above shows that $d(a, q_3^*) < \max\{d(a, q_1^*), d(a, q_2^*)\} \leqslant r_0$ for any point $a \in A$ and hence $r(q_3^*) < r_0$ by the compactness of A, which contradicts the definition of r_0. ∎

2.3. *Convexity properties.*   Manifolds $\tilde{M}$ also share many of the same convexity properties of Euclidean spaces with the usual inner product. We discuss only a few that are needed for this chapter. Proofs and other examples may be found in [**BO**, Sections 2–4].

Convexity of a subset $A \subseteq \tilde{M}$ is defined exactly as in Euclidean space since there is a unique geodesic joining any two distinct points of $\tilde{M}$. A continuous function $f: \tilde{M} \to \mathbb{R}$ is said to be *convex* if for any geodesic γ: $\mathbb{R} \to \tilde{M}$ the function $f \circ \gamma$: $\mathbb{R} \to \mathbb{R}$ is a convex function on $\mathbb{R}$. A continuous function f: $\mathbb{R} \to \mathbb{R}$ is convex if $f[\lambda a + (1 - \lambda)b] \leqslant \lambda f(a) + (1 - \lambda)f(b)$ for all $\lambda \in [0,1]$ and numbers a, b with $a < b$. If f: $\mathbb{R} \to \mathbb{R}$ is C^2, then f is *convex* if and only if $f''(x) \geqslant 0$ for all $x \in \mathbb{R}$. A function f: $\tilde{M} \to \mathbb{R}$ or f: $\mathbb{R} \to \mathbb{R}$ is *strictly convex* if strict inequality holds in all inequalities stated above. We note that if f: $\tilde{M} \to \mathbb{R}$ is a convex function, then $\tilde{M}^a = \{ p \in \tilde{M}: f(p) \leqslant a \}$ is a closed convex subset of $\tilde{M}$ for all real numbers a.

The following three results are basic to the study of manifolds of nonpositive sectional curvature.

PROPOSITION 2.3A. *Let A be a closed convex subset of a space $\tilde{M}$. Then for each point p of $\tilde{M}$ there exists a unique point $P(p)$ in A such that $d(p,P(p)) \leqslant d(p,q)$ for all $q \in A$.*

The point $P(p)$ is called the foot point of p on A and P: $\tilde{M} \to A$ is called the orthogonal projection onto A.

PROPOSITION 2.3B. *Let A be a closed convex subset of a space $\tilde{M}$. Then the function f_A: $\tilde{M} \to \mathbb{R}$ given by $f_A(p) = d^2(p, P(p)) = d^2(p, A)$ is a continuous convex function. Moreover f_A is strictly convex on $\tilde{M} - A$ if the sectional curvature in $\tilde{M}$ is negative.*

PROPOSITION 2.3C. *If ϕ is any isometry of a space $\tilde{M}$, then the function d_ϕ^2: $\tilde{M} \to \mathbb{R}$ given by $d_\phi^2(p) = d^2(p, \phi p)$ is a C^∞ convex function on $\tilde{M}$. The function d_ϕ^2 has a positive minimum value at a point p of $\tilde{M}$ if and only if $(\phi \circ \gamma)(t) = \gamma(t + \omega)$ for all $t \in \mathbb{R}$, where $\omega = d(p, \phi p)$ and $\gamma(t)$ is the unique geodesic with $\gamma(0) = p$ and $\gamma(\omega) = \phi(p)$. (In this case we say that ϕ translates γ by an amount ω.)*

2.4. *Examples.* Before proceeding further we describe some examples of complete, simply-connected spaces $\tilde{M}$ with sectional curvature $K \leqslant 0$.

1. Euclidean space $\mathbb{R}^n$ with the usual inner product. Here $K \equiv 0$. The geodesics are the straight lines and the isometries are all products $R \circ T$, where $R \in O(n)$, the orthogonal group, and T is a

translation of $\mathbb{R}^n$. The group $O(n)$ is a normal subgroup of the isometry group $I(\mathbb{R}^n)$.

2. Hyperbolic space H^n with $K \equiv -1$. We describe two models, both of which have useful features.

a) *Upper half-space model.* Let $H^n = \{(x_1, \ldots, x_n) \in \mathbb{R}^n: x_n > 0\}$. The inner product on $T_p H^n$ at a point $p = (p_1, \ldots, p_n)$ is the usual Euclidean inner product multiplied by $(1/p_n^2)$. The geodesics of H^n are either the "vertical" lines $t \to (x_1, x_2, \ldots, x_{n-1}, e^t x_n)$ or Euclidean circular arcs in $\mathbb{R}^n$ that are orthogonal to the hyperplane $x_n = 0$ and have unit speed in the hyperbolic metric.

We describe the isometries of H^n in this model in the case $n = 2$, but for $n \geqslant 3$ the description of isometries is more convenient in the other model of H^n described below. If $n = 2$, then $\mathrm{SL}(2, \mathbb{R})$ acts on H^2 by fractional linear transformations: identifying a point (x, y) with the complex number $z = x + iy$ we define

$$\begin{pmatrix} a & b \\ c & d \end{pmatrix}(z) = \frac{az + b}{cz + d}$$

where a, b, c, d are real numbers such that $ad - bc = 1$. The matrices A and $-A$ in $\mathrm{SL}(2, \mathbb{R})$ induce the same action on the hyperbolic plane H^2 and one may show that $I_0(H^2)$, the connected component of the isometry group of H^2 that contains the identity, equals $\mathrm{PSL}(2, \mathbb{R})$, the quotient of $\mathrm{SL}(2, \mathbb{R})$ by the two element subgroup $\{\pm I\}$. For a nice description of the geometry of H^2 in terms of the elementary theory of functions of one complex variable see [S]. See also [Bea].

b) *Hyperboloid model.* Let

$$H^n = \left\{ (x_1, \ldots, x_{n+1}) \in \mathbb{R}^{n+1} : \left(\sum_{i=1}^{n} x_i^2 \right) - x_{n+1}^2 \right.$$

$$\left. = -1 \text{ and } x_{n+1} > 0 \right\}.$$

Let $SO(n,1)$ denote the subgroup of $SL(n,\mathbb{R})$ which leaves invariant the bilinear form ϕ:

$$(x, y) \rightarrow \left(\sum_{i=1}^{n} x_i y_i \right) - x_{n+1} y_{n+1}, \qquad \text{where} \quad x = (x_1, \ldots, x_{n+1})$$

and $y = (y_1, \ldots, y_{n+1})$ are arbitrary elements of $\mathbb{R}^{n+1}$. The restriction of ϕ to the tangent spaces of H^n is positive definite and hence a Riemannian inner product on H^n. Moreover $SO(n,1) = I_0(H_n)$. The geodesics of H^n that start at $e_{n+1} = (0, \ldots, 0, 1)$ are unit speed parameterizations relative to ϕ of the curves obtained by intersecting H^n with 2-planes in $\mathbb{R}^{n+1}$ that contain the vector e_{n+1}. All other geodesics of H^n are images of the geodesics just described under the group $SO(n,1)$, which acts transitively on H^n.

3. The space $\tilde{M}_n$ consisting of all positive definite symmetric $n \times n$ matrices of determinant one. This example will be described below in Section 5. See also Section 3 of [**Mo**].

2.5. *Points at infinity for $\tilde{M}$.* Two unit speed geodesics γ, σ of a space $\tilde{M}$ are said to be *asymptotic* if there exists $c > 0$ such that $d(\gamma t, \sigma t) \leqslant c$ for all $t \geqslant 0$. The relation of being asymptotic is an equivalence relation on the unit speed geodesics of $\tilde{M}$. An equivalence class of unit speed geodesics of $\tilde{M}$ is called a *point at infinity* for $\tilde{M}$. If γ is a geodesic of $\tilde{M}$ then $\gamma(\infty)$ will denote the equivalence class of γ and $\gamma(-\infty)$ will denote the equivalence class of the geodesic γ^{-1}: $t \rightarrow \gamma(-t)$.

Example 1. If $\tilde{M} = \mathbb{R}^n$ with the usual inner product, then two geodesics are asymptotic if and only if they are parallel. The points of $\tilde{M}(\infty)$ are in one-one correspondence with the pencils of oriented parallel lines in $\mathbb{R}^n$ and each pencil is represented by a unique oriented straight line through the origin.

Example 2. If $\tilde{M} = H^n$ in the upper half-space model, then any two vertical geodesics $\gamma(t) = (x_1, \ldots, x_{n-1}, x_n e^t)$ and $\sigma(t) = (y_1, \ldots, y_{n-1}, y_n e^t)$ are asymptotic and the corresponding point at infinity is denoted by $\{\infty\}$. All other points at infinity for H^n are in one-one correspondence with the points of the hyperplane $x_n = 0$; if $\gamma(t), \sigma(t)$ are nonvertical asymptotic geodesics of H^n then $\gamma(t)$ and $\sigma(t)$ converge to the same point of $x_n = 0$ as $t \to +\infty$, where convergence is measured relative to the extrinsic Euclidean norm of $\mathbb{R}^{n+1}$. Hence $H^n(\infty)$ may be identified with $\{x_n = 0\} \cup \{\infty\}$, which is topologically an $(n-1)$-sphere.

The next result shows that the points of $\tilde{M}(\infty)$ may be identified with the unit vectors at a fixed point p, and hence $\tilde{M}(\infty)$ may be identified with an $(n-1)$-sphere, as we have already seen in the two examples above.

PROPOSITION 2.5A. *Let γ be a unit speed geodesic of a space $\tilde{M}$, and let p be any point of $\tilde{M}$. Then there exists a unique unit speed geodesic σ such that $\sigma(0) = p$ and σ is asymptotic to γ.*

Proof. (Existence). For each positive integer n let σ_n be the unique geodesic of $\tilde{M}$ such that $\sigma_n(0) = p$ and $\sigma_n(t_n) = \gamma(n)$, where $t_n = d(p, \gamma(n))$. If $A = \gamma(\mathbb{R})$, then by the convexity of the function $p \to d^2(p, A)$ (Proposition 2.3B) and the fact that $\sigma_n(t_n) \in A$ it follows that $d(\sigma_n(t), A) \leqslant d(p, A)$ for $0 \leqslant t \leqslant t_n$. If σ is a unit speed geodesic such that $\sigma(0) = p$ and $\sigma'(0)$ is a cluster point of the sequence of unit vectors $\{\sigma_n'(0)\}$, then $d(\sigma t, A) \leqslant d(p, A)$ for all $t \geqslant 0$ by continuity. It now follows easily that σ is asymptotic to γ.

(Uniqueness). Let $\sigma_1(t), \sigma_2(t)$ be unit speed geodesics such that $\sigma_1(0) = \sigma_2(0) = p$ and σ_i is asymptotic to γ for $i = 1, 2$. It follows that σ_1 and σ_2 are asymptotic and hence $d(\sigma_1(t), \sigma_2(t)) \leqslant c$ for all $t \geqslant 0$ and some positive constant c. By the Law of Cosines (2.2B) this is only possible if $\sigma_1 = \sigma_2$. ∎

2.5B. *Notation.* We extend the notation introduced in (2.1). Given a point x in $\tilde{M}(\infty)$ we let γ_{px} denote the unique geodesic of $\tilde{M}$ such that $\gamma_{px}(0) = p$ and $\gamma_{px}(\infty) = x$; that is, γ_{px} starts at p and belongs to the asymptote class x. We let $V(p, x)$ denote $\gamma_{px}'(0)$, the initial velocity of γ_{px}. Finally if p is any point of $\tilde{M}$ and if r, q are points of $\tilde{M}$ or $\tilde{M}(\infty)$ that are distinct from p, then we define $\sphericalangle_p(q, r)$, the angle subtended at p by q and r, to be the angle at p between $V(p, q)$ and $V(p, r)$.

2.5C. *Cone topology.* Let $M^* = \tilde{M} \cup \tilde{M}(\infty)$. We describe a topology that makes M^* a closed n-disk. See [**EO**, Section 2] for proofs and further details. Given points $p \in \tilde{M}$, $x \in \tilde{M}(\infty)$ and positive numbers ϵ, R we define

$$C(p, \epsilon, R, x) = \{ q \in M^*: d(p, q) > R \quad \text{and} \quad \sphericalangle_p(q, x) < \epsilon \}.$$

By convention the condition $d(p, q) > R$ applies only to points q in $\tilde{M}$. The set $C(p, \epsilon, R, x)$ is called the R-truncated cone with vertex p, angle ϵ and axis x. For fixed points p, x the truncated cones $C(p, \epsilon, R, x)$ as ϵ and R vary over all positive numbers form a neighborhood basis at x in $M^* = \tilde{M} \cup \tilde{M}(\infty)$ and the resulting topology for M^* is called the *cone or sphere topology.*

With respect to the cone topology on M^* and its induced topology on $\tilde{M}(\infty)$ one has the following useful facts.

PROPOSITION 2.5D.   *For a fixed point $p \in \tilde{M}$ let $S_p\tilde{M}$ denote the unit tangent vectors at p. Then the map $V: \tilde{M} \times \tilde{M}(\infty) \to S\tilde{M}$ given by $(p, x) \to V(p, x) = \gamma'_{px}(0)$ is a homeomorphism with respect to the product topology in $\tilde{M} \times \tilde{M}(\infty)$.*

From Proposition 2.5A we saw already that the restriction of V to $\{ p \} \times \tilde{M}(\infty)$ is a bijection onto $S_p\tilde{M}$, the unit vectors at p, for every point p of $\tilde{M}$.

PROPOSITION 2.5E.   *Let $A = \{(p, q, r) \in \tilde{M} \times M^* \times M^*: p \neq q, p \neq r\}$. Then the map $\sphericalangle: (p, q, r) \to \sphericalangle_p(q, r) = \sphericalangle(V(p, q), V(p, r))$ is continuous.*

COROLLARY 2.5F.   *Let γ be a unit speed geodesic of $\tilde{M}$. Let $\{ \gamma_n \}$ be a sequence of geodesics of $\tilde{M}$ such that $\gamma'_n(0) \to \gamma'(0)$ as $n \to \infty$, and let $\{t_n\} \subseteq \mathbb{R}$ be a sequence such that $t_n \to +\infty$ as $n \to \infty$. Then*
1) $\gamma_n(\infty) \to \gamma(\infty)$ *and* $\gamma_n(-\infty) \to \gamma(-\infty)$ *as* $n \to \infty$
2) $\gamma_n(t_n) \to \gamma(\infty)$ *as* $n \to +\infty$.

2.5G. *Joining points at infinity.*

DEFINITION.   We say that distinct points x, y in $\tilde{M}(\infty)$ can be *joined by a geodesic of $\tilde{M}$* if there exists a geodesic γ of $\tilde{M}$ such that $\gamma(\infty) = x$ and $\gamma(-\infty) = y$.

Example 1. $\tilde{M} = \mathbb{R}^n$. Here each point x in $\tilde{M}(\infty)$ can be joined to exactly one point y in $\tilde{M}(\infty)$. The joining geodesic is not unique.

Example 2. $\tilde{M} = H^n$, hyperbolic n-space. Here any two distinct points x, y of $\tilde{M}(\infty)$ can be joined by a unique geodesic of $\tilde{M}$.

Example 3. $\tilde{M}$ has sectional curvature $K \leqslant c < 0$ for some negative constant c. Here also any two distinct points x, y of $\tilde{M}(\infty)$ can be joined by a unique geodesic γ of $\tilde{M}$. One constructs γ as follows: fix a point p in $\tilde{M}$ and let σ_n denote the geodesic joining $\gamma_{px}(n)$ and $\gamma_{py}(n)$ for any positive integer n. Then the condition $K \leqslant c < 0$ implies that $d(p, \sigma_n)$ is uniformly bounded above. If σ_n is parameterized so that $d(p, \sigma_n(0))$ is uniformly bounded above, then $\{\sigma_n'(0)\}$ converges to a vector v as $n \to \infty$, and the geodesic σ with initial velocity v joins x to y. The uniqueness of σ follows from the strict convexity of $p \to d^2(p, \sigma)$ on $\tilde{M} - \sigma$ given by Proposition 2.3B. See [**BO**, Lemma 9.10] for further details.

Examples 1 and 3 represent the two extreme cases. For an example of intermediate behavior see the discussion of the symmetric space $\tilde{M}_n$ in Section 5.

2.5H. *Action of isometries on $\tilde{M}(\infty)$.*   If x is any point of $\tilde{M}(\infty)$ and if ϕ is any isometry of $\tilde{M}$, then we define $\phi(x) = (\phi \circ \gamma)(\infty)$ where γ is any geodesic of $\tilde{M}$ such that $\gamma(\infty) = x$. By the definition of a point at infinity it follows that this definition of $\phi(x)$ is independent of the choice of the geodesic γ that represents x. It is easy to verify that the isometries of $\tilde{M}$ act as homeomorphisms of $\tilde{M}(\infty)$ with the cone topology, and moreover the map $(\phi, x) \to \phi(x)$ of $I(\tilde{M}) \times \tilde{M}(\infty) \to \tilde{M}(\infty)$ is continuous with respect to the product topologies.

3.   NONSIMPLY-CONNECTED MANIFOLDS OF NONPOSITIVE SECTIONAL CURVATURE

3.1. *Fundamental group as a group of isometries.*   An isometry ϕ of $\tilde{M}$ is called *elliptic* if ϕ fixes some point of $\tilde{M}$. A group Γ of isometries of $\tilde{M}$ is called a *deckgroup* if 1) Γ has no elliptic elements except the identity and 2) Γ is *discrete*; that is, if C is any compact subset of $\tilde{M}$ then $\phi(C) \cap C$ is nonempty for only finitely many elements of Γ. If one identifies those points of $\tilde{M}$ that lie on the same orbit of a deckgroup Γ, then the resulting quotient space $\tilde{M}/\Gamma$ becomes a Riemannian manifold of the same dimension as $\tilde{M}$ such that the projection π: $\tilde{M} \to \tilde{M}/\Gamma$ is locally an isometry. The theory of covering spaces says that any complete, nonsimply-connected manifold M of nonpositive sectional curvature can be expressed as such a quotient space $\tilde{M}/\Gamma$ for a suitable complete, simply-connected space $\tilde{M}$ of nonpositive sectional curvature and a suitable deckgroup Γ of isometries of $\tilde{M}$. Moreover, the fundamental group of M is isomorphic to the group Γ.

A basic principle is that one can study the geometry of $M = \tilde{M}/\Gamma$ and the algebra of the fundamental group of M by studying the action of the isometry group Γ on $\tilde{M}$ and especially on $\tilde{M}(\infty)$. This principle is the theme of this chapter. We give some elementary examples of this principle in this section and other more complicated examples in Section 6.

3.2. Lattices.

Let $\tilde{M}$ be any complete, simply-connected space with nonpositive sectional curvature. A discrete group $\Gamma \subseteq I(\tilde{M})$, possibly with elliptic elements, is called a *lattice* if there exists a positive constant A such that if $0 \subseteq \tilde{M}$ is any open set having the property that $\phi(0)$ is disjoint from 0 for all nonidentity elements ϕ in Γ, then the volume of 0 is $\leqslant A$. A lattice $\Gamma \subseteq I(\tilde{M})$ is called *uniform* if the quotient space $\tilde{M}/\Gamma$ is compact and *nonuniform* if $\tilde{M}/\Gamma$ is noncompact.

3.3. Examples.

1) If $\{v_1, \ldots, v_n\}$ are linearly independent vectors in $\mathbb{R}^n$, then the elements $\{T_1, \ldots, T_n\}$, where T_i is a translation in $\mathbb{R}^n$ by v_i, generate a free abelian group Γ of rank n and the quotient space $\mathbb{R}^n/\Gamma$ is a flat n-torus. All flat n-tori are obtained in this manner. If M is any compact flat n-manifold, then M is a quotient space $\mathbb{R}^n/\Gamma^*$, where Γ^* is a group of rigid motions of $\mathbb{R}^n$ that contains some translation group Γ as defined above as a finite index subgroup. See [W] for a further discussion of complete manifolds with $K \equiv 0$.

2) If M is any compact orientable surface of genus $g \geqslant 2$, then there exists a uniform lattice Γ in $\mathrm{PSL}(2, \mathbb{R})$, the connected isometry group of the hyperbolic plane, such that M is diffeomorphic to the quotient space H^2/Γ. In particular M admits a complete Riemannian metric with Gaussian curvature $K \equiv -1$. The set of such lattices Γ forms a $(6g - 6)$-dimensional space called the Teichmüller space for M.

3) The group $\mathrm{SL}(2, \mathbb{Z})$ consisting of those 2×2 matrices with determinant 1 and integer entries is a nonuniform lattice (the modular group) in $\mathrm{PSL}(2, \mathbb{R}) = I_0(H^2)$.

4) If $\tilde{M}$ is any symmetric space of noncompact type (see the next section), then A. Borel has shown in [**Bo**] that $I_0(\tilde{M})$, the connected isometry group of $\tilde{M}$, admits both uniform and nonuniform lattices. In particular $\mathrm{SL}(n, \mathbb{Z})$, the group of $n \times n$ matrices with determinant 1 and integer entries is a nonuniform lattice in $I_0(\tilde{M}_n)$, where the symmetric space $\tilde{M}_n = \mathrm{SL}(n, \mathbb{R})/\mathrm{SO}(n, \mathbb{R})$ is discussed below in Section 5.

3.4. *Elementary properties of the fundamental group.* Let M be a complete manifold with nonpositive sectional curvature, and represent M as a quotient space $\tilde{M}/\Gamma$, where $\tilde{M}$ is a complete, simply-connected manifold of nonpositive sectional curvature and Γ is a discrete group of isometries of $\tilde{M}$ that contains no elliptic elements. Since $\tilde{M}$ is diffeomorphic to a Euclidean space the homotopy groups $\pi_k(M)$ are zero for $k \geqslant 2$ by covering space theory. In a certain sense this means that the topology of M is concentrated in the fundamental group $\pi_1(M)$, which is isomorphic to Γ as remarked above. We now present two results on the fundamental group of a complete manifold M with $K \leqslant 0$.

THEOREM 3.4A. *Let M be a complete manifold with nonpositive sectional curvature. Then no element of $\pi_1(M)$ different from the identity has finite order.*

THEOREM 3.4B. *(Preissmann [**P**]). Let M be a compact manifold with strictly negative sectional curvature. Then every abelian subgroup of $\pi_1(M)$ is infinite cyclic.*

For generalizations of Theorem 3.4B to compact manifolds of nonpositive sectional curvature see [**CE**, Chapter 9], [**GW**], and [**LY**].

Proof of Theorem 3.4A. We express M as a quotient $\tilde{M}/\Gamma$, where $\tilde{M}$ is simply connected and $\Gamma \subseteq I(\tilde{M})$ is discrete and has no elliptic elements. Identifying $\pi_1(M)$ with Γ we suppose that some element $\phi \neq 1$ of Γ has finite order and we let Γ^* denote the finite cyclic subgroup of Γ generated by ϕ. By the Cartan fixed-point theorem (2.2E) the elements of Γ^* and ϕ in particular fix some point p in $\tilde{M}$, but this contradicts the fact that Γ has no elliptic elements except the identity.

Proof of Theorem 3.4B. We again express M as a quotient space $\tilde{M}/\Gamma$ and identify $\pi_1(M)$ with Γ. Let $A \subseteq \Gamma$ be an abelian subgroup.

We show first that every element ϕ of Γ translates some geodesic γ of $\tilde{M}$; that is, $(\phi \circ \gamma)(t) = \gamma(t + \omega)$ for all $t \in \mathbb{R}$ and some $\omega > 0$. By Proposition 2.3C it suffices to show that the function d_ϕ: $p \to d(p, \phi p)$ has a positive minimum value on $\tilde{M}$. Given $\phi \neq 1$ in Γ we let $\{p_n\} \subseteq \tilde{M}$ be a sequence such that $d_\phi(p_n) \to a = \inf(d_\phi)$. If $\{p_n\}$ has a cluster point p in $\tilde{M}$, then d_ϕ assumes a positive minimum value at p since Γ has no elliptic elements. Suppose that $\{p_n\}$ has no cluster points in $\tilde{M}$ and fix a point p in $\tilde{M}$. By the compactness of $M = \tilde{M}/\Gamma$ there exists a positive constant R and a sequence $\{\phi_n\} \subseteq \Gamma$ such that $d(p_n, \phi_n p) \leqslant R$ for every positive integer n. If $q_n = \phi_n^{-1}(p_n)$, then the sequence $\{q_n\}$ is bounded in $\tilde{M}$ and has a cluster point q. If $\psi_n = \phi_n^{-1}\phi\phi_n$, then $d_{\psi_n}(q_n) = d_\phi(p_n) \to a$ as $n \to \infty$. Since $\{q_n\}$ is bounded and Γ is discrete it follows that only finitely many of the elements $\{\psi_n\}$ are distinct, and hence by passing to a subsequence we may assume that $\psi_n = \psi$ for all n, where $\psi = \xi\phi\xi^{-1}$ for some element ξ in Γ. By continuity and by passing to a subsequence if necessary we have $d_\psi(q) = \lim_{n \to \infty} d_{\psi_n}(q_n) = \lim_{n \to \infty} d_\phi(p_n) = a$. Hence, $d_\phi(\xi^{-1}q) = d_\psi(q) = a$, which proves that d_ϕ assumes a positive minimum value in $\tilde{M}$.

Now let $\phi \neq 1$ be any element of the abelian subgroup $A \subseteq \Gamma$, and let γ be a geodesic of $\tilde{M}$ such that $(\phi \circ \gamma)(t) = \gamma(t + \omega)$ for all $t \in \mathbb{R}$ and some $\omega > 0$. If $F: \tilde{M} \to \mathbb{R}$ is the function defined by

$F(p) = d^2(p, \gamma(\mathbb{R}))$, then by Proposition 2.3B F is convex on $\tilde{M}$ and strictly convex on $\tilde{M} - \gamma(\mathbb{R})$ since the curvature of $\tilde{M}$ is strictly negative by hypothesis. If $\psi \neq 1$ is any element of A and if $\sigma(t) = (\psi \circ \gamma)(t)$, then $(\phi \circ \sigma)(t) = (\phi \circ \psi \circ \gamma)(t) = (\psi \circ \phi \circ \gamma)(t) = (\psi \circ \gamma)(t + \omega) = \sigma(t + \omega)$ for all $t \in \mathbb{R}$. Since $\gamma(\mathbb{R})$ is invariant under ϕ it follows that $F \circ \phi = F$ and hence $(F \circ \sigma)(t) = (F \circ \sigma)(t + \omega)$ for all $t \in \mathbb{R}$. In particular $F \circ \sigma$ is a bounded convex function on $\mathbb{R}$ and hence must be constant by elementary properties of convex functions. Since F is strictly convex on $\tilde{M} - \gamma(\mathbb{R})$ it follows that $\sigma = \gamma$.

The argument above shows that ϕ translates a unique geodesic γ of $\tilde{M}$ and every element ψ of A leaves γ invariant. Hence $(\psi \circ \gamma)(t) = \gamma(t + \omega_\psi)$ for all $t \in \mathbb{R}$ and some $\omega_\psi \neq 0$. If we restrict A to γ and regard A as a discrete group of translations $\{\omega_\psi\}$ of $\mathbb{R}$, then it is clear that A must be infinite cyclic and generated by an element ψ such that $|\omega_\psi|$ has minimum value. ∎

4. SYMMETRIC SPACES OF NONCOMPACT TYPE

4.1. *Symmetry diffeomorphisms.* For each point p in a simply-connected space $\tilde{M}$ one may define a symmetry diffeomorphism s_p: $\tilde{M} \to \tilde{M}$ by

$$s_p = \exp_p \circ S \circ \exp_p^{-1}$$

where $S(v) = -v$ for all vectors v in $T_p\tilde{M}$. Equivalently, $s_p\gamma(t) = \gamma(-t)$ for all $t \in \mathbb{R}$ and all geodesics γ such that $\gamma(0) = p$. We let G^* denote the group of diffeomorphisms of $\tilde{M}$ generated by the symmetry diffeomorphisms s_p, $p \in \tilde{M}$. Since each s_p is its own inverse, G^* is the set of all finite products $s_{p_1} \circ s_{p_2} \circ \cdots \circ s_{p_m}$, where m is any positive integer and $(p_1, \ldots, p_m)$ are arbitrary points of $\tilde{M}$, not necessarily all distinct.

4.2. *Definition of symmetric space.* For an expanded presentation of the results in this section see the appendix of [E5]. A complete simply-connected space $\tilde{M}$ is said to be a *symmetric space* if $G^* \subseteq I(\tilde{M})$ or equivalently if every symmetry diffeomorphism s_p,

$p \in \tilde{M}$, is an isometry of $\tilde{M}$. A symmetric space $\tilde{M}$ is said to be of *noncompact type* if $\tilde{M}$ cannot be written as the Riemannian product of a Euclidean space and another manifold. If $\tilde{M}$ is a symmetric space of noncompact type, then it can be shown that $I_0(\tilde{M}) \subseteq G^*$ $\subseteq I(\tilde{M})$ (see [L, p. 88]). Moreover, $I_0(\tilde{M})$ is a semisimple Lie group with trivial center and no compact normal subgroups except the identity. (See for example Lemma 3.1 of [E3].)

In general, a Riemannian manifold M is said to be a symmetric space if for every point p of M there exists an isometry s_p of M that fixes p and has order 2. Any symmetric space $\tilde{M}$ of noncompact type has a dual space M of *compact type* with sectional curvature $K \geqslant 0$. See [H] for an extensive discussion.

4.3. *Examples.* A diffeomorphism s_p of order 2 and with fixed point p in a Riemannian manifold M is an isometry of M if for any pair of distinct points q, r, both distinct from p, one has $d(q, r) = d(s_p(q), s_p(r))$. Comparing the geodesic triangles with vertices $\{p, q, r\}$ and $\{p, s_p(q), s_p(r)\}$ it becomes apparent that any Riemannian manifold M with a side-angle-side congruence axiom for geodesic triangles must be a symmetric space. In particular:

1. Any Euclidean space $\mathbb{R}^n$ is a symmetric space with sectional curvature $K \equiv 0$.

2. Any hyperbolic space H^n is a symmetric space of noncompact type with sectional curvature $K \equiv -1$.

3. Any sphere $S^{n-1} \subseteq \mathbb{R}^n$ of radius 1 is a symmetric space of compact type with sectional curvature $K \equiv 1$.

4. If G is any compact Lie group and $\langle \, , \, \rangle$ is any inner product on G that is invariant under left and right translations by elements of G, then G is a symmetric space with sectional curvature $K \geqslant 0$. See [Mi, pp. 109–115] for details.

5. Let G be a connected semisimple Lie group with finite center and no compact normal subgroup except the identity. Let K be any maximal compact subgroup of G. Then the space of left cosets G/K together with an appropriate G-left invariant metric becomes a symmetric space of noncompact type. See [H, pp. 178–179] for details. See also the corollary in Section 4.5 below.

6. The space $\tilde{M}_n$ of noncompact type consisting of all $n \times n$ symmetric, positive definite matrices with determinant 1. See Section 5.

For the remainder of this section $\tilde{M}$ will denote a complete, simply-connected symmetric space of nonpositive sectional curvature but not necessarily of noncompact type. We now describe some of the geometry of such a space $\tilde{M}$.

4.4. *Transvections in* $I(\tilde{M})$.

4.4A. DEFINITION.   Fix a point p in $\tilde{M}$. An isometry ϕ of $\tilde{M}$ is a *transvection at* p if there exists a unit speed geodesic γ of $\tilde{M}$ such that 1) $\gamma(0) = p$; 2) $(\phi \circ \gamma)(t) = \gamma(t + \omega)$ for all $t \in \mathbb{R}$ and some $\omega \in \mathbb{R}$; 3) $d\phi$: $T_{\gamma(s)}\tilde{M} \to T_{\gamma(s+\omega)}\tilde{M}$ is parallel translation along γ from $\gamma(s)$ to $\gamma(s + \omega)$.

DEFINITION.   A vector field X on $\tilde{M}$ is an *infinitesimal transvection at* p if its flow transformations $\{e^{tX}\}$ are transvections at p.

PROPOSITION 4.4B.   *Let p be a point of $\tilde{M}$, and let γ be a geodesic of $\tilde{M}$, not necessarily of unit speed, such that $\gamma(0) = p$. Then there exists an infinitesimal transvection X at p such that $e^{tX}(p) = \gamma(t)$ for all $t \in \mathbb{R}$, where $\{e^{tX}\}$ denotes the flow transformations of X, and de^{tX}: $T_{\gamma(s)}\tilde{M} \to T_{\gamma(s+t)}\tilde{M}$ is a parallel translation along γ for all $s, t \in \mathbb{R}$.*

Proof.   For each $t \in \mathbb{R}$ let s_t denote the geodesic symmetry $s_{\gamma(t)}$ at $\gamma(t)$, and define $p_t = s_{t/2} \circ s_0$. It is routine to verify that $s_t \gamma(s) = \gamma(-s + 2t)$ for all $s, t \in \mathbb{R}$, and hence $p_t \gamma(s) = \gamma(s + t)$ for all s, $t \in \mathbb{R}$. This implies a) $p_t(p) = \gamma(t)$ for all $t \in \mathbb{R}$. Moreover, one can also show b) dp_t: $T_{\gamma(s)}\tilde{M} \to T_{\gamma(s+t)}\tilde{M}$ equals parallel translation along γ from $\gamma(s)$ to $\gamma(s + t)$ for all $s, t \in \mathbb{R}$. See Lemma (8.1.2) of [W, p. 232] for details. Since $p_t \circ p_s$ and p_{s+t} have the same value and differential map at p (or any other point of γ) we conclude that $p_t \circ p_s = p_{s+t}$ for all $s, t \in \mathbb{R}$. This shows that the transformations $\{p_t\}$ are the flow transformations of a vector field X on $\tilde{M}$.

Assertion b) and the discussion above show that X is an infinitesimal transvection at p with the properties stated in the proposition.
∎

COROLLARY. $I_0(\tilde{M})$ *acts transitively on* $\tilde{M}$.

Proof. Fix a point p in $\tilde{M}$. By the theorem of Hopf-Rinow the space $\tilde{M}$ is the union of all geodesics of $\tilde{M}$ that start at p. The result now follows from Proposition 4.4B since the flow transformations of an infinitesimal transvection X at p lie in $I_0(\tilde{M})$. ∎

4.5. *Maximal compact subgroups of* $I_0(\tilde{M})$. If $\tilde{M}$ is a symmetric space we let G denote $I_0(\tilde{M})$ and for each point $p \in \tilde{M}$ we let $G_p = \{ g \in G: g(p) = p \}$. The subgroup G_p is a compact subgroup of $I_0(\tilde{M})$ in the induced topology (this follows from Theorem 2.2 of [**H**, p. 167]). If $q \in \tilde{M}$ is any other point, then clearly $G_q = gG_pg^{-1}$ if $g \in I_0(\tilde{M})$ is any element such that $g(p) = q$. On the other hand if $K \subseteq G$ is any compact group, then $K \subseteq G_p$ for some point p in $\tilde{M}$ by the Cartan fixed-point theorem (2.2E). This discussion proves the following.

THEOREM. *Let* $\tilde{M}$ *be a symmetric space, and let* $G = I_0(\tilde{M})$. *Then every maximal compact subgroup* K *of* G *equals* G_p *for some point* p *in* $\tilde{M}$, *and all maximal compact subgroups of* G *are conjugate in* G.

COROLLARY. *Let* $\tilde{M}$ *be a symmetric space, and let* $G = I_0(\tilde{M})$. *Let* K *be a maximal compact subgroup of* G, *and let* $p \in \tilde{M}$ *be a point fixed by* K. *Then the map from the left coset space* G/K *to* $\tilde{M}$ *given by* $gK \to g(p)$ *is a bijection.*

The following result, see for example [**ChE**, Proposition 4.4], is also useful.

PROPOSITION. *Let* $\tilde{M}$ *be a symmetric space and let* K *be a maximal compact subgroup of* $G = I_0(\tilde{M})$. *Then* $G(x) = K(x)$ *for every point* $x \in \tilde{M}(\infty)$. *In particular every* G-*orbit in* $\tilde{M}(\infty)$ *is closed in* $\tilde{M}(\infty)$.

4.6. *Tangent space of $\tilde{M}$ at p.* Fix a point p in $\tilde{M}$ and let $\not p = \{X: X$ is an infinitesimal transvection at $p\}$. From the discussion in Section 4.4 it follows that $\not p$ is isomorphic to $T_p\tilde{M}$ under the map $X \to X(p)$, $X \in \not p$. In the sequel we identify $\not p$ with $T_p\tilde{M}$.

4.7. *Curvature tensor of $\tilde{M}$ at p.* Let $p \in \tilde{M}$ be any point and identify $T_p\tilde{M}$ with the vector space $\not p$ of infinitesimal transvections at p. If X, Y are elements of $\not p$, then the skew symmetric curvature transformation $R(X, Y): \not p \to \not p$ (compare Section 1) is given by $R(X, Y)Z = -ad[X, Y](Z) = -[[X, Y], Z]$, where $[X, Y]$ denotes the Lie bracket of the vector fields X and Y. See Theorem 4.2 of [**H**, p. 180]. In particular if $[X, Y] = 0$, then $R(X, Y) \equiv 0$.

4.8. *Holonomy of $\tilde{M}$ at p.*

PROPOSITION. *Let $\tilde{M}$ be a symmetric space of noncompact type. Let p be a point of $\tilde{M}$, and let $K = \{g \in G: g(p) = p\}$. Then the holonomy group Φ_p at p equals $dK = \{d\phi: \phi \in K\}$.*

For a discussion with references see [**H**, p. 162].

4.9. *Rank of a symmetric space $\tilde{M}$.* A *k-flat* in a symmetric space $\tilde{M}$ is by definition a complete, totally geodesic k-dimensional submanifold of $\tilde{M}$ with zero sectional curvature. The *rank* of a symmetric space $\tilde{M}$ is the maximum dimension k of a k-flat in $\tilde{M}$. If k is the rank of a symmetric space $\tilde{M}$, then every geodesic of $\tilde{M}$ is contained in at least one k-flat of $\tilde{M}$.

The k-flats in a symmetric space $\tilde{M}$ can be described algebraically as follows. Fix a point p in $\tilde{M}$, and let $\not p$ denote the vector space of infinitesimal transvections at p (see Section 4.4). The k-flats of $\tilde{M}$ that contain p are precisely expressible as orbits $A(p)$, where $A = e^a = \{e^X: X \in a\}$ and $a \subseteq \not p$ is a vector subspace of dimension k such that $[X, Y] = 0$ for all $X, Y \in a$. The fact that $A(p)$ is flat follows from the discussion in 4.7.

4.10. *Regular and singular points at infinity.* A geodesic γ in a symmetric space $\tilde{M}$ of rank k is said to be *regular* if γ is contained in exactly one k-flat of $\tilde{M}$. A geodesic γ of $\tilde{M}$ is *singular* if it is not regular. The unit vectors in $\tilde{M}$ tangent to regular geodesics of $\tilde{M}$ form a dense open subset of the unit tangent bundle of $\tilde{M}$.

If $G_x = \{ g \in G = I_0(\tilde{M})$: $g(x) = x \}$ for an arbitrary point $x \in \tilde{M}(\infty)$, then it can be shown that G_x acts transitively on $\tilde{M}$ for every point x in $\tilde{M}(\infty)$. Hence if a geodesic γ of $\tilde{M}$ is regular (respectively, singular), then any geodesic of $\tilde{M}$ asymptotic to γ is also regular (respectively, singular). We may now define a point x in $\tilde{M}(\infty)$ to be *regular* (respectively, *singular*) if x admits a representative geodesic γ that is regular (respectively, singular). The set of regular points at infinity, denoted $R(\infty)$, is a dense open subset of $\tilde{M}(\infty)$ that is invariant under $I(\tilde{M})$. If $\tilde{M}$ has rank $k \geqslant 2$, then the set of singular points at infinity is a closed, nowhere dense subset of $\tilde{M}(\infty)$ that is invariant under $I(\tilde{M})$.

5. THE SYMMETRIC SPACE $\tilde{M}_n = \mathrm{SL}(n, \mathbb{R})/\mathrm{SO}(n, \mathbb{R})$

Let $n \geqslant 2$ be any positive integer and let $\tilde{M}_n$ denote the space of positive definite, symmetric $n \times n$ matrices with real coefficients and determinant 1. With a suitable inner product $\tilde{M}_n$ becomes a symmetric space of noncompact type and rank $n - 1$, and moreover if $\tilde{M}$ is a symmetric space of noncompact type with $n = \dim I_0(\tilde{M})$, then after rescaling the metric of $\tilde{M}$ by constants on the irreducible Riemannian factors there exists an isometric totally geodesic imbedding of $\tilde{M}$ into $\tilde{M}_n$. See the appendix of [E5] for details. See also [Mo, Section 3] and [K] for further discussion of $\tilde{M}_n$. The space $\tilde{M}_2$ is another model for the hyperbolic plane.

The space $\tilde{M}_n$ is diffeomorphic to a Euclidean space of dimension $-1 + \frac{1}{2}n(n + 1)$ and in fact the matrix exponential map Exp is such that $\mathrm{Exp}(A) = \sum_{n=0}^{\infty}(A^n/n!)$ gives an explicit diffeomorphism between $\not{p} = \{$symmetric $n \times n$ matrices of trace zero$\}$ and $\tilde{M}_n$. The group $\mathrm{SL}(n, \mathbb{R})$ acts transitively on $\tilde{M}_n$ by diffeomorphisms if one defines $\tau_g(P) = gPg^t$ for all $g \in \mathrm{SL}(n, \mathbb{R})$ and $P \in \tilde{M}_n$. The kernel of the map τ: $\mathrm{SL}(n, \mathbb{R}) \to \mathrm{Diff}(\tilde{M}_n)$ is the center of $\mathrm{SL}(n, \mathbb{R})$, the finite group of diagonal matrices with entries $+1$ or -1. Note that $\mathrm{SO}(n, \mathbb{R}) = \{ g \in \mathrm{SL}(n, \mathbb{R})$: $g(I) = I \}$ and hence $\tilde{M}_n$ may be identified with the coset space $\mathrm{SL}(n, \mathbb{R})/\mathrm{SO}(n, \mathbb{R})$.

5.1. *Metric of $\tilde{M}_n$.* Using the matrix exponential map Exp we identify $T_I \tilde{M}_n$ with $\not{p}$, the set of symmetric $n \times n$ matrices of trace zero. We define $\langle \, , \, \rangle$ on $\not p$ by $\langle A, B \rangle_I = \text{Trace}(AB^t) = \text{Trace}(AB)$, the Euclidean inner product in $\mathbb{R}^{n^2}$.

If $P \in \tilde{M}_n$ is any point and if $g \in \text{SL}(n, \mathbb{R})$ is any element such that $g(I) = gg^t = P$, then we define an inner product on $T_p \tilde{M}_n$ by $\langle dg(A), dg(B) \rangle_p = \langle A, B \rangle_I$ for any $A, B \in \not p$. Although g is not uniquely determined, if g_1, g_2 are any two elements with $g_1(I) = g_2(I) = P$, then $g_2 = g_1 k$ for some $k \in \text{SO}(n, \mathbb{R})$ and hence the inner product at P is well defined.

It is not difficult to show that $\tau(\text{SL}(n, \mathbb{R})) = I_0(\tilde{M}_n)$.

5.2. *Geodesics and symmetries of $\tilde{M}_n$.* Since $I_0(\tilde{M}_n)$ acts transitively on $\tilde{M}_n$ it suffices to describe the unit speed geodesics of $\tilde{M}_n$ that start at the identity I. These are precisely the curves $\gamma_X(t) = \text{Exp}(tX)(I) = \text{Exp}(2tX)$, where $X \in \not p$ and $\|X\|^2 = \text{Trace}(X^2) = 1$.

The geodesic symmetry s_I at the identity is the map $A \to A^{-1}$ for $A \in \tilde{M}_n$ as one can see from the description of the geodesics through I. In general the geodesic symmetry at $P = g(I)$ is given by $s_p = g \circ s_I \circ g^{-1}$ and a computation shows that $s_P(Q) = PQ^{-1}P$ for all P, Q in $\tilde{M}_n$.

5.3. *Curvature of $\tilde{M}_n$.* Given a unit vector $X \in \not p$ (i.e. Trace $(X^2) = 1$) the corresponding unit speed geodesic $\gamma_X(t) = \text{Exp}(tX)(I) = \text{Exp}(2tX)$ induces transvections $p_t = s_{t/2} \circ s_0 = \text{Exp}(tX)$ by the discussion of geodesic symmetries above. (See Section 4.4 for notation.) Hence $\not p$ can be identified with the infinitesimal transvections of $\tilde{M}_n$ at I by the discussion in 4.4.

If X, Y are nonzero elements of $\not p$ then the curvature transformation $R(X, Y)$ is given by

$$R(X, Y)Z = -[[X, Y], Z]$$

for all $Z \in \not p$, where $[X, Y] = XY - YX$, (matrix bracket). Compare to 4.7.

5.4. *Holonomy of $\tilde{M}_n$ at I.* We identify $T_I \tilde{M}_n$ with $\not p$ and observe that $K = \tau(\text{SO}(n, \mathbb{R}))$ is the maximal compact subgroup of $I_0(\tilde{M}_n)$ that fixes I. Any element $g \in \text{SO}(n, \mathbb{R})$ satisfies $g^t = g^{-1}$ and hence $g(\text{Exp}(X)) = g(\text{Exp}(X))g^t = g(\text{Exp}(X))g^{-1} =$

$\mathrm{Exp}(gXg^{-1})$ for all $X \in \not{p}$. From the discussion in 4.8 we now obtain the following:

PROPOSITION.   *Two unit vectors X, Y in $\not{p}$ lie in the same orbit of the holonomy group Φ_I at I if and only if $Y = gXg^{-1}$ for some $g \in \mathrm{SO}(n, \mathbb{R})$. Therefore the orbits of the holonomy group are in one–one correspondence with the n-tuples $(\lambda_1, \ldots, \lambda_n)$ such that $\lambda_1 \geqslant \lambda_2 \geqslant \cdots \geqslant \lambda_n$, $\sum_{i=1}^n \lambda_i = 0$ and $\sum_{i=1}^n \lambda_i^2 = 1$, where the $\{\lambda_i\}$ are the eigenvalues of X in $\not{p}$.*

5.5. *Rank and flats in $\tilde{M}_n$.*   From the discussion in 5.3 and 4.9 one can show that for any integer $k \geqslant 1$ the k-flats in $\tilde{M}_n$ that contain the identity I have the form $\mathrm{Exp}(S) = \{\mathrm{Exp}(X)\colon X \in S\}$, where S is a k-dimensional subspace of $\not{p}$ such that $XY = YX$ for all $X, Y \in S$. Since commuting symmetric matrices can be diagonalized simultaneously it follows that gSg^{-1} consists of diagonal matrices for a suitable $g \in \mathrm{SO}(n, \mathbb{R})$. In particular, if S_0 denotes the diagonal matrices in $\not{p}$ and F_0 denotes $\mathrm{Exp}(S_0)$, then for every k-flat $F = \mathrm{Exp}(S)$ that contains I there exists $g \in \mathrm{SO}(n, \mathbb{R})$ such that $g(F) = gFg^t = gFg^{-1} = \mathrm{Exp}(gSg^{-1}) \subseteq \mathrm{Exp}(S_0) = F_0$. In particular $\tilde{M}_n$ has rank $n - 1 = \dim(F_0)$ and all $(n-1)$ flats containing I have the form $g(F_0) = gF_0g^{-1}$ for $g \in \mathrm{SO}(n, \mathbb{R})$.

5.6. *Regular geodesics of $\tilde{M}_n$.*   If $X \in \not{p} = T_I\tilde{M}_n$ is a unit vector and if $\gamma_X(t) = \mathrm{Exp}(tX)(I) = \mathrm{Exp}(2tX)$ is the corresponding geodesic, then the discussion in 5.5 and the definition in Section 4.10 show that γ_X is regular if and only if the eigenvalues of X are all distinct.

5.7. *Points at infinity for $\tilde{M}_n$.*

5.7A. *Flags in $\mathbb{R}^n$.*   To describe the space $\tilde{M}_n(\infty)$ we recall the notion of a *flag in $\mathbb{R}^n$*. A *flag* F in $\mathbb{R}^n$ is a collection of nonzero vector subspaces $V_1, \ldots, V_k$ of $\mathbb{R}^n$ such that $V_k = \mathbb{R}^n$ and V_i is a proper subspace of V_{i+1} for every i. Of necessity one has $k \leqslant n$ and we say that a flag $F = (V_1, \ldots, V_k)$ is *regular* if $k = n$. In this case V_i is a hyperplane in V_{i+1} for every i. The group $\mathrm{SL}(n, \mathbb{R})$ acts on flags in a natural way: given an element $g \in \mathrm{SL}(n, \mathbb{R})$ and a flag $F = (V_1, \ldots, V_k)$ in $\mathbb{R}^n$ we define $g(F) = (g(V_1), \ldots, g(V_k))$.

Two flags $F = (V_1, \ldots, V_k)$ and $F^* = (V_1^*, \ldots, V_r^*)$ are said to be *in opposition* if $k = r$ and if $\mathbb{R}^n$ is the direct sum of V_i and V_{k-i}^* for every i. Every flag $F = (V_1, \ldots, V_k)$ determines an inverse flag $F^{-1} = (V_1^*, \ldots, V_k^*)$, where V_i^* is the orthogonal complement of V_{k-i} in $\mathbb{R}^n$ for each i. It is not difficult to show that flags F_1, F_2 are in opposition if and only if there exists an element g in $\mathrm{SL}(n, \mathbb{R})$ such that $g(F_1) = F_1$ and $g(F_1^{-1}) = F_2$. If F_1 is a regular flag, then it follows from the definition of being in opposition that F_1 is in opposition to the flags in a dense open subset of $\mathscr{F}_0$, the space of regular flags equipped with a natural topology relative to which $\mathrm{SL}(n, \mathbb{R})$ acts continuously.

5.7B. *Eigenvalue-flag pairs for a point in* $\tilde{M}_n(\infty)$.   Given a point x in $\tilde{M}_n(\infty)$ we associate a flag $F(x) = (V_1(x), \ldots, V_k(x))$ in $\mathbb{R}^n$ and a vector $\lambda(x) = (\lambda_1(x), \ldots, \lambda_k(x))$ such that

a) $\lambda_1(x) > \lambda_2(x) > \cdots > \lambda_k(x)$,

b) $\displaystyle\sum_{i=1}^{k} m_i \lambda_i(x) = 0$ and

c) $\displaystyle\sum_{i=1}^{R} m_i \lambda_i^2(x) = 1$,   where $m_i = \dim V_i(x) - \dim V_{i-1}(x)$.

The construction will show that given a flag $F = (V_1, \ldots, V_k)$ in $\mathbb{R}^n$ and a vector $\lambda = (\lambda_1, \ldots, \lambda_k)$ in $\mathbb{R}^k$ that satisfies a), b), and c), then there exists a unique point x in $\tilde{M}_n(\infty)$ such that $\lambda(x) = \lambda$ and $F(x) = F$. Hence $\tilde{M}_n(\infty)$ can be regarded as the collection of pairs (λ, F), where F is a flag of k subspaces in $\mathbb{R}^n$ for some integer k with $2 \leqslant k \leqslant n$, and λ is a vector with k components that satisfies a), b), and c). The discussion will also show that $x \in \tilde{M}_n(\infty)$ is a regular point at infinity if and only if $F(x)$ is a regular flag in $\mathbb{R}^n$.

We saw in Proposition 2.5A that $\tilde{M}_n(\infty)$ can be identified with the unit vectors in $T_I \tilde{M}_n$, which can in turn be identified with $\not p_1 = \{\text{symmetric } n \times n \text{ matrices } X \text{ with } \mathrm{Trace}(X) = 0 \text{ and } \mathrm{Trace}(X^2) = 1\}$. Specifically, $x \in \tilde{M}_n(\infty)$ is associated with the element $X \in \not p_1$ such that $\gamma_X(\infty) = x$, where $\gamma_X(t) = \mathrm{Exp}(tX)(I) = \mathrm{Exp}(2tX)$.

Given a vector X in $\not p_1$ we let $\{\lambda_i(X)\}_{i=1}^{k}$ be the distinct eigenvalues of X arranged so that $\lambda_1(X) > \cdots > \lambda_k(X)$. Now let

$E_i(X)$ be the eigenspace of X associated to $\lambda_i(X)$ and let $V_i(X)$ be the orthogonal direct sum of the eigenspaces $\{E_j(X): 1 \leqslant j \leqslant i\}$. The symmetric matrix X in $\not{p}_1$ is completely determined by the vector $\lambda(X) = (\lambda_1(X), \ldots, \lambda_k(X))$ and the flag $F(X) = (V_1(X), \ldots, V_k(X))$. We now assign to the point $x \in \tilde{M}_n(\infty)$ the eigenvalue-flag pair $(\lambda(x), F(x))$, where $\lambda(x) = \lambda(X)$, $F(x) = F(X)$ and $X \in \not{p}_1$ is the vector such that $\lambda_X(\infty) = x$. Clearly the eigenvalues $\{\lambda_i(X)\}$ satisfy a), b), and c) above.

5.7C. *Joining points in $\tilde{M}_n(\infty)$.* The discussion above allows us to describe precisely those points x, y in $\tilde{M}_n(\infty)$ that can be joined by a geodesic of $\tilde{M}_n$ (compare to 2.5G).

PROPOSITION. *Let x, y be distinct points of $\tilde{M}_n(\infty)$. Let $(\lambda(x), F(x))$ and $(\lambda(y), F(y))$ be the corresponding eigenvalue-flag pairs. Then there exists a geodesic γ of $\tilde{M}_n$ such that $\gamma(\infty) = x$ and $\gamma(-\infty) = y$ if and only if 1) $F(x)$ and $F(y)$ are in opposition and 2) $\lambda_i(x) = -\lambda_{k-i+1}(y)$ for $1 \leqslant i \leqslant k$ where $\lambda(x) = (\lambda_1(x), \ldots, \lambda_k(x))$ and $\lambda(y) = (\lambda_1(y), \ldots, \lambda_k(y))$.*

If $n = 2$, (hyperbolic plane), then any two distinct points of $\tilde{M}_n(\infty)$ can be joined by a geodesic of $\tilde{M}_n$, but this is far from being true if $n \geqslant 3$ as the result above shows.

5.8. *Action of $I_0(\tilde{M}_n)$ on $\tilde{M}_n(\infty)$.*

PROPOSITION. *Let $g \in \mathrm{SL}(n, \mathbb{R})$ and $x \in \tilde{M}_n(\infty)$ be given. Let $(\lambda(x), F(x))$ be the eigenvalue-flag pair associated to x. Then $(\lambda(x), gF(x))$ is the eigenvalue-flag pair associated to $g(x)$ or equivalently 1) $\lambda(gx) = \lambda(x)$ and 2) $F(gx) = gF(x)$.*

This result shows in particular that $\mathrm{SL}(n, \mathbb{R}) = I_0(\tilde{M}_n)$ never acts transitively on $\tilde{M}_n(\infty)$ for $n \geqslant 3$, and more generally gives a complete description of the orbits of $\mathrm{SL}(n, \mathbb{R})$ in $\tilde{M}_n(\infty)$.

Sketch of proof. We consider only the case that x is a regular point at infinity. Since $\mathrm{SL}(n, \mathbb{R})$ acts transitively on the space $\mathscr{F}_0$ of regular flags we may reduce to the case that $F(x) =$

$(V_1, V_2, \ldots, V_n)$, where $V_i = \text{span}\{e_1, \ldots, e_i\}$ for all i and $\{e_1, \ldots, e_n\}$ is the natural orthonormal basis of $\mathbb{R}^n$. In this case one shows by direct computation that $g(x) = x$ if and only if g is upper triangular. On the other hand the upper triangular $n \times n$ matrices are precisely those that fix $F(x)$. Next, if $x = \gamma_X(\infty)$ for $X \in \not{h}_1$, then $\phi(x) = (\phi \circ \gamma_X)(\infty) = \gamma_{\phi X \phi^{-1}}(\infty)$ for every $\phi \in \text{SO}(n, \mathbb{R})$ since $(\phi \circ \gamma_X)(s) = \phi \gamma_X(s) \phi^t = \phi \text{Exp}(2sX) \phi^{-1} = \text{Exp}(2s(\phi X \phi^{-1})) = \gamma_{\phi X \phi^{-1}}(s)$ for all s in $\mathbb{R}$. Assertions 1) and 2) of the proposition are now clear for $\phi \in \text{SO}(n, \mathbb{R})$ and x as above. Finally the discussion above shows that 1) and 2) hold for all elements of $\text{SL}(n, \mathbb{R})$ since any element $h \in \text{SL}(n, \mathbb{R})$ can be written as $h = \phi g$, where $g \in \text{SL}(n, \mathbb{R})$ is upper triangular and $\phi \in \text{SO}(n, \mathbb{R})$. ∎

COROLLARY. *An element g in $\text{SL}(n, \mathbb{R})$ fixes a point x in $\tilde{M}_n(\infty)$ if and only if g leaves invariant the associated flag $F(x)$.*

6. ACTION OF ISOMETRIES ON $\tilde{M}(\infty)$

In this section $\tilde{M}$ will denote an arbitrary complete, simply-connected manifold of nonpositive sectional curvature, not necessarily a symmetric space. If the isometry group of $\tilde{M}$ is large in a certain sense ($I(\tilde{M})$ satisfies the duality condition—see below) then much of the geometry of $\tilde{M}$ is reflected in the topological action of $I(\tilde{M})$ on $\tilde{M}(\infty)$. Moreover, if $M = \tilde{M}/\Gamma$ is a quotient manifold of $\tilde{M}$ with finite Riemannian volume, then much of the geometry of M is reflected in the action of the fundamental group Γ on $\tilde{M}(\infty)$. We illustrate this principle by considering the geodesic flow.

6.1. *Duality Condition.* Let $\Gamma \subseteq I(\tilde{M})$ be an arbitrary group of isometries of $\tilde{M}$. Points x, y in $\tilde{M}(\infty)$ are said to be Γ-*dual* if there exists a sequence $\{\phi_n\} \subseteq \Gamma$ such that $\phi_n(p) \to x$ and $\phi_n^{-1}(p) \to y$ as $n \to \infty$ for any point p of $\tilde{M}$. The following useful result is Lemma 2.4a of [ChE].

PROPOSITION 6.1A. *Let $\Gamma \subseteq I(\tilde{M})$ be any group, and let x, y be points in $\tilde{M}(\infty)$ that are Γ-dual. If $z \in \tilde{M}(\infty)$ is any point of $\tilde{M}(\infty)$ that can be joined to x by a geodesic of $\tilde{M}$, then $y \in \overline{\Gamma(z)}$.*

6.1B. DEFINITION. A group $\Gamma \subset I(\tilde{M})$, not necessarily discrete, satisfies the *duality condition* if the points $\gamma(\infty)$ and $\gamma(-\infty)$ are Γ-dual for any geodesic γ of $\tilde{M}$.

REMARK. It follows from the discussion in Proposition 3.7 of [E1] that a subgroup Γ of $I(\tilde{M})$ satisfies the duality condition if and only if every vector in the unit tangent bundle $S\tilde{M}$ of $\tilde{M}$ is *nonwandering modulo* Γ with respect to the geodesic flow $\{g^t\}$ in $S\tilde{M}$; that is, given an open set O in $S\tilde{M}$ there exist sequences $\{t_n\} \subseteq \mathbb{R}$ and $\{\phi_n\} \subseteq \Gamma$ such that $t_n \to +\infty$ and $(d\phi_n \circ g^{t_n})(O) \cap O$ is nonempty for every n.

Example 1. Let $\Gamma \subseteq I(\tilde{M})$ be a lattice with no elliptic elements; that is, the quotient space $\tilde{M}/\Gamma$ is a smooth manifold of finite volume and the same dimension as $\tilde{M}$. Then Γ satisfies the duality condition by the Poincaré recurrence lemma and the remark above.

Example 2. Let $\Gamma = I_0(\tilde{M})$, where $\tilde{M}$ is a symmetric space. Then Γ satisfies the duality condition by the discussion of transvections in 4.4.

PROPOSITION 6.1C. *Let* $\Gamma \subseteq I(\tilde{M})$ *satisfy the duality condition. Then for any two points* x, y *in* $\tilde{M}(\infty)$ *the closed sets* $\overline{\Gamma(x)}$ *and* $\overline{\Gamma(y)}$ *are either disjoint or identical.*

This result is Lemma 2.9 of [**BBE**].

PROPOSITION 6.1D. *Let* $\tilde{M}$ *have sectional curvature satisfying* $K \leqslant c < 0$, *and let* $\Gamma \subseteq I(\tilde{M})$ *be a group that satisfies the duality condition. Then* $\overline{\Gamma(x)} = \tilde{M}(\infty)$ *for every point* $x \in \tilde{M}(\infty)$.

Proof. Let x, y be any points in $\tilde{M}(\infty)$ and let z be a point in $\tilde{M}(\infty)$ distinct from both x and y. Since $K \leqslant c < 0$ it follows from the discussion in Section 2.5G that there exist geodesics γ_1 and γ_2 of $\tilde{M}$ that join z to x and z to y respectively. The point z is Γ-dual to both x and y since Γ satisfies the duality condition, and it now follows from Proposition 6.1A that $y \in \overline{\Gamma(x)}$. ∎

6.2. *Geodesic flows.* We consider the geodesic flow $\{g^t\}$ on the unit tangent bundle SM of a complete manifold M of nonpositive sectional curvature, and relate the action of $\{g^t\}$ on SM to the action of the fundamental group Γ of M on $\tilde{M}(\infty)$, where $M = \tilde{M}/\Gamma$.

PROPOSITION 6.2A. *Let $M = \tilde{M}/\Gamma$ be a complete manifold of nonpositive sectional curvature and finite volume, where $\tilde{M}$ is simply connected and Γ is a discrete group of isometries of $\tilde{M}$ without elliptic elements. Then the following properties are equivalent:*

1) The geodesic flow $\{g^t\}$ has a dense orbit in the unit tangent bundle SM.

2) The group Γ has a dense orbit in $\tilde{M}(\infty)$.

3) Every orbit of Γ in $\tilde{M}(\infty)$ is dense in $\tilde{M}(\infty)$.

Proof. This result is contained in Theorem 4.14 of [**E2**] and the discussion in Section 6.1. ∎

COROLLARY 6.2B. *Let M be a complete manifold with finite volume and sectional curvature satisfying $K \leqslant c < 0$. Then the geodesic flow has a dense orbit in the unit tangent bundle SM.*

Proof. This is an immediate consequence of Propositions 6.1D and 6.2A. ∎

COROLLARY 6.2C. *Let M be a compact manifold with $K \equiv 0$. Then the geodesic flow does not have a dense orbit in SM.*

Proof. We may write $M = \mathbb{R}^n/\Gamma$, where $n = \dim M$ and Γ is a uniform lattice in $I(\mathbb{R}^n)$. A theorem of Bieberbach (compare [**W**, p. 100]) says that Γ contains a normal free abelian subgroup Γ_0 with rank n and finite index in Γ, and moreover Γ_0 consists of translations of $\mathbb{R}^n$. Hence any orbit of Γ in $\mathbb{R}^n(\infty)$ is finite since the translations in $\mathbb{R}^n$ fix every point of $\mathbb{R}^n(\infty)$. Now apply Proposition 6.2A. ∎

COROLLARY 6.2D. *Let $n \geqslant 3$ be any integer and let $M = \tilde{M}_n/\Gamma$ be a compact quotient manifold of $\tilde{M}_n$. Then the geodesic flow in SM has no dense orbit.*

Proof. By the discussion in 5.8 the full isometry group $I(\tilde{M}_n)$ has no dense orbits in $\tilde{M}_n(\infty)$ and hence no subgroup Γ of $I(\tilde{M}_n)$ can have a dense orbit in $\tilde{M}_n(\infty)$; we use the fact that $I_0(\tilde{M}_n)$ has finite index in $I(\tilde{M}_n)$ and $I_0(\tilde{M}_n)$ orbits in $\tilde{M}_n(\infty)$ are closed by 4.5. ∎

6.2E. REMARK. More generally, if $\tilde{M}$ is any symmetric space of noncompact type and rank $k \geqslant 2$ and if M is any finite volume quotient manifold, then the geodesic flow does not have a dense orbit in the unit tangent bundle SM. Since $I(\tilde{M})$ satisfies the duality condition it suffices by Proposition 6.2A to show that $I(\tilde{M})$ (and hence any subgroup Γ) has a proper closed invariant subset A in $\tilde{M}(\infty)$. The set A consisting of the singular points in $\tilde{M}(\infty)$ satisfies these properties by the discussion in 4.10.

7. ACTION OF GEODESIC SYMMETRIES ON $\tilde{M}(\infty)$

Given a point $p \in \tilde{M}$ we define the geodesic symmetry s_p on $\tilde{M}(\infty)$ by

$$s_p(x) = \gamma_{px}(-\infty)$$

for all $x \in \tilde{M}(\infty)$. By the discussion in 2.5 the function s_p is a homeomorphism of $\tilde{M}(\infty)$ and extends continuously the geodesic symmetry $s_p \colon \tilde{M} \to \tilde{M}$. The action of the symmetry diffeomorphism group G^* on $\tilde{M}(\infty)$ is closely related to the action of the holonomy group Φ_p acting on the unit vectors of $T_p\tilde{M}$ for an arbitrary point $p \in \tilde{M}$. This relationship is provided by the following result of [E7].

THEOREM 7.1. *Let* $p \in \tilde{M}$ *and* $x \in \tilde{M}(\infty)$ *be arbitrarily given points. Let* G^* *denote the symmetry diffeomorphism group acting by homeomorphisms on* $\tilde{M}(\infty)$, *and let* Φ_p *denote the holonomy group at* p. *Then* $y \in \overline{G^*(x)}$ *if* $V(p, y)$ *lies in the* Φ_p *orbit of* $V(p, x)$. *(See 2.5B for notation.)*

THEOREM 7.2. *Let* $\tilde{M}$ *be irreducible and suppose that* G^* *admits a proper closed invariant subset in* $\tilde{M}(\infty)$. *Then* $\tilde{M}$ *is isometric to a symmetric space of noncompact type and rank* $k \geqslant 2$.

Proof. By Theorem 7.1 the holonomy group Φ_p has a proper, closed invariant subset in the sphere of unit vectors in $T_p\tilde{M}$ for any point p of $\tilde{M}$. The conclusion of the corollary is now a consequence of the following special case of a result of M. Berger [**Be**]. ∎

THEOREM. *Let N be an irreducible complete simply-connected Riemannian manifold, and let p be any point of N. If the holonomy group Φ_p has a proper closed invariant subset in the sphere of unit vectors in T_pN, then N is isometric to a symmetric space with rank at least 2.*

Finding proper, closed G^*-invariant subsets in $\tilde{M}(\infty)$ is closely related to finding proper, closed Γ-invariant subsets in $\tilde{M}(\infty)$ if Γ is a subgroup of $I(\tilde{M})$ that satisfies the duality condition.

PROPOSITION 7.3. *Let $X \subseteq \tilde{M}(\infty)$ be a proper closed subset, and let $\Gamma \subseteq I(\tilde{M})$ be a group that satisfies the duality condition and leaves X invariant. Then for any point p of $\tilde{M}$ the sets $X \cap s_p(X)$ and $X \cup s_p(X)$ are G^*-invariant and one of them is a proper closed subset of X.*

Proof. We show that X is invariant under the index 2 subgroup G_e^* consisting of those elements of G^* that are products of an even number of geodesic symmetries s_p, $p \in \tilde{M}$. Since G_e^* is normal in G^* it will then follow immediately that $s_p(X)$ is invariant under G_e^* and hence that $X \cap s_p(X)$ and $X \cup s_p(X)$ are both invariant under G^* for any point p of $\tilde{M}$. If $X \cup s_p(X) = \tilde{M}(\infty)$, then $X \cap \tilde{s}_p(X)$ must be nonempty since $\tilde{M}(\infty)$ is connected, which completes the proof.

To show that X is invariant under G_e^* it suffices to consider elements of G_e^* of the form $\phi = s_p \circ s_q$, where p, q are arbitrary points of $\tilde{M}$. Given a point $x \in X$ let $y = s_q(x)$ and $z = s_p(y) = \phi(x)$. The points y and z are Γ-dual since Γ satisfies the duality condition and hence $\phi(x) \in \overline{\Gamma(x)} \subseteq X$ by Proposition 6.1A. ∎

THEOREM 7.4. *Let $\tilde{M}$ be an irreducible, complete, simply-connected manifold of nonpositive sectional curvature whose isometry* group $I(\tilde{M})$ satisfies the duality condition. Then $\tilde{M}$ is isometric to a symmetric space of noncompact type and rank $k \geqslant 2$ if and only if $\tilde{M}(\infty)$ admits a proper closed subset invariant under $I(\tilde{M})$.

Proof. If $\tilde{M}$ is a symmetric space of noncompact type and rank $k \geqslant 2$, then by Remark 6.2E the group $I(\tilde{M})$ leaves invariant the set of singular points in $\tilde{M}(\infty)$, which is a closed nowhere-dense subset of $\tilde{M}(\infty)$. Conversely, suppose that $I(\tilde{M})$ leaves invariant some proper closed subset X of $\tilde{M}(\infty)$. Then the symmetry diffeomorphism group G^* leaves invariant a proper closed subset of $\tilde{M}(\infty)$ by Proposition 7.3, and hence $\tilde{M}$ is isometric to a symmetric space of noncompact type and rank $k \geqslant 2$ by Corollary 7.2. ∎

We conclude this section with the following geodesic flow characterization of finite volume, locally symmetric spaces of higher rank.

THEOREM 7.5. *Let $\tilde{M}$ be an irreducible, complete, simply-connected manifold of nonpositive sectional curvature, and let $M = \tilde{M}/\Gamma$ be a quotient manifold of finite volume and dimension equal to that of $\tilde{M}$. Then $\tilde{M}$ is isometric to a symmetric space of noncompact type and rank $k \geqslant 2$ if and only if the geodesic flow in the unit tangent bundle SM does not have a dense orbit.*

Proof. This is an immediate consequence of Theorem 7.2, Proposition 7.3 and Proposition 6.2A since the group Γ satisfies the duality condition by Example 1 of 6.1. ∎

8.   RANK OF A MANIFOLD OF NONPOSITIVE CURVATURE

To any complete manifold M of nonpositive sectional curvature one may assign an integer rank(M), with $1 \leqslant \text{rank}(M) \leqslant \dim M$, that measures the "flatness" of M. To each unit vector v tangent to M let $r(v)$ denote the dimension of the vector space $J(\gamma_v)$ consisting of all perpendicular parallel vector fields $E(t)$ along the geodesic γ_v with initial velocity v such that the sectional curvatures $K(E, \gamma_v')(t)$ are zero for all t. Now define rank(M) to be 1 plus the smallest of the integers $r(v)$, as v ranges over SM. If M is a symmetric space of noncompact type or a quotient manifold of such a space, then rank(M) agrees with the other definition of rank given above in 4.9. If M has strictly negative sectional curvature, then rank(M) = 1 while if M is flat ($K \equiv 0$), then rank(M) = dim(M).

In [Ba] and [BS] the following striking result of W. Ballmann and K. Burns–R. Spatzier is proved. A somewhat simplified proof based on Theorem 7.2 may be found in [E7].

THEOREM 8.1. *Let $\tilde{M}$ be a complete, irreducible manifold of nonpositive sectional curvature with rank($\tilde{M}$) = $k \geqslant 2$ and sectional curvature satisfying $K \geqslant -a^2$ for some positive constant a. If $I(\tilde{M})$ admits a lattice Γ, then $\tilde{M}$ is isometric to a symmetric space of noncompact type and rank $k \geqslant 2$.*

As a consequence of the result above one obtains the following rigidity theorem of Gromov [BGS]. See also [E4] for a special case.

THEOREM 8.2. *Let $\tilde{M}$ be a symmetric space of noncompact type and rank $k \geqslant 2$, and let $M^* = \tilde{M}/\Gamma$ be a compact quotient manifold such that no finite cover of M^* splits as a Riemannian product. Let M be any compact manifold of nonpositive sectional curvature whose fundamental group is isomorphic to the fundamental group of M^*. Then M is isometric to M^* if the metric of M^* is rescaled by constants on the local de Rham factors of M^*.*

REMARKS. 1. The Mostow Rigidity Theorem proves the result above in the special case that the universal cover M' of M is a symmetric space of noncompact type. The proof of Theorem 8.2 either involves methods similar to the methods used by Mostow (see [BGS]) or a direct application of the Mostow Rigidity Theorem after using Theorem 7.2 to prove that M' is a symmetric space (see [E4] and [E7]).

2. Theorem 8.2 implies in particular that on a compact manifold M^* as described in the theorem there exist no metrics of nonpositive sectional curvature except those obtained from the original metric by rescaling the metric by constants on local de Rham factors of M^*.

Rank of the fundamental group.   We remarked earlier in Section 3.4 that the fundamental group of a complete manifold M of nonpositive sectional curvature carries all of the homotopy information of M. It is reasonable to ask how much of the geometry of M is expressible in terms of algebraic properties of the fundamental group of M, and not unreasonable to expect that a considerable amount of the geometry can be so expressed. We give one example from [**BE**].

Given an abstract group Γ and an integer $k \geqslant 1$ we let $A_k(\Gamma) = \{\phi \in \Gamma: Z(\phi)$ contains a free abelian subgroup of rank $r \leqslant k$ as a subgroup of finite index$\}$, where $r \leqslant k$ is arbitrary and $Z(\phi) = \{\psi \in \Gamma: \psi\phi = \phi\psi\}$. Define

$$r(\Gamma) = \inf\{k \geqslant 0: \Gamma = \phi_1 A_k(\Gamma) \cup \cdots \cup \phi_m A_k(\Gamma),$$

$$\text{where } \phi_1, \ldots, \phi_m \in \Gamma\}.$$

We then define

$$\text{rank}(\Gamma) = \sup\{r(\Gamma^*): \Gamma^* \text{ is a finite index subgroup of } \Gamma\}. \quad (8.3)$$

In [**BE**] the following result is proved.

THEOREM 8.4. *Let M be a complete manifold of finite volume whose sectional curvature satisfies $-a^2 \leqslant K \leqslant 0$ for some positive constant a. Then rank(M) equals the rank of the fundamental group of M.*

REMARK.   The definition of $r(\Gamma)$ is taken from work of Prasad and Raghunathan [**PR**] who proved that $r(\Gamma) = \text{rank}(M)$ if $M = \tilde{M}/\Gamma$, where $\tilde{M}$ is a symmetric space of noncompact type and rank $k \geqslant 2$ and Γ is a lattice contained in $I_0(\tilde{M})$.

From Theorems 8.4 and 8.1 we then obtain the following characterization of symmetric spaces of noncompact type and rank at least two (see [**BE**]).

THEOREM 8.5.   *Let M be a complete manifold of finite volume that admits no finite Riemannian cover that splits as a Riemannian* product. *Assume that the sectional curvature satisfies* $-a^2 \leqslant K \leqslant 0$ *for some positive constant a. Then the following conditions are equivalent*:

1) *The universal Riemannian covering manifold $\tilde{M}$ of M is a symmetric space of noncompact type and rank $k \geqslant 2$.*

2) a) *The fundamental group of M is finitely generated.*

   b) *No finite index subgroup of the fundamental group of M is a direct product $A \times B$.*

   c) *The fundamental group of M has rank $k \geqslant 2$ in the sense of (8.3).*

REFERENCES

[Ba] W. Ballmann, "Nonpositively curved manifolds of higher rank," *Ann. of Math.* 122 (1985), 597–609.

[BBE] W. Ballmann, M. Brin and P. Eberlein, "Structure of manifolds of nonpositive curvature, I," *Ann. of Math.* 122 (1985), 171–203.

[BBS] W. Ballmann, M. Brin and R. Spatzier, "Structure of manifolds of nonpositive curvature, II," *Ann. of Math.* 122 (1985), 205–235.

[BE] W. Ballmann and P. Eberlein, "Fundamental groups of manifolds of nonpositive curvature," *J. Diff. Geom.* 25 (1987), 1–22.

[Bea] A. Beardon, *The geometry of discrete groups*, Springer, New York, 1983.

[Be] M. Berger, "Sur les groupes d'holonomie homogène des variétés à connexion affine et des variétés Riemanniennes," *Bull. Soc. Math. France* 83 (1953), 279–330.

[BGS] W. Ballmann, M. Gromov and V. Schroeder, *Manifolds of nonpositive curvature*, Birkhäuser, Boston, 1985.

[BC] R. Bishop and R. Crittenden, *Geometry of manifolds*, Academic Press, New York, 1964.

[BO] R. Bishop and B. O'Neill, "Manifolds of negative curvature," *Trans. Amer. Math. Soc.* 145 (1969), 1–49.

[Bo] A. Borel, "Compact Clifford-Klein forms of symmetric spaces," *Topology* 2 (1963), 111–122.

[BS] K. Burns and R. Spatzier, "Manifolds of nonpositive curvature and their buildings," *Publ. IHES* no. 65, pp. 35–59.

[CE] J. Cheeger and D. Ebin, *Comparison theorems in Riemannian geometry*, North Holland, Amsterdam, 1975.

[ChE] S. Chen and P. Eberlein, "Isometry groups of simply connected manifolds of nonpositive curvature," *Ill. J. Math.* 24 (1980), 73–103.

[E1] P. Eberlein, "Geodesic flows on negatively curved manifolds, I," *Ann. of Math.* 95 (1972), 492–510.

[E2] ____, "Geodesic flows on negatively curved manifolds, II," *Trans. Amer. Math. Soc.* 178 (1973), 57–82.

[E3]   ____, "Isometry groups of simply connected manifolds of nonpositive curvature, II," *Acta Math.* 149 (1982), 41–69.

[E4]   ____, "Rigidity of lattices of nonpositive curvature," *Erg. Th. Dyn. Syst.* 3 (1983), 47–85.

[E5]   ____, "Structure of manifolds of nonpositive curvature," in *Global differential geometry and global analysis* 1984, Lecture Notes in Mathematics, vol. 1156, Springer, New York, pp. 86–153.

[E6]   ____, "Symmetry diffeomorphism group of a manifold of nonpositive curvature, I," to appear in *Trans. Amer. Math. Soc.*

[E7]   ____, "Symmetry diffeomorphism group of a manifold of nonpositive curvature, II," to appear in *Indiana Univ. Math. J.*

[EO]   P. Eberlein and B. O'Neill, "Visibility manifolds," *Pac. J. Math.* 46 (1973), 45–109.

[GW]   D. Gromoll and J. Wolf, "Some relations between the metric structure and the algebraic structure of the fundamental group in manifolds of nonpositive curvature," *Bull. Amer. Math. Soc.* 77 (1971), 545–552.

[H]   S. Helgason, *Differential geometry and symmetric spaces*, Academic Press, New York, 1962.

[K]   F. I. Karpelevic, "The geometry of geodesics and the eigenfunctions of the Beltrami-Laplace operator on symmetric spaces," *Trans. Moscow Math. Soc.* (AMS Translation) Tom 14 (1965), 51–199.

[LY]   H. B. Lawson and S.-T. Yau, "Compact manifolds of nonpositive curvature," *J. Diff. Geom.* 7 (1972), 211–228.

[L]   O. Loos, *Symmetric spaces*, vol. 1, W. A. Benjamin, New York, 1969.

[Mi]   J. Milnor, *Morse Theory*, Annals of Math Studies Number 51, Princeton University Press, Princeton, 1963.

[Mo]   G. D. Mostow, *Strong rigidity of locally symmetric spaces*, Annals of Math Studies Number 78, Princeton University Press, Princeton, 1973.

[P]   A. Preissmann, "Quelques propriétés globales des espaces de Riemann," *Comm. Math. Helv.* 15 (1942–43), 175–216.

[PR]   G. Prasad and M. A. Raghunathan, "Cartan subgroups and lattices in semisimple Lie groups," *Ann. of Math.* 96 (1972), 296–317.

[S]   C. L. Siegel, *Topics in complex function theory*, vol. II, J. Wiley and Sons, New York, 1971, pp. 14–30.

[W]   J. Wolf, *Spaces of constant curvature*, Third Edition, Publish or Perish, Boston, 1974.

WHAT IS ANALYSIS IN THE LARGE?

Marston Morse

1. INTRODUCTION

All mathematics is more or less "in the large" or "in the small." It is highly improbable that any definition of these terms could be given that would be satisfactory to all mathematicians. Nor does it seem necessary or even desirable that hard and fast definitions be given. The German terms *im Grossen* and *im Kleinen* have been used for some time with varying meanings. It will perhaps be interesting and useful to the reader to approach the subject historically by way of examples.

No proofs are given. In attempting to give the reader a conception of analysis in the large two ways are open. The first is to attempt an elementary exposition of the fundamental techniques. Unfortunately, this method of exposition is attempted much too often. The explanations given are fragmentary and give an exaggerated notion of the importance of some special technique, and no adequate notion of the subject as a whole. In a new and comprehensive field possibly the only way to give the beginner a stimulating and adequate notion of what the subject is about is to give examples and results which are themselves relatively complete. The cooperative reader can readily imagine the variety of techniques that might be used to obtain the stated results, and may himself invent new techniques, but in the presence of significant results he is less apt to be concerned with trivialities and subjective bypaths.

2. AN EXAMPLE FROM
DIFFERENTIAL GEOMETRY

Most of classical differential geometry is "in the small," that is, most theorems are proved merely in the neighborhood of a point. It is proved, for example, that in the neighborhood of a point P of a surface Σ, Σ can be referred to isothermic parameters so that neighboring P,

$$(2.1) \qquad ds^2 = \lambda(u, v)[du^2 + dv^2]$$

with $\lambda(u, v) \neq 0$.† The question in the large as to what sort of closed surfaces can be represented as a whole with parameters (u, v) and ds^2 of the form (2.1) has been asked and answered in general only in recent years. It is required that there be just one curve $u = $ const. and just one curve $v = $ const. through each point. Among two-sided or orientable surfaces which admit such parameters, those of the topological type‡ of the torus are the only possibilities.

One could continue by asking a more general question. What sort of closed surfaces S admit a representation in terms of parameters (u, v) in such a manner that there is one and only one curve $u = $ const. and one and only one curve $v = $ const. through each point? Such a representation of S would in particular imply the existence at each point P of S of a vector tangent to the curve $u = $ const. through P. There would thus exist a field of vectors,

---

† Appropriate hypotheses as to the regularity of the representation of the surface must be made.

‡ A surface is of the topological type of the torus if it is the (1–1) continuous image of the torus.

one for each point of P, tangent to S and P and varying continuously with P. For such a field to exist S must be the topological type of the torus.

Thus, in questions as to the existence of parameter nets without singularities, the controlling factors are those of topology. One can see why analysis or geometry in the large depends so heavily on topology.

3. AN EXAMPLE FROM THE THEORY OF FUNCTIONS OF A COMPLEX VARIABLE

The theorem that a function $f(z)$ of a complex variable z which has no singularities in the extended plane other than poles is a rational function of z, is a theorem in the large the proof of which illustrates some of the salient characteristics of analysis in the large. One begins by representing $f(z)$ neighboring $z = z_0$ as the sum of the "principal part" of $f(z)$ at z_0 and a function analytic at z_0. This is the preliminary analysis in the small.

Upon subtracting the principal parts of $f(z)$ at each pole from $f(z)$ one obtains a function $\phi(z)$ bounded in absolute value and with at most removable singularities. According to Liouville, $\phi(z)$ is a constant. The theorem follows.

The analysis in the large comes in the proper definition of the extended plane and the proof of the Liouville theorem. Details will not be given but it will be of interest to state that the theorem of Liouville can be reduced to a theorem of topological character on the nature of vector fields.

4. DIFFERENTIAL EQUATIONS IN THE LARGE. AN EXAMPLE FROM THE WORKS OF HENRI POINCARÉ

It is no mere coincidence that Poincaré was the first to comprehend fully the possibilities of analysis in the large, and at the same time was the father of modern topology. Poincaré was not satisfied with the classical theory of differential equations. He wished to know something concerning the system of trajectories as a whole. He was greatly interested in the movements of the planets but found insufficient generality and completeness in the classical theory. His interest in celestial mechanics is in the background of all of his papers on differential equations.

Poincaré's first papers on differential equations are not pretentious in their generality, but in method they are most novel. Poincaré is concerned with an ordinary first order differential equation† defined at each point of a 2-sphere. In terms of any system of local coordinates (u, v) representing the neighborhood of a point (u_0, v_0) on the sphere the differential conditions have the form

$$\frac{du}{U(u, v)} = \frac{dv}{V(u, v)}.$$

The functions U and V are supposed real and analytic in (u, v) neighboring (u_0, v_0). Points (u_0, v_0) at which both U and V vanish are termed "singular points." These points are supposed finite in number on the sphere.

Poincaré makes certain assumptions concerning the singular points (u_0, v_0). To state these conditions we shall take (u_0, v_0) as the origin. Then U and V have developments of the form

$$U = au + bv + \cdots,$$
$$V = cu + dv + \cdots$$

neighboring the origin. Poincaré assumes in most of his work that the roots λ_1 and λ_2 of the equation

$$\begin{vmatrix} a - \lambda, & b \\ c & d - \lambda \end{vmatrix} = 0$$

are distinct, different from 0, never pure imaginary, and that neither $\lambda_1|\lambda_2$ nor $\lambda_2|\lambda_1$ is a positive integer. These conditions will be satisfied by most analytic examples.

Curves on the sphere which satisfy the differential equation are termed *characteristics*. In general, characteristics are without singularity except at most when they pass through a singular point

---

† Poincaré, "Sur les courbes définies par les équations différentielles," *Journal de Liouville*, 1881, 1882.

of the differential equation. Typical of analysis in the large, Poincaré's work permits a subdivision into three parts as follows:

(a) *a study of characteristics neighboring a singular point;*

(b) *the assignment of an index* ±1 *to each singular point and the establishment of a relation between these indices (this part of the analysis would now be regarded as an essay in combinatorial topology)*;

(c) *a description of the characteristics in the large with particular reference to recurrence and limiting trajectories [results* (a) *and* (b) *are preliminary to* (c)].

Part (a). In his study of characteristics neighboring a singular point, Poincaré shows that there are three principal kinds of singular points as follows:

"*Noeud.*" Neighboring a noeud (u_0, v_0) each characteristic tends to (u_0, v_0) with a definite limiting direction. For example, the differential equation

$$\frac{du}{u} = \frac{dv}{2v}$$

has a noeud at the origin. In this example, the characteristics have the form $kv = hu^2$ where h and k are constants.

"*Foyer.*" The characteristics approach such a singular point in the form of spirals, with the arc length becoming infinite. For example, the differential equation

$$\frac{du}{u - v} = \frac{dv}{u + v}$$

has a foyer at the origin with logarithmic spirals as characteristics.

"*Col.*" There are just two characteristics which tend to a col as a limiting point. For example, the equation

$$\frac{du}{u} = \frac{dv}{-v}$$

has a col at the origin. The characteristics $uv =$ const. include the two characteristics $u = 0$ and $v = 0$ passing through the origin.

Part (b). In the development (b), Poincaré assigns an index 1 to each noeud and to each foyer, and an index -1 to each col.

Poincaré shows that *the sum of the indices of the singular points on the sphere equals* 2. Thus, there must exist at least two singular points.

A closed characteristic without a multiple point is called a *cycle*. If a characteristic tends to a noeud or a foyer as a limit point there is in general no natural way to continue the characteristic, and it is agreed that in such cases the characteristic shall end at the noeud or foyer. If a characteristic g tends to a col the convention is made that g may be continued turning either to the right or left and departing from the col on a characteristic. By virtue of this convention, the notion of a cycle is enlarged. With this understood we see that a cycle can have no singularity other than those occurring at a col.

Part (c). Poincaré ends with a relatively complete description of the characteristics. He shows that *a characteristic continued without limit in a given sense either terminates at a noeud, or is a cycle, or is asymptotic to a cycle.* A foyer is to be regarded as a degenerate cycle to which the neighboring spirals are asymptotic.

The reader is asked to observe the fundamental difference between the modes of analysis required in Parts (a), (b), and (c), and then to note how (a) and (b) are preliminary to (c) and make (c) possible. The index theorem of Poincaré has its topological generalization in the fixed point theorems of Brouwer, Alexander, Lefschetz, and H. Hopf. The analysis of characteristics in (c) is the predecessor of the modern study of recurrence and transitivity which G. D. Birkhoff has developed so fully and to which Hedlund, Morse, von Neumann, Koopman, E. Hopf and others have made significant contributions.

5. ELEMENTARY EXAMPLES IN EQUILIBRIUM THEORY IN THE LARGE

Equilibrium theory in the large makes an extensive use of topology. The principles of analysis brought out in the previous examples appear here again. Briefly summed we have seen in these examples that analysis in the large has involved (a) a preliminary analysis in the small, (b) a local determination of indices, and (c) an integration of this local analysis by various means (including topology) into the final theorems in the large. The examples which we shall now present will show how various problems which from a local point of view appear most diverse, from a topological point of view are essentially the same.

We begin with certain results concerning a function f of a point on a closed bounded n-manifold Σ lying in a euclidean space of sufficiently high dimension. We suppose throughout that Σ is locally represented in terms of n parameters (u) with convenient conditions of differentiability and regularity. In terms of the local parameters (u) f shall be a function $F(u)$ at least three times continuously differentiable. A *critical* or *equilibrium point* of f is a point at which each partial derivative of F is null.

For the purposes of this exposition we shall make an assumption which is in general fulfilled. We shall suppose that each critical point is *nondegenerate* in the sense that the terms F_2 of the second order in the Taylor's formula for F about the critical point is a nondegenerate quadratic form. Then, as in the elementary theory of conic sections, it is possible to make a real nonsingular linear transformation from the variables (u) to the variables (v) such that F_2 takes the form

$$F_2 = -v_1^2 - \cdots - v_k^2 + v_{k+1}^2 + \cdots + v_n^2.$$

The number k is called the *index* of the critical point.

A manifold such as Σ possesses an ith Betti number R_i ($i = 1, \cdots, n$). This is the maximum number of independent nonbounding i-cycles† on Σ. For example, if Σ is a torus then $R_0 = 1$, $R_1 = 2, R_2 = 1$. We shall be concerned with a 3-dimensional torus T_3. Such a manifold can be obtained by starting with a 2-dimensional torus T_2 and a 2-plane π_2 lying in a euclidean 3-plane, with π_2 not intersecting T_2. To obtain T_3 we revolve T_2 about π_2 in a 4-plane containing our 3-plane. Such a T_3 is sometimes called a product of three circles. For T_3 one has $R_0 = 1$, $R_1 = 3$, $R_2 = 3$,

---

† For details see Seifert-Threlfall, *Lehrbuch der Topologie.* Leipzig: 1934, Chap. III.

$R_3 = 1$. These numbers are the binomial coefficients when $n = 3$. We can obtain a 1–1 continuous image of an ordinary torus by identifying opposite sides of a square. Similarly one can obtain a 1–1 continuous image of T_3 by identifying opposite faces of a cube. With this identification, three mutually perpendicular edges of the cube represent three independent nonbounding 1-cycles, as can be shown. Similarly, three mutually perpendicular faces of the cube represent three independent nonbounding 2-cycles. A point is a 0-cycle and T_3 itself is a 3-cycle. In this way one intuitively accounts for the fact that $R_0 = 1$, $R_1 = 3$, $R_2 = 3$, $R_3 = 1$. A 3-dimensional manifold which is a 1–1 continuous image of T_3 will be called a *topological* 3-torus.

The theorem which will be used in what follows is that on Σ the number M_i of critical points of f of index i satisfies the fundamental relation†

$$(5.1) \qquad\qquad M_i \geqq R_i.$$

Thus on a topological 3-torus one can infer the existence of at least $1 + 3 + 3 + 1 = 8$ critical points.

Examples

1. *Triangles of light.* Let there be given three nonintersecting, simple, closed, nonsingular, analytic curves C_1, C_2, C_3 all lying in a 2-plane. We shall be concerned with triangles with vertices p_1, p_2, p_3 on C_1, C_2, C_3, respectively. Such a triangle will be called a *triangle of light* if a ray of light following this triangle is reflected at p_i as if C_i were a mirror, or if the angle in the triangle at p_i is π. How many triangles of light can we affirm to exist?

Let f be the sum of the lengths of the sides of the triangle $p_1p_2p_3$. We can refer C_i to a parameter u_i which is proportional to the arc length and varies from 0 to 2π. Then f becomes a function $f(u_1, u_2, u_3)$. The domain of definition of f is clearly a topological 3-torus. As a matter of analysis in the small, one proves by ele-

---

† See Morse, "Calculus of variations in the large." Colloquium lectures, *American Mathematical Society* (1934), Chap. VI. Also, Seifert-Threlfall, *Variationsrechnung im Grossen*. Leipzig: 1938.

mentary methods that f has a critical point if and only if the corresponding triangle is a triangle of light.

These triangles of light can then be classified according to the index of the corresponding critical point. According to relation (5.1) in the general theory of critical points there are at least $8 = 1 + 3 + 3 + 1$ of these triangles of light.

2. *Normals from a point to a topological 3-torus.* Let Σ_3 be a topological 3-torus in a euclidean 4-space. Let p be a fixed point not on Σ_3. We seek normals from p to Σ_3. To obtain these we let f be the distance from p to Σ_3 regarding f as a function of the point (u) of Σ_3. It can be shown that except for a subset of special points p the critical points of f are nondegenerate. Moreover, one then shows by a local analysis that f has a critical point (u) if and only if the line segment from p to (u) is normal to Σ_3 at (u). According to our general theorem there are then at least eight normals from p to Σ_3. These normals can be classified and it can be shown that the index of a nondegenerate critical point is the number of centers of principal curvature of Σ_3 between p and (u) on the given normal. Similar theorems hold for a topological 2-torus. Here the number of normals is at least 4.

3. *Three-planes passing through a fixed 2-plane and tangent to the preceding topological 3-torus Σ_3.* We suppose that the fixed 2-plane π_2 does not pass through a hole in Σ_3, that is, we suppose that π_2 can be moved indefinitely away from Σ_3 without intersecting Σ_3. We can then show that if π_2 is nonspecialized there are at least eight 3-planes through π_2 tangent to Σ_3.

4. *Heavy chain in equilibrium, with ends free to move on a topological 2-torus and on a closed curve C, respectively.* We suppose that the curve C and the topological 2-torus Σ_2 lie in euclidean 3-space but that no point of C and Σ_2 lie on the same plumb line. We suppose that a chain is provided which is larger than the maximum distance from a point of C to a point of Σ_2. The end points of the chain are supposed free to move on Σ_2 and C respectively and the chain is permitted to pass through Σ_2 or C. If the position of C is nonspecialized relative to Σ_2, then there are at least eight positions of equilibrium of the chain, seven of which are unstable. The function the critical points of which are sought gives the height of the center of gravity of the chain as a function of the end points of the chain.

The examples of this section are unified by the fact that the function involved is defined in each case on a topological 3-torus. The examples belong equally well to mechanics, geometry, or the calculus of variations. The restrictions as to nonspecialized positions of the configurations involved can all be removed by replacing the definition of a critical point in terms of derivatives by a topological definition of a critical point, and by replacing the classification of critical points according to their indices by a topological classification of a group theoretic character. This type of generalization both in its form and genesis is characteristic of analysis in the large.

It is possible that analysis in the large may eventually reduce to topology, but not until topology has been greatly broadened. It is equally conceivable that the apparently less general situations which arise with such frequency in problems in analysis in the large may form the canonical cases about which the topology of the future can be built.

Analysis is full of difficult but significant unsolved problems in the large. We mention only one example. How does the topological structure of the contour manifolds of the Jacobi least action integral J in the problem of three or more celestial bodies vary with the value of J? The independent variable in J is a closed path. The solution of this problem may disclose that the planetary orbits exist for essentially topological reasons. On the purely topological side, the number of problems the solution of which is necessary for a rapid advance of analysis in the large is very great, presenting a field that is virtually untouched.

REFERENCES

The reader wishing to enter this field should read the current introductions to topology and differential geometry. In the latter field, Reference 1 is recommended. Reference 2 introduces the reader to critical point theory and may well be followed by Reference 3. A description of the three different levels at which critical point theory proceeds is found in the fourth reference and others. The advanced reader can now turn to the fifth reference with its Section 7, "Application au calcul des variations." In Reference 6, the Lefschetz theorem is proved with the aid of the Morse theory of normals and focal points and the relation (5.1) in its appropriate form. Reference 7 contains one of the deepest applications of the critical point theory. In Reference 8, the analysis in the large of functions on a 3-torus, as applied in Examples 1 through 4, finds another kindred application.

1. Georges de Rham, *Variétés différentiables*. Paris: Hermann et Cie., 1955.

2. M. Morse and G. B. van Schaack, "The critical point theory under general boundary conditions," *Annals of Mathematics*, 35 (1934).

3. H. Seifert and W. Threlfall, *Variationsrechnung im Grossen*. Leipzig und Berlin: Teubner, 1938.

4. M. Morse, "Recent advances in variational theory in the large," *Proceedings of the International Congress of Mathematics*, (1950), pp. 143–55.

5. J.-P. Serre, "Homologie singulière des espaces fibrés," *Annals of Mathematics*, 54 (1951), pp. 425–505.

6. A. Andreotti and T. Frankel, "The Lefschetz theorem on hyperplane sections," *Annals of Mathematics*, 69 (1959), pp. 713–17.

7. R. Bott and H. Samelson, "Applications of the theory of Morse to symmetric spaces," *American Journal of Mathematics*, 70 (1958), pp. 964–1029.

8. Leon van Hove, "The occurrence of singularities in the elastic frequency distribution of a crystal," *Physical Review*, 89 (1953), pp. 1189–93.

9. John Milnor, "Morse theory" (based on lecture notes by M. Spivak and R. Wells), *Annals of Mathematics Series*, No. 51. Princeton, N.J.: Princeton University Press, 1963.

10. *Differential and Combinatorial Topology, A symposium in honor of Marston Morse*. Princeton Mathematical Series, Princeton, N.J.: Princeton University Press, 1965.

SURFACE AREA

Lamberto Cesari

There is a growing interest in a general theory concerning the analytical properties of transformations, or mappings

$$(T, A): \ p = p(w), \quad w \in A \subset X, \quad p \in Y, \quad \text{or} \quad T: A \to Y, \quad (1)$$

from a set A of a "space X" into a "space Y." Let us say explicitly that T is meant to be single-valued but not necessarily one-one—that is, each $w \in A$ is mapped into one and only one $p = p(w) \in Y$ (image of w), all these $p = p(w)$, $w \in A$, form a set $T(A) \subset Y$ (graph) of T, but each $p \in T(A)$ may be the image of more than one $w \in A$, even infinitely many $w \in A$ (counter images of p). Real functions of one real variable, parametric and nonparametric curves, surfaces, etc., are examples of such mappings, and the analytical entities attached to (T, A) may be called total variation, length, area, etc., or, more generally, line integrals, surface integrals, etc.

The last concepts are usually introduced under very restrictive conditions on T, but recently it has been recognized that length, area, etc., can be introduced under the mere hypothesis of continuity of T (and even this may not be required), and that the finiteness of the length (area, etc.) assures the existence of a line integral (surface integral, etc.).

If X and Y are Euclidean spaces E_k and E_N, respectively, of k and N dimensions, and A is a nondegenerate interval of X, or any open set of X, then we may say that T is a parametric k-variety in E_N (a parametric curve for $k = 1$, a parametric surface for $k = 2$). The theory for $k = 1$ and $k = 2$ has reached some degree of completeness and has been extended in a number of directions in the last decades. Hopefully, a theory adequate in scope, depth, and generality for k-varieties will be developed for $k > 2$. Several books [3a], [17] are dedicated to the theory. A series of articles [4]–[6] in the *American Mathematical Monthly* illustrate briefly some of the results. The present exposition draws upon these articles; we shall discuss some of the questions concerning surfaces because surfaces entail a deeper connection with topology and measure theory than curves.

A completely abstract, or axiomatic approach has been initiated in reference [12], where the underlying structure, or formal theory alone has been emphasized for general transformations. Nevertheless, we shall discuss in Sections 1–10 only actual continuous mappings (precisely surfaces, parametric and nonparametric) and their geometrical, analytical, and topological properties. In Section 11 we shall report on work on discontinuous nonparametric surfaces. Many different viewpoints, Almgren' varifolds, Fleming's integral currents, Young's generalized curves and surfaces, discontinuous parametric surfaces are not discussed in the present elementary exposition.

1. THE CONCEPT OF CONTINUOUS SURFACES

We shall denote by E_2 the real Euclidean w-plane, $w = (u, v)$, by E_N any real Euclidean space [for $N = 3$ let E_3 be the p-space with $p = (x, y, z)$], by $\overline{M}$, M^*, and M^0 the closure, the boundary, and the set of the interior points of a set M in any such space, by $|p| = (x^2 + y^2 + z^2)^{1/2}$ the Euclidean norm, and by $|p - q|$ the Euclidean distance of two points p, q.

By a surface S we shall mean a mapping (1) from some set $A \subset E_2$ into E_3 (or E_N), where A might be, for example, a square, a circle, a polygonal region, or a (closed) simple Jordan region J. Or, A could be, more generally, a (closed) Jordan region of finite connectivity $\nu \geqslant 0$, say $J = J_0 - (J_1 + \cdots + J_\nu)^0$, $J_i \subset J_0^0$, $J_i J_j = 0$, $i \neq j$, $i, j = 1, \ldots, \nu$, where all J_i and J_0 are closed simple Jordan regions. It has been found convenient to take for A any "admissible" set, that is, either any Jordan region J as above, or a finite sum of disjoint Jordan regions, or any set $G \subset E_2$, open in E_2, or any set $G \subset J$, open in J (further generalizations have been, and are being, studied). We will suppose $N = 3$. Thus (T, A) is defined by a continuous vector function $T(w)$, $w \in A$, say,

$$S = (T, A): \ p = T(w), \quad w \in A, \quad T(w) = [x(w), y(w), z(w)],$$

or   (1.1)

$$S = (T, A): \ x = x(u, v), \quad y = y(u, v),$$

$$z = z(u, v), \quad (u, v) \in A.$$

The set $T(A) \subset E_3$ is the graph of $S = (T, A)$, but it in no way defines (T, A). For example, the mappings T: $x = u$, $y = v$, $z = 0$, $(u, v) \in Q$, and T': $x = \sin^2 k\pi u$, $y = v$, $z = 0$, $(u, v) \in Q$, where Q is the square $Q = [0 \leqslant u, v \leqslant 1]$ and $k > 1$ an integer, have both the same graph $T(Q) = T'(Q) = U$, where U is the unit square $[0 \leqslant x, y \leqslant 1, \ z = 0]$ in the x, y-plane. However T covers U just once, while T' covers U exactly k times. Finally, if P: $x = \phi(t)$, $y = \psi(t)$, $0 \leqslant t \leqslant 1$, denotes the well-known Peano curve covering U, then T'': $x = \phi(u)$, $y = \psi(u)$, $z = 0$, $(u, v) \in Q$, has the same graph as T' and T, but is a completely different surface. Indeed, we shall mention that the "areas" of T, T', T'' are different numbers, namely 1, k, and 0, respectively.

Actually there are cases where our intuition associates to different mappings (or surfaces) the same entity (as, for example, T and T' with $k = 1$). Indeed, various concepts of "equivalence" have been taken into consideration, such as the Lebesgue and Fréchet equivalences (for example, T and T' above for $k = 1$ are Lebesgue equivalent). Classes of equivalent mappings are then denoted as

Lebesgue surfaces and Fréchet surfaces. If A, A' are topological equivalent, and (T, A), (T', A') are continuous maps into the same space E_n, then (T, A), (T', A') are said to be comparable. If there is a homeomorphism h: $A \to A'$, or $w' = h(w)$, from A onto A' such that $T'(h(w)) = T(w)$ for all $w \in A$, then (T, A) and (T', A') are said to be Lebesgue equivalent. If for every $\epsilon > 0$ there is a homeomorphism h_ϵ: $A \to A'$, or $w' = h_\epsilon(w)$, from A onto A' such that

$$|T'(h_\epsilon(w)) - T(w)| \leqslant \epsilon \qquad (1.2)$$

for all $w \in A$, then (T, A) and (T', A') are said to be Fréchet equivalent. In general, the infimum of all numbers $\epsilon > 0$ for which (1.2) holds for some homeomorphism h_ϵ is said to be the Fréchet distance $\|T, T'\|$ of (T, A) and (T', A'). It can be shown that $\|T, T'\| \geqslant 0$, $\|T, T'\| = \|T', T\|$, and $\|T, T'\| \leqslant \|T, T''\| + \|T'', T\|$, hence, Fréchet equivalent comparable maps form a metric space. These definitions have been discussed in references [3] and [17], and for curves also in [5]. Below, the book [3] will often be quoted as [SA] followed by indications of page and section.

Surfaces, or mappings $S = (T, A)$ for which each point $p \in T(A)$ has only one counterimage are said to be simple [and if T^{-1} is continuous, then T is a homeomorphism between A and $T(A)$]. If for every $p \in T(A)$ the set $T^{-1}(p) \subset A$ is connected, then T is said to be monotone; if $T^{-1}(p) \subset A$ is totally disconnected, then T is said to be light. For example, surfaces of the type

$$T: x = u, \quad y = v, \quad z = z(u, v), \quad (u, v) \in Q,$$

that is, T: $z = z(x, y)$, $(x, y) \in Q$, are said, somewhat improperly, to be "nonparametric surfaces," and they are certainly simple. The mapping T: $x = u$, $y = v$, $z = (1 - u^2 - v^2)^{1/2}$, $(u, v) \in Q = (u^2 + v^2 \leqslant 1)$, the graph of which is a "halfsphere," is nonparametric and simple. The mapping T: $x = 2(r - r^2)^{1/2}\cos\theta$, $y = 2(r - r^2)^{1/2}\sin\theta$, $z = 2r - 1$, where $(u, v) \in Q$, $r\cos\theta = u$, $r\sin\theta = v$, the graph of which is the whole sphere $x^2 + y^2 + z^2 = 1$, is monotone, but not light since the point $p = (0, 0, 1)$ is the image of all points $(u, v) \in Q^*$. The mapping T: $x = 1 - u^2$, $y = u - u^3$,

$z = v$, $(u, v) \in A = (-2 \leqslant u \leqq 2, 0 \leqslant v \leqslant 1)$, the graph of which is a portion of a cylinder with generatrices parallel to the z-axis, is light, but not monotone.

Of particular interest are the "flat surfaces" or "plane mappings," that is, those mappings with a graph contained in a plane. For example, if τ_r, $r = 1, 2, 3$, denotes the projections of E_3 onto the yz, zx, xy coordinate planes, say E_{21}, E_{22}, E_{23}, then $T_r = \tau_r T$, $r = 1, 2, 3$, are plane mappings, namely,

$$T_1: x = 0, \qquad y = y(u, v), \quad z = z(u, v), \quad (u, v) \in A,$$
$$T_2: x = x(u, v), \quad y = 0, \qquad z = z(u, v), \quad (u, v) \in A, \quad (1.3)$$
$$T_3: x = x(u, v), \quad y = y(u, v), \quad z = 0, \qquad (u, v) \in A.$$

It was pointed out already by S. Banach and G. Vitali that properties of a mapping T as a parametric continuous surface (for example, the finiteness of the "area") have no, or very little bearing on the properties of the single functions $x(u, v)$, $y(u, v)$, $z(u, v)$, but they have an essential bearing on the properties of the pairs of functions (y, z), (z, x), (x, y), that is, on the plane mappings T_r, $r = 1, 2, 3$. For example, no matter which continuous function $\phi(u, v)$, $(u, v) \in A$, we consider, the mapping T: $x = y = z = \phi(u, v)$, $(u, v) \in A$, has a graph completely contained in the straight line $x = y = z$ in E_3, and its "area" is zero. For nonparametric surfaces T: $x = u$, $y = v$, $z = z(u, v)$, all properties of T are, of course, reflected into properties of the single function $z(u, v)$.

2. LEBESGUE AREA

For parametric surfaces (mappings) $S = (T, A)$: $p = p(w)$, $w \in A$, $p = (x, y, z)$, $w = (u, v)$, defined by functions $x(u, v)$, $y(u, v)$, $z(u, v)$, which are continuous in A with their first partial derivatives, it is usual to assume for the area of S the value of the integral (area integral)

$$I(T, A) = (A^0) \int |J| \, du \, dv = (A^0) \int \left(J_1^2 + J_2^2 + J_3^2 \right)^{1/2} du \, dv, \quad (2.1)$$

and for the surface integral the value of

$$I(T, A, f) = (A^0) \int\!\!\int f[p(w), J(w)] \, du \, dv, \qquad (2.2)$$

where $J = J(w)$ is the vector Jacobian $J = (J_1, J_2, J_3)$, $J_1 = y_u z_v - z_u y_v$, etc., and where $f(p, t)$ is any given function of (p, t) continuous in (p, t) for all $p = (x, y, z) \in T(A)$, and all $t = (t_1, t_2, t_3)$. To assure that $I(T, A, f)$ is invariant with respect to Lebesgue or Fréchet equivalences, we add another condition

(h): $f(p, kt) = kf(p, t)$ for all $k \geqslant 0$, $t \in E_3$ and $p \in T(A)$.

Thus, for $f = |t|$, $I(T, A, f)$ is the area integral $I(T, A)$ and condition (h) is satisfied. The definitions (2.1) and (2.2) are adequate under the restrictive conditions mentioned but inadequate generally. To see this, let us denote by $\phi(t)$, $0 \leqslant t \leqslant 1$, $\phi(0) = 0$, $\phi(1) = 1$, the well-known monotone nondecreasing, continuous function which is constant on each complementary interval I_j of the ternary Cantor set in $(0,1)$. Thus, $\phi'(t) = 0$ almost everywhere in $(0,1)$ [15, p. 368]. Then the nonparametric light mapping (surface) S': $x = u$, $y = v$, $z = \phi(u)$, $(u, v) \in Q = (0, 1, 0, 1)$ should have "area" greater than, or equal to, $\sqrt{2}$, while the integral $I(T, A)$ has the value 1. Analogously, the monotone mapping (surface) S'': $x = \phi(u)$, $y = v$, $z = 0$ (a monotone mapping from Q into the square $U = [0 \leqslant x, y \leqslant 1, z = 0]$) should have "area" 1 (or at least 1), while the integral $I(T, A)$ has the value 0.

A definition shown to be adequate was proposed by Lebesgue in 1900. Simply, the Lebesgue area of a surface S is the "lower limit" of the elementary areas of the polyhedral surfaces approaching S. To make this definition precise, we need a few more words.

We shall denote as a figure F any finite sum of disjoint closed polygonal regions in E_2 (for example, a square, a polyhedral region). A mapping (P, F): $x = p(w)$, $w \in F$, from a figure F is said to be quasilinear, or a polyhedral surface if (1) $p(w)$ is single valued and continuous in F, and (2) there exists some subdivision D of F into nonoverlapping triangles t such that each component $x(u, v), y(u, v), z(u, v)$ of $p(w)$ is linear in each $t \in D$—that is, of the form $au + bv + c$, a, b, and c constants for each t. Then P maps each $t \in D$ into a triangle $\Delta \subset E_3$, which may be degenerated into a segment, or a single point. Then by the elementary area $a(P, F)$ of (P, F) is meant the sum $a(P, F) = \Sigma a(\Delta)$, where $a(\Delta)$ is the usual area of the triangle $\Delta = P(t)$ and Σ ranges over all $t \in D$. We shall say that a sequence $[A_n]$ of admissible sets A_n invades an admissible set A, if $A_n \subset A$, $A_n \subset A_{n+1}$, $A_n^0 \to A^0$ as $n \to \infty$. Finally, a sequence (T_n, A_n), $n = 1, 2, \ldots$, is said to be convergent toward (T, A) if (1) A_n invades A, and (2) $d_n \to 0$ as $n \to \infty$ where $d_n = \sup|T_n(w) - T(w)|$ for all $w \in A_n$. Thus, if $A_n = A$, $n = 1, 2, \ldots$, the convergence of (T_n, A) toward (T, A) is the uniform convergence in A of $T_n(w)$ toward $T(w)$, $w \in A$.

Suppose now that (T, A) is a given continuous mapping from an admissible set A and denote by γ the class of all sequences $[(P_n, F_n), n = 1, 2, \ldots]$ of quasilinear mappings convergent toward (T, A). Then the Lebesgue area $L(T, A)$ of (T, A) is defined by

$$L(S) = L(T, A) = \underset{\gamma}{\text{Inf}} \underset{n \to \infty}{\liminf} a(P_n, F_n). \qquad (2.3)$$

Of course some reader may ask why this definition is chosen instead of considering simply "the supremum, or the limit of the elementary areas of the polyhedral surfaces inscribed in S." H. Schwarz and G. Peano discovered in 1890 that such a supremum may be $+\infty$ and such a limit may not exist even with as simple a surface as a portion of "circular cylinder" [**SA**, 4.2, p. 24]. Definition (2.3), proposed by Lebesgue in 1900, is the analogue of one of the alternative definitions of Jordan length for a curve [**5**]. It can be proved [**SA**, 5.9, p. 37] that there exists some sequence (P_n, F_n), $n = 1, 2, \ldots$, convergent toward (T, A) with $\lim a_n(P_n, F_n) = L(T, A)$ as $n \to \infty$.

In case A is itself a figure (for example, a square) it is not restrictive to suppose [**SA**, 6.2, p. 61] that $F_n = A$ for all n. Then γ is the class of all sequences $[(P_n, A), n = 1, 2, \ldots]$ of quasilinear mappings convergent uniformly in A toward (T, A).

In case A is a figure and T is quasilinear, then $L(T, F) = a(T, F)$, that is, the Lebesgue area coincides with the elementary area. The proof of this fact, intuitive as it may appear, is difficult (see [**SA**, p. 108]).

3. THE LOWER SEMICONTINUITY PROPERTY OF LEBESGUE AREA

Definition (2.3) has, among others, the advantage that it makes it easy to prove the lower semicontinuity of $L(S)$, namely, the property which can be expressed by the following statement.

THEOREM 3.1. *If (T_n, A_n), $n = 1, 2, \ldots$, is convergent toward (T, A), then $L(T, A) \leqslant \liminf L(T_n, A_n)$ as $n \to \infty$.*

Proof. We shall prove (3.1) in the case A is a given figure, assuming in Definition (2.3) $F_n = A$ for all n, according to the remark at the end of Section 2. We may also assume $\lambda = \liminf L(T_n, A_n) < +\infty$, and $L(T_n, A) < +\infty$ for all n. Then, if $d_n = \max|T_n(w) - T(w)|$, we have $d_n \to 0$ as $n \to +\infty$. By the definition of $L(T, A)$ there exists some sequence (P_{nm}, A), $m = 1, 2, \ldots$, of quasilinear mappings convergent toward (T_n, A) as $m \to \infty$ with $a(P_{nm}, A) \to L(T_n, A)$. Thus we have $\delta_{nm} \to 0$ as $m \to \infty$ where $\delta_{nm} = \max|P_{nm}(w) - T_n(w)|$ for all $w \in A$. For each n, there exists an integer $m = m(n)$ such that $\delta_{nm} < 1/n$, $|a(P_{nm}, A) - L(T_n, A)| < 1/n$. Now we have $|P_{nm}(w) - T(w)| \leqslant |P_{nm} - T_n| + |T_n - T| \leqslant \delta_{nm} + d_n < d_n + 1/n$ for all $w \in A$. If $P'_n = P_{n, m(n)}$, then the sequence (P'_n, A), $n = 1, 2, \ldots$ converges toward (T, A), that is, belongs to the class γ relative to (T, A). Thus, by Definition (2.3) we have

$$L(T, A) \leqslant \liminf_{n \to \infty} a(P'_n, A) \leqslant \liminf_{n \to \infty} \left[L(T_n, A) + \frac{1}{n} \right] = \lambda. \ \blacksquare$$

We shall denote as before by L and λ the numbers $L = L(T, A)$ and $\lambda = \liminf L(T_n, A_n)$. Even for polyhedral surfaces we may have $L < \lambda \leqslant +\infty$ (in particular, we may have $L < +\infty$, $\lambda = +\infty$). We give the following example. Suppose $A = F$ is the unit square $A = (0 \leqslant u \leqslant 1, 0 \leqslant v \leqslant 1)$, and T is the identity mapping T: $x = u$, $y = v$, $z = 0$. Thus, $S = (T, A)$ is the unit square in the x, y-plane. Suppose $A_n = A$ and define $S_n = (T_n, A)$ as follows: $x_n = u$, $y_n = v$, $z_n = h_n(u - (i/n))$ for $i/n \leqslant u \leqslant (i/n) + (1/2n)$, $z_n = h_n((i/n) + (1/n) - u)$ for $(i/n) + (1/2n) \leqslant u \leqslant (i/n) + (1/n)$, where $i = 0, 1, \ldots, n - 1$, $0 \leqslant u \leqslant 1$, $0 \leqslant v \leqslant 1$ and h_n, $n = 1, 2, \ldots$, are arbitrary positive numbers. As n describes the sequence $n = 1, 2, \ldots$, we have a sequence of polyhedral surfaces S_n, $n = 1, 2, \ldots$. Each S_n consists of $2n$ rectangular strips of side lengths $\rho = 1$ and $\rho' = (1/2n)(1 + h_n^2)^{1/2}$. Each strip has a side of length 1 on the x, y-plane, and the opposite side is above and parallel to this plane at a height of $h_n/2n$. If we suppose all numbers h_n equal to 1, then the strips are all at 45-degree angles with the xy-plane, and we have $h_n/2n = 1/2n \to 0$ as $n \to \infty$; if we suppose $h_n = n^{1/2}$, the angle of the $2n$ strips with the xy-plane becomes closer and closer to 90° as n increases, but still we have $h_n/2n = 2^{-1}n^{-1/2} \to 0$ as $n \to \infty$. Because $x_n = x$, $y_n = y$, $|z_n - z| = h_n/2n$, we have $T_n \to T$ as $n \to \infty$ (uniformly on A). On the other hand, we have $L = L(T, A) = a(T, A) = 1$, and $L(T_n, A) = a(T_n, A) = (1 + h_n^2)^{1/2}$. Thus, for $h_n = 1$, $L(T_n, A) = 2^{1/2}$, $n = 1, 2, \ldots$, and $L = 1$, $\lambda = 2^{1/2} > 1$. For $h_n = n^{1/2}$, $L(T_n, A) = (1 + n)^{1/2}$, and $L = 1$, $\lambda = +\infty$.

In a slightly more general form, statement 3.1 can be worded to say that the Lebesgue area of continuous maps is lower semicontinuous in the Fréchet metric.

We shall return to the concept of area in Section 6. We mention here that an axiomatic approach to area has long been sought. For simplicity, suppose that A is a fixed figure F, and consider the class of all continuous mappings (T, A) from A into E_3. If a functional $Z(T, A)$ defined in this class must be considered an area, it should satisfy certain axioms, a number of which have been proposed. Of these, we list only a few.

1. $Z(T, A)$ *is lower semicontinuous, that is, it satisfies Theorem (3.1) (with L replaced by Z).*
2. $Z(P, A) = a(P, A)$ *for every quasilinear mapping (P, A).*
3. *There exists a sequence (P_n, A) of quasilinear mappings convergent toward (T, A) with $a(P_n, A) \to Z(T, A)$ as $n \to \infty$.*

Obviously the Lebesgue area satisfies all three axioms.

THEOREM 3.2. *Any functional $Z(T, A)$ satisfying Axioms 1 and 2 is less than, or equal to, $L(T, A)$. Thus, the Lebesgue area is the maximum of all functionals satisfying Axioms 1 and 2.*

Proof. Indeed, for every sequence (P_n, A) convergent toward (T, A) with $a(P_n, A) \to L(T, A)$, we have, by Axioms 1 and 2,

$$Z(T, A) \leqslant \liminf Z(P_n, A) = \liminf a(P_n, A) = L(T, A),$$

that is, $Z \leqslant L$, and Theorem 3.2 is proved.

THEOREM 3.3. *Any functional $Z(T, A)$ satisfying Axioms 1, 2, and 3 coincides with $L(T, A)$.*

Proof. Indeed, if (P_n, A) is the sequence mentioned in Axiom 3 for the mapping (T, A), we have, by Axioms 3 and 2, and by the definition of Lebesgue area,

$$Z(T, A) = \liminf Z(P_n, A) = \liminf a(P_n, A) \geqslant L(T, A),$$

and thus $Z \geqslant L$ and $Z \leqq L$, that is, $Z = L$. Thereby Theorem 3.3 is proved.

4. **PLANE MAPPINGS: THEIR TOTAL VARIATION AND ABSOLUTE CONTINUITY**

Let (T, A): $p = T(w)$, $w \in A$, $w = (u, v)$, $p = (x, y)$, be any continuous mapping from an admissible set A of the oriented u, v-plane E_2 into the oriented, x, y-plane E_2', that is,

$$(T, A): x = x(u, v), \qquad y = y(u, v), \qquad (u, v) \in A.$$

For every simple closed polygonal region $\pi \subset A$, let us consider the oriented boundary π^* and the image $C: (T, \pi^*)$ of π^*, that is, the continuous mapping (T, π^*) defined by T on π^* (restriction of T on π^*). It is a closed oriented curve C of the x, y-plane E_2. For each point $p_0 = (x_0, y_0)$ not on the graph (C) of C, we can define the topological index $O(p_0; C)$ of C with respect to the point $p_0 \in E_2' - (C)$. Roughly speaking, $O(p_0; C)$ is the integral number of times $(\leqq 0)$ in which C "links" the point p_0 in the positive direction. Suppose we assume on π^* a parameter, say s, $0 \leqslant s \leqslant a$. Suppose also we use polar coordinates (ρ, θ) of center p_0 in E_2.

Then as s describes $(0, a)$, that is, w describes π^* once in the positive sense, $p = T(w)$ describes C. The modulus $\rho = \rho(s)$ of p, that is, $\rho = \rho(s) = |p - p_0|$, is a single-valued continuous function of s on $(0, a)$ and $\rho(a) = \rho(0)$. The argument $\theta(s)$ of p is defined only as mod 2π. If we fix any one of its values, say $\bar{\theta} = \theta(0)$, for $s = 0$, then by continuity, $\theta(s)$ is defined on $(0, a)$ as a single-valued continuous function of s, and $\theta(a) \equiv \theta(0) = \bar{\theta}(\text{mod } 2\pi)$. Then, by definition, $O(p_0; C) = (2\pi)^{-1}[\theta(a) - \theta(0)]$, and it is easy to prove that $O(p_0; C)$ does not depend on the parametrization of π^*, and on the choice of $\bar{\theta}$. This purely analytical definition [1, p. 462] of $O(p_0; C)$ is certainly the most elementary one. Other equivalent and purely topological definitions are given in topology. If C is rectifiable, and we think of E_2' as the plane of the complex variable $\zeta = x + iy$, with $\zeta_0 = x_0 + iy_0$, then $O(p_0; C)$ is given by the line integral

$$O(p_0; C) = (2\pi i)^{-1} \int_C (\zeta - \zeta_0)^{-1} d\zeta.$$

The topological index $O(p; C)$ has a number of analytical and topological properties [**SA**, 8.3–8.11, p. 85]. Let us mention here that $O(p; C)$, $p = (x, y) \in E_2' - (C)$, is always finite (but not necessarily bounded in E_2'); $O(p; C)$ is constant on each complementary component of the graph (C) of C and is 0 in the unbounded one. If we assume $O(p, C) = 0$ at all points $p \in (C)$, then $O(p; C)$, $p \in E_2'$, is a single-valued, integral-valued ($\gtreqless 0$) function of p in E_2', and $O(p; C)$ is B-measurable. Thus, the L-integral

$$v(T; \pi) = (E_2') \int |O(p; C)| \, dx \, dy$$

exists (finite, or $+\infty$). Analogously, we may consider the numbers

$$v_+(T, \pi) = (E_2') \int O^+(p; C) \, dx \, dy \geqslant 0,$$

$$v_-(T, \pi) = (E_2') \int O^-(p; C) \, dx \, dy \geqslant 0,$$

where $O^+ = \frac{1}{2}[|O| + O]$, $O^- = \frac{1}{2}[|O| - O]$, and $v_+ + v_- = v$. Finally, if $O(p; C)$ is L-integrable, that is, $v(T, \pi) < +\infty$, then the number

$$u(T, \pi) = (E_2') \int O(p; C) \, dx \, dy \gtreqqless 0,$$

also exists, and $|u| \leqslant v$.

If D denotes any finite system of closed, nonoverlapping, simple, polygonal regions $\pi \subset A$, let

$$V(T, A) = \sup_D \sum_{\pi \in D} v(T, \pi),$$

and for every point $p \in E_2'$,

$$N(p; T, A) = \sup_D \sum_{\pi \in D} |O(p; C)|.$$

Then $N(p; T, A)$, $0 \leqslant N \leqslant +\infty$, is an integral-valued, single-valued function of p in E_2', and $N(p)$ is lower semicontinuous in E_2, that is, $N(p_0) \leqslant \liminf N(p)$ as $p \to p_0$, for every $p_0 \in E_2'$. The function $N(p, T, A)$, $p \in E_2'$, is said to be the multiplicity function of (T, A). (It is similar to the corrected multiplicity of a real function of one real variable considered in [4, p. 328].) Finally, let

$$W(T, A) = (E_2') \int N(p; T, A) \, dx \, dy.$$

Both V and W can be considered as total variations of (T, A). Indeed, V is of the Jordan total variation type, and W of the Banach total variation of a real function of one real variable type [4, 321]. The following basic theorem extends the Banach theorem for real functions [4, 321].

THEOREM 4.1. *For every continuous plane mapping* (T, A) *we have* $V(T, A) = W(T, A) = L(T, A)$.

For a proof see [SA, pp. 186 and 390]. Then the common value (finite, or $+\infty$) of V, W, L can be assumed as the total variation of the plane mapping (T, A), and (T, A) is said to be of bounded variation if $V(T, A) = W(T, A) < +\infty$. Analogously, let us write

$$V_+(T, A) = \sup_D \sum_{\pi \in D} v_+(T, \pi),$$

$$V_-(T, A) = \sup_D \sum_{\pi \in D} v_-(T, \pi),$$

$$N_+(p; T, A) = \sup_D \sum_{\pi \in D} O^+(p; C),$$

$$N_-(p; T, A) = \sup_D \sum_{\pi \in D} O^-(p; C),$$

$$W_+(T, A) = (E_2') \int N_+(p; T, A) \, dx \, dy,$$

$$W_-(T, A) = (E_2') \int N_-(p; T, A) \, dx \, dy.$$

THEOREM 4.2. *For every continuous plane mapping (T, A) we have* [SA, p. 187]

$$V_+(T, A) = W_+(T, A), \qquad V_-(T, A) = W_-(T, A),$$

$$V_+ + V_- = V, \qquad W_+ + W_- = W.$$

Thus, the common values $V_+ = W_+$, $V_- = W_-$ can be assumed as the positive and negative total variations of (T, A), and the difference $\mathscr{B}(T, A) = V_+(T, A) - V_-(T, A)$ as the signed, or relative, total variation of (T, A).

Finally, if (T, A) is of bounded variation, that is, $V(T, A) < +\infty$, $v(T, \pi) < +\infty$ for every $\pi \in A$, and thus also is the following number defined:

$$U(T, A) = \sup_D \sum_{\pi \in D} |u(T, \pi)|.$$

THEOREM 4.3. *For every continuous bounded variation plane mapping (T, A) we have $U(T, A) = V(T, A)$.*

Let us observe finally that V is "overadditive"; that is, if (A') is any finite subdivision of A into nonoverlapping admissible sets, or, more generally, any finite system of nonoverlapping admissible sets $A' \subset A$, then $V(T, A) \geqslant \Sigma V(T, A')$, and the sign $>$ may hold even in the apparently elementary case where A' and A are polygonal regions.

These considerations show how the concept of "plane mapping of bounded variation" can be founded on topological and measure theoretical considerations. On the same basis, we can introduce the corresponding concept of absolute continuity.

A continuous (plane) mapping (T, A): $p = T(w)$, $w \in A$, $w = (u, v)$, $p = (x, y)$, from the uv-plane E_2 into the xy-plane E_2', is said to be absolutely continuous if both the following conditions hold.

1. *Given $\epsilon > 0$, there is a $\delta = \delta(T, A, \epsilon) > 0$ such that for each finite system $D = (\pi)$ of nonoverlapping simple closed polygonal regions $\pi \subset A$ with $\Sigma|\pi| < \delta$ we have $\Sigma v(T, \pi) < \epsilon$.*
2. *For every simple closed polygonal region $\pi \subset A$ and finite subdivision (π') of π into nonoverlapping simple polygonal regions π' we have $V(T, \pi) = \Sigma V(T, \pi')$.*

In (1), $|\pi|$ denotes the Lebesgue measure of the set π in E_2 (area). Condition (1) is familiar, and essentially requires that v (and thus V) is "an absolutely continuous set function." Condition (2) simply requires that V is "additive" (at least in the class of the simple polygonal regions $\pi \subset A$). Conditions (1) and (2) are independent, as examples have shown, and it is merely their logical union $(1) \cup (2)$ which we assume here as a definition of absolute continuity. There are many interesting properties, each of which is a necessary and sufficient condition for a mapping (T, A) to be absolutely continuous.

5. A CHARACTERIZATION OF SURFACES WITH FINITE AREA

The basic concepts of area of a mapping (T, A), $T: A \to E_3$ (Section 2), and of total variation of the plane mappings (T_r, A) defined by Equation (1.3) in Section 4, are connected by a basic theorem, which extends formally to Lebesgue area, a known Jordan theorem for Jordan length [5, p. 489].

THEOREM 5.1. *For every continuous mapping* (T, A): $p = T(w)$, $w \in A$, $w = (u, v)$, $p = (x, y, z)$, *we have*

$$V(T_r, A) \leqslant L(T, A) \leqslant V(T_1, A) + V(T_2, A) + V(T_3, A),$$

$$r = 1, 2, 3. \tag{5.1}$$

Thus, $L < +\infty$ *if and only if all plane mappings* (T_r, A), $r = 1, 2, 3$, *are of bounded variation* (Cesari, 1942).

This theorem, despite its analogy with the elementary Jordan theorem for curves (see [5]), has been shown to have a deep topological and measure theoretical basis. The proof is given in [**SA**, p. 295], and consists of the process of stretching and smoothing the continuous surface $S = (T, A)$ into continuous polyhedral surfaces $S_n = (P_n, F_n)$, $n = 1, 2, \ldots$ with $P_n \to T$ as $n \to \infty$, and $a(P_n, F_n) \leqslant V_1 + V_2 + V_3$, $V_r = V(T_r, A)$, $r = 1, 2, 3$.

6. PEANO AND GEÖCZE AREAS

Let (T, A): $p = p(w)$, $w \in A$, be any continuous mapping from $A \subset E_2$ into E_3, and let us denote by T_1, T_2, T_3 the plane mappings of Equation (1.2) which are the projections $T_r = \tau_r T$ of T on the (y, z), (z, x), (x, y) coordinate planes E_{21}, E_{22}, E_{23}. If π denotes any simple polygonal region $\pi \subset A$, π^* the oriented boundary of π, let $C: (T, \pi^*)$, $C_r: (T_r, \pi^*)$, $r = 1, 2, 3$, be the continuous oriented closed curves which are the images of π^* under T and T_r. Thus, $(C) \in E_3$, $(C_r) \subset E_{2r}$, and C_r is the "projection" of C on E_{2r}.

According to Section 4, we put

$$v_r = v(T_r, \pi) = (E_{2r}) \int |O(p; C_r)| \, dp, \qquad r = 1, 2, 3,$$

$$v = (v_1, v_2, v_3), \qquad |v| = (v_1^2 + v_2^2 + v_3^2)^{1/2},$$

$$V(T, A) = \sup_D \sum_{\pi \in D} |v(T, \pi)|,$$

where the supremum is taken with respect to all finite systems D of nonoverlapping simple polygonal regions $\pi \subset A$. Thus, $v(T_r, \pi) \leqslant v(T, \pi)$, $V(T_r, A) \leqslant V(T, A) \leqslant +\infty$, $r = 1, 2, 3$. The number V can be thought of as an "area" of the surface $S = (T, A)$. A variant of this definition follows. Denote by α any plane in E_3, τ_α the projections of E_3 onto α, $T_\alpha = \tau_\alpha T$ the projection of T on α, C_α: (T_α, π^*) the oriented continuous closed curve which is the image of π^* under T_α, $(C_\alpha) \subset \alpha$. Then put $v(T_\alpha, \pi) = (\alpha) \int |O(p; C_\alpha)| \, dp$, $v^*(T, \pi) = \sup_\alpha v(T_\alpha, \pi)$, and

$$P(T, A) = \sup_D \sum_{\pi \in D} v^*(T, \pi) = \sup_D \sum_{\pi \in D} \sup_\alpha v(T_\alpha, \pi).$$

Also, $P(T, A)$, like $V(T, A)$ and $L(T, A)$, is an "area" of T, in the sense that all three definitions (just as many others) are precise mathematical formulations of concepts which all strongly appeal to our intuition as areas. In fact, the following theorem has been proved.

THEOREM 6.1. *For every continuous mapping (surface) (T, A) we have $L(T, A) = V(T, A) = P(T, A)$* (C. B. Morrey, T. Rado, L. Cesari, J. Cecconi).

The common value of the three numbers L, V, P, is defined as the area of (T, A), and is usually called the Lebesgue area of (T, A). The present definitions of V and P are the final result of successive refinements of a concept first proposed by Peano in 1890. This process of successive refinements is necessary for us to reach the basic identity in Theorem 6.1, which did not hold for the previous somewhat crude concepts. A number of authors, among them Z. de Geöcze, S. Banach, L. Tonelli, R. Caccioppoli, T. Rado, E. J. McShane, C. B. Morrey, J. Cecconi, and L. Cesari, have contributed to these refinements. By common agreement, the area V is usually designated as the Geöcze area and P as the Peano area of $S = (T, A)$, to honor two mathematicians who proved to have such deep insight into the forthcoming theory. For a direct proof of Theorem 6.1, see [**SA**, p. 390]. If $V(T, A) < +\infty$, then $V_r(T, A) < +\infty$, $r = 1, 2, 3$, and for every $\pi \subset A$, the following numbers also exist

$$u_r = u(T_r, \pi) = (E_{2r}) \int O(p; C_r)\, dp, \qquad r = 1, 2, 3,$$

$$u = (u_1, u_2, u_3), \qquad |u| = \left(u_1^2 + u_2^2 + u_3^2\right)^{1/2},$$

$$U(T, A) = \sup_D \sum_{\pi \in D} |u(T, \pi)|,$$

where the preceding conventions are used. Then we have, obviously,

$$0 \leqslant |u(T, \pi)| \leqslant v(T, \pi) < +\infty,$$

$$0 \leqslant U(T, A) \leqslant V(T, A) \leqslant +\infty,$$

and it is clear that for some π and T, we may well have $|u| < |v|$. Nevertheless, the following theorem holds.

THEOREM 6.2. *For every mapping (T, A) with $V(T, A) < +\infty$, we have $U(T, A) = V(T, A)$* (Cesari).

7. A WEIERSTRASS-TYPE INTEGRAL

By using the same blend of topological and analytical considerations with which we defined Geöcze and Peano areas, we may now define the concept of parametric surface integral through a Weierstrass-type limit process.

Let $S = (T, A)$: $p = T(w)$, $w \in A$, $w = (u, v)$, $p = (x, y, z)$, be a given mapping of finite area, and, for simplicity, let us suppose that A is a closed finitely connected Jordan region. Let $f(p, t)$ be a continuous function of (p, t), $p = (x, y, z)$, $t = (t_1, t_2, t_3)$, for all $p \in T(A)$ and real vector t satisfying the usual condition $f(p, kt) = kf(p, t)$ for all $k \geqslant 0$, $p \in T(A)$, and t. For every simple polygonal region $\pi \subset A$ let us consider the vector

$$u = u(T, \pi) = (u_1, u_2, u_3), \qquad u_r = u(T_r, \pi), \qquad r = 1, 2, 3,$$

of norm $|u| = (u_1^2 + u_2^2 + u_3^2)^{1/2}$. If $|u| > 0$, then the unit vector

$$a = a(T, \pi) = (a_1, a_2, a_3), \qquad a_r = \frac{u_r}{|u|},$$

$$r = 1, 2, 3, \qquad a_1^2 + a_2^2 + a_3^2 = 1$$

can be thought of as the vector of the direction cosines of an "average normal" n to the piece of the surface S defined by T on π. If $u = 0$ let us take for $a = (a_1, a_2, a_3)$ any vector with $|a| = 1$. If D is any finite system $D = (\pi)$ of nonoverlapping simple polygonal regions $\pi \subset A$, and for each $\pi \in D$, we denote by $\tilde{p}$ any point of $T(\pi)$, we may consider the sum

$$M = \sum_{\pi \in D} f(\tilde{p}, u) = \sum_{\pi \in D} f(\tilde{p}, a)|u(\pi, T)|$$

as an "approximate expression" of a Weierstrass-type integral $E(T, A, f)$ of f on the surface $S = (T, A)$. An index $\delta = \delta(D)$ measuring the "fineness" of D can be introduced in various ways. Then it can be proved that the following limit exists and is finite

$$E(T, A, f) = \lim_{\delta \to 0} M = \lim_{\delta \to 0} \sum_{\pi \in D} f(\tilde{p}, a)|u(\pi, T)|,$$

for every continuous mapping (T, A) of finite area (Cesari, [8]). The integral E is invariant with respect to both Lebesgue and Fréchet equivalences; that is, $E(T, A, f)$ has the same value in correspondence with Lebesgue or Fréchet equivalent mappings.

An index δ of fineness of a system D is, for example, defined by the maximum of all following numbers: $\operatorname{diam} T(\pi)$ for $\pi \in D$; $|\Sigma T_r(\pi^*)|$, $r = 1, 2, 3$, and $U(T, A) - \Sigma_{\pi \in D}|u(T, \pi)|$, $U(T_r, A) - \Sigma_{\pi \in D}|u(T_r, \pi)|$, $r = 1, 2, 3$.

8. RELATION BETWEEN AREA AND AREA INTEGRAL

What relation exists between area and the area integral of Section 2, and between the Weierstrass-type integral E and the classical surface integral I of Section 2? As mentioned in Section 2, the finiteness of the area does not imply the existence of the partial derivatives of $x(u, v)$, $y(u, v)$, $z(u, v)$, and, hence, of the ordinary Jacobians $J_1 = y_u z_v - y_v z_u, \ldots$. The example $S = (T, A)$: $x = y = z = \phi(u, v)$, where $\phi(u, v)$ is any continuous function with partial derivatives at no point, is typical, because $L(S) = 0$ is certainly finite here. Thus, it is clear that $L(S) < +\infty$ implies the existence of the Weierstrass-type surface integral $E(T, A, f)$, but neither area integral $I(T, A)$ nor the classical integral $I(T, A, f)$ need exist.

Nevertheless, under the sole hypothesis of finiteness of the area, certain "generalized Jacobians" $\mathcal{J}_r(w)$, $w \in A^0$, can be defined almost everywhere in A^0, and for which we have $\mathcal{J}_r(w) = J_r(w)$, $r = 1, 2, 3$, almost everywhere in A^0, whenever the functions x, y, z have ordinary partial derivatives almost everywhere in A^0. Of the various equivalent definitions of generalized Jacobians, the following is certainly the simplest and holds for almost all $w_0 \in A^0$:

$$\mathcal{J}_r(w_0) = \lim_{\sigma \to 0} \frac{\mathcal{B}(T_r, q)}{|q|}, \qquad r = 1, 2, 3,$$

where $w_0 \in A^0$, q denotes any square with sides parallel to the u and v axes, with $w_0 \in q$, $q \subset A^0$, and $\sigma = \operatorname{diam} q$. With these definitions, we can prove the following theorem, which corresponds formally to a Tonelli's theorem for curves ([5], p. 492).

THEOREM 8.1. *For any continuous mapping* $S = (T, A)$ *with* $L(T, A) < +\infty$, *we have*

$$L(T, A) \geqslant (A^0) \int |\mathcal{J}| \, du \, dv, \tag{8.1}$$

where $\mathcal{T} = (\mathcal{T}_1, \mathcal{T}_2, \mathcal{T}_3)$. *The equality sign holds if and only if the plane mappings* (T_r, A), $r = 1, 2, 3$, *are absolutely continuous.*

Suppose now that $S = (T, A)$ and $f(p, t)$ are given as in Section 7. Then we have the following theorem.

THEOREM 8.2. *If* $L(S) < +\infty$ *and the plane mappings* (T_r, A), $r = 1, 2, 3$, *are absolutely continuous, then*

$$E(T, A, f) = (A^0) \int f[p(w), \mathcal{T}(w)] \, du \, dv. \qquad (8.2)$$

In other words, the integral E is given as an ordinary surface integral (with generalized Jacobians) whenever the area is given by the corresponding area integral.

In both Theorems 8.1 and 8.2 the integrals are L-integrals.

The question finally presents itself, whether any continuous surface $S = (T, A)$: $x = x(u, v)$, $y = y(u, v)$, $z = z(u, v)$, $(u, v) \in A$, has a "representation" (T^*, A): $x = X(u, v)$, $y = Y(u, v)$, $z = Z(u, v)$, $(u, v) \in A$, for which the partial derivatives $X_u, \ldots, Z_v$ exist almost everywhere in A^0, and for which the area is given by the classical area integral. In other words, we ask whether another continuous mapping (T^*, A) exists which is Lebesgue, or Fréchet equivalent to (T, A), for which the plane mappings (T_r^*, A), $r = 1, 2, 3$, are absolutely continuous, and for which $X_u, \ldots, Z_v$ exist almost everywhere in A^0. The answer is affirmative when Fréchet equivalence is considered (Cesari).

On the other hand, it was proved from an abstract and more general viewpoint (Cesari; see [12]) that for any continuous Fréchet surface S of finite Lebesgue area and any representation (T, A) of S [that is, for any element (T, A) of the equivalence class defining S], there is a suitable measure function μ (defined in a suitable algebra $\mathcal{A}$ of subsets M of A) such that

$$L(T, A) = (A^0) \int d\mu,$$

$$E(T, A, f) = (A^0) \int f[p(w), \theta(w)] \, d\mu,$$

where $\theta(w)$ is a unit vector, $|\theta(w)| = 1$, defined μ almost everywhere in A^0 and representing a generalized normal vector to the surface S. In other words, every representation (T, A) of a surface S of finite Lebesgue area can be used for the computation of the area of S and of the surface integrals on S, provided we express these as Lebesgue-Stieltjes integrals in terms of a suitable measure function (area measure), and a suitable generalized normal vector function (Radon-Nikodym derivatives). As stated previously, we refer to [12] for this more general viewpoint, and we shall return in Section 10 to analytical and topological properties of continuous surfaces.

9. THE INTEGRAL $E(T, A, f)$ IN THE CALCULUS OF VARIATIONS

Under the sole conditions of continuity of T and finiteness of the area, the Weierstrass type integral $E(T, A, f)$ has been shown to have the expected properties for parametric problems of the calculus of variations. For instance in [9] and [10] Cesari proved necessary and sufficient conditions for the lower semicontinuity of $E(T, A, f)$. For a parametric integrand $f(p, t)$, continuous for $p \in E_3$, $t \in E_3$, we say that $E(T, A, f)$ is positive definite [semidefinite] if $f(p, t) > 0$ for all $p \in E_3$, $t \in E_3$, $t \neq 0$ [$f(p, t) \geq 0$]. If f is of class C^1 for $p \in E_3$, $t \in E_3$, $t \neq 0$, then $E(T, A, f)$ is said to be positive regular [semiregular] provided $W(p, t, \bar{t}) > 0$ for all $p \in E_3$, $t, \bar{t} \in E_3$, $t \neq \bar{t}$ [$W(p, t, t) \geq 0$], where W is the usual Weierstrass function $W(p, t, \bar{t}) = f(p, \bar{t}) - \sum_i \bar{t}_i f_{t_i}(p, t)$, $p \in E_3$, $t, \bar{t} \in E_3$. For instance, Cesari [9] proved that any integral E which is positive definite and positive semiregular [or positive semidefinite and positive regular] is lower semicontinuous in the Fréchet metric. On the basis of these results Cesari then applied the integral E to the study of the Plateau problem in E_3 for parametric continuous integrals. The Plateau problem for area had been studied by Douglas and Rado in the years 1929–30. Cesari proved the following in [11].

THEOREM 9.1. *Let M be a closed bounded convex set in E_3, and C a closed simple continuous curve in M. Let Ω be the family of all* *oriented Fréchet surfaces S in A as continuous maps from a disk, whose Lebesgue area is finite and whose boundary curve is C, and let us suppose that Ω is not empty. Then every positive definite and semiregular integral $E(T, A, f)$ has an absolute minimum in Ω.*

In the proof of this theorem one encounters minimizing sequences $[S_k]$ whose Lebesgue areas are equibounded, but this does not endow them with any compactness property. Thus a smoothing process had to be devised [11] so as to obtain a modified subsequence which is compact in the Fréchet metric. Extensions of Theorem 9.1 are known.

As mentioned before, Cesari [12] has extended the Weierstrass-type integral E to abstract situations where the area-related functions u, v are replaced by general quasi-additive set functions with values in E_n and under suitable assumptions. Further extensions for quasi-additive set functions with values in a Banach space have been considered by Warner [18] and later by Brandi and Salvadori [2] with applications to problems of the calculus of variations. It is relevant here to note that while most arguments in E_n depend on the existence of the usual inner product, the corresponding results in Banach spaces are based on martingale theory.

10. FINE-CYCLIC ELEMENTS

For simplicity, we shall suppose, as in Section 7, that A is a closed, finitely connected, Jordan region, say $A = J_0 - (J_1 + \cdots + J_\nu)^0$ (Section 1), where $0 \leqslant \nu < +\infty$ is the order of connectivity of A. Thus, for $\nu = 0$, A is called a disk; for $\nu = 1$, A is called an annular region. We have already mentioned in Section 1 the concepts of monotone mappings and of light mappings. We should now recall a precise characterization of the topological structure of any continuous mapping. For the purpose, let us mention here that if we have a mapping f: $y = f(x)$ from a "set" A onto a set B, and a mapping g: $z = g(y)$ from B into a space C, then the composition mapping, $F = gf$: $z = g[f(x)]$ from A into C is said to be the product of f and g (in this order) and denoted by

gf. Also, $F = gf$ is said to be a factorization of F into the two factors f and g. With this convention, we may state at once the factorization theorem of analytic topology, namely, that every continuous mapping, say (T, A): $p = p(w)$, $w \in A$, has a factorization $T = lm$ into two factors, a monotone mapping m, followed by a light mapping l (monotone-light factorization).

To understand this statement in the terms necessary here, let us consider for every point $p \in T(A)$ the inverse set $T^{-1}(p) \subset A$. Since A is compact and T is continuous, the set $T^{-1}(p)$ is compact; hence, its components g are continua $g \subset A$. By a "continuum" is meant, as usual, a bounded, closed, connected set, and hence even single points may be considered continua. Note that, if T is monotone (Section 1), then, for every $p \in T(A)$, the inverse set $g = T^{-1}(p) \subset A$ is just one continuum; if T is light (Section 1), then, for every $p \in T(A)$, all components g of $T^{-1}(p) \subset A$ are single points. In any case, for any continuous mapping (T, A), the collection $\Gamma = \{g\}$ of all components g of the set $T^{-1}(p)$, $p \in T(A)$, is a decomposition of A into disjoint continua $g \subset A$ (which may well be all single points of A). Γ is the collection of all maximal continua of constancy for T in A.

We may also consider the family $\mathscr{G}_0$ of all sets $G \subset A$ which are open in A and are the union of continua $g \in \Gamma$, say $G = \bigcup g$ (that is, have the property that $g \in \Gamma$, $gG \neq 0$ implies $g \subset G$).

We may now consider the elements g of Γ as "points," say $\tilde{g}$, and Γ as the "space," say $\tilde{\Gamma}$, the points of which are $\tilde{g}$. To consider $\tilde{\Gamma}$ as a space, we actually should define a topology on $\tilde{\Gamma}$, which turns out to be the equivalent of defining in $\tilde{\Gamma}$ the collection $\tilde{\mathscr{G}}$ of the open sets $\tilde{G}$. We can do so easily by considering each set $G = \bigcup g$, $G \in \mathscr{G}_0$, as the union $\tilde{G} = \bigcup \tilde{g}$ of the corresponding elements $\tilde{g}$. Then $\tilde{\mathscr{G}}$ is simply the family of all sets $\tilde{G}$.

By these natural definitions, $\tilde{\Gamma}$ can be proved to be not only a topological space, but also a "Peano space" (in particular, compact, connected, locally connected). Now, if m: $\tilde{g} = m(w)$, $w \in A$, is the mapping from A onto $\tilde{\Gamma}$, which maps each point $w \in g$, $g \in \Gamma$ into the point $\tilde{g} \in \tilde{\Gamma}$, then m is obviously monotone, because for each $\tilde{g} \in \tilde{\Gamma}$, the set $m^{-1}(\tilde{g}) = g$ is exactly the continuum $g \in \Gamma$, $g \subset A$. Finally, if l: $p = p(\tilde{g})$, $\tilde{g} \in \tilde{\Gamma}$, is the mapping from $\tilde{\Gamma}$ onto $T(A) \subset E_3$ defined by $l = Tm^{-1}$, then l is "light," because for each

$p \in T(A)$, the components g of the set $l^{-1}(p)$ are all single points $\tilde{g} \in \tilde{\Gamma}$. While we refer to the usual expositions for more formal proofs, we note that $T = lm$ is a monotone-light decomposition of T, that $m: A \to \tilde{\Gamma}$, $l: \tilde{\Gamma} \to T(A)$, and that $\tilde{\Gamma} = m(A)$ is the middle space, or hyperspace, of the decomposition [21].

Obviously any "space" M which is homeomorphic to $\tilde{\Gamma}$ may be considered a middle space, or hyperspace, for T, because, if h is a homeomorphism of $\tilde{\Gamma}$ onto M, and $m' = hm$, $l' = lh^{-1}$, then $T = l'm'$, $m': A \to M$, $l': M \to T(A)$, and m' is monotone and l' is light. Thus, M is the middle space and can be called a model of $\tilde{\Gamma}$. Also, for every monotone-light factorization $T = lm$ of T, $m: A \to M$, $l: M \to T(A)$, M is homeomorphic to $\tilde{\Gamma}$, that is, M is a model of $\tilde{\Gamma}$ [21].

If for some readers the previous considerations appear somewhat abstract, the following remark may be of help: A model M can be built in the Euclidean space E_3. If T is light, we may take for m the identity mapping, and $M = A$ is the middle space; if T is monotone, we may take for l the identity mapping, and $M = T(A)$, that is, the graph of (T, A) is the middle space M.

For example, in the monotone mapping T: $x = u$, $y = 0$, $z = 0$, $(u, v) \in A = (0 \leqslant u, v \leqslant 1)$, M is a segment (an arc); in the monotone mapping T: $x = \sin \pi r \cos \theta$, $y = \sin \pi r \sin \theta$, $z = \cos \pi r$, $(u, v) \in A = (u^2 + v^2 \leqslant 1)$, where $r \cos \theta = u$, $r \sin \theta = v$, $M = T(A)$ is the unit sphere in E_3; in the light mapping T: $x = u$, $y = v$, $z = z(u, v)$, $(u, v) \in A$, where $z(u, v)$ is a continuous function constant on no proper subcontinuum of A, the middle space is $M = A$.

Let (T, A): $p = p(w)$, $w \in A = (u^2 + v^2 \leqslant 1)$ be the monotone mapping from the disk A into E_3 defined by $x = (2r - 1)\cos \theta$, $y = (2r - 1)\sin \theta$, $z = 0$, if $\frac{1}{2} \leqslant r \leqslant 1$, and by $x = y = 0$, $z = 1 - 2r$, if $0 \leqslant r \leqslant \frac{1}{2}$, where $r \cos \theta = u$, $r \sin \theta = v$. Then Γ is the collection of all circles $u^2 + v^2 = r^2$ with $0 < r \leqslant \frac{1}{2}$, and of all single points $(u, v) \in A$, with $u = v = 0$, or $\frac{1}{4} < u^2 + v^2 \leqslant 1$. The middle space $M = T(A)$ is made up of the unit circle of the x-y plane and of the unit segment of the z-axis issuing from its center (a disk and a thread).

Let us consider now mappings (T, A): $p = T(w)$, $w \in A = (0 \leqq u, v \leqslant 1)$, from the unit square into E_3, where either Γ is the collection of all segments $g = (0 \leqslant v \leqslant 1, u = t)$, $0 \leqslant t \leqslant 1$ [surfaces $S_1 = $

(T, A)], or Γ is the collection of the boundaries $g = [\max(|2u - 1|,$ $|2v - 1|) = t]$, $0 \leqslant t \leqslant 1$, of all squares contained in A, concentric and similar to A [surfaces $S_2 = (T, A)$]. In either case, we may assume $T(w) = f(t)$, where $f(t)$ is a continuous function of t in $0 \leqslant t \leqslant 1$, constant on no subinterval of $(0, 1)$. If we suppose that $f(t)$ never takes twice the same value $f = (x, y, z)$, then T is monotone, and $M = T(A)$ is an arc PQ. The two types of surfaces, S_1 and S_2, apparently identical in E_3, are different. Surfaces S_1 can be thought of as limit cases of thin strips, and surfaces S_2, as limit cases of thin cones. The two surfaces S_1 and S_2, even defined by the same vector function f, are neither Lebesgue nor Fréchet equivalent.

Mappings (T, A) with $L(T, A) = 0$ can be characterized topologically, namely, M is a space of dimension less than, or equal to, 1 (T. Rado, 1945, for $\nu = 0$; R. F. Williams, 1958, for $\nu \geqslant 0$). In the case where A is a disk ($\nu = 0$), M has been further characterized (as a dendrite of analytic topology). Simply stated, we may expect M to be a ramified system of threads.

Mappings (T, A): $p = T(w)$, $w \in A$, with $L(T, A) > 0$, must therefore possess a middle space M with some two-dimensional parts (see, for example, the preceding example where M consists of a disk and a thread). These parts are important and, where A is a disk ($\nu = 0$), they are the cyclic elements Σ of M. A subset Σ of M is said to be a cyclic element of M if Σ is a proper continuum and is not disconnected by suppressing any one of the points of Σ. Any two cyclic elements Σ of M are not overlapping (they may have in common at most one point), and the collection $\{\Sigma\}$ of all cyclic elements of M is at most countable. Thus, in the preceding example, the only cyclic element Σ of M is the disk. Again, for $\nu = 0$, the cyclic elements Σ of M have been fully characterized: each is either a disk or a sphere [21].

For $\nu > 0$, a further decomposition of M may be necessary. The parts σ of M of dimension 2 may actually be finer than the cyclic elements Σ of M, and they are denoted as the fine-cyclic elements σ of M. A subset σ of M is said to be a fine-cyclic element of M if σ is a proper continuum and is not disconnected by suppressing any finite system of points of σ [13], [16]. Any two fine-cyclic elements σ of M are not overlapping (they may have in common at most finitely many points), and the collection $\{\sigma\}$ of all fine-cyclic elements σ of M is at most countable. For example, suppose that $\nu = 1$, A is the annular region $1 \le u^2 + v^2 \le 4$, and (T, A) is the monotone mapping defined by $x = \varphi(u)$, $y = v$, $z = 0$, where $\varphi(u) = u$ for $-1 \le u \le 1$, $= 1$ for $u \ge 1$, $= -1$ for $u \le -1$. Then, the surface $S = (T, A)$ is made up of the two Jordan regions, $R_1 = [1 \le x^2 + y^2 \le 4$, $-1 \le x \le 1$, $y \ge 0]$ and $R_2 = [1 \le x^2 + y^2 \le 4$, $-1 \le x \le 1$, $y \le 0]$, having in common the two points $(-1, 0)$ and $(1, 0)$. We may take $M = T(A) = R_1 + R_2$, and M presents two fine-cyclic elements, R_1 and R_2, but only one cyclic element, M itself.

11. AREA OF DISCONTINUOUS NONPARAMETRIC SURFACES AND RELATED INTEGRALS OF THE CALCULUS OF VARIATIONS

To simplify the exposition let us consider here nonparametric nonnecessarily continuous surfaces S: $z = g(x, y)$, $(x, y) \in G$, where g is a real-valued function of class $L_1(G)$ and G is any bounded open subset of E_2. To define Lebesgue area in the same spirit as (2.3), we consider here the class γ of all sequences $[(P_n, F_n), n = 1, 2, \ldots]$ "convergent" to g. That is, F_n is a sequence of figures $F_n \subset G$, $F_n \subset F_{n+1}$, invading G, or $\lim F_n^0 = G$, P_n is quasilinear on F_n, and $\int_{F_n} |P_n(x, y) - g(x, y)| \, dx \, dy \to 0$ as $n \to \infty$. Then, in analogy with (2.3), for (generalized) Lebesgue area of S: $z = g(x, y)$, $(x, y) \in G$, we take

$$L(S) = \inf_{\gamma} \liminf_{n \to \infty} a(P_n, F_n), \qquad 0 \le L(S) \le \infty.$$

To obtain a characterization of the surfaces S as above whose generalized Lebesgue area is finite, Cesari [22] in 1936 proposed a definition of functions g of bounded variation (BV) which we state here, for the sake of simplicity, only for the case G is an interval, $G = [a < x < b, c < y < d]$. Then g is said to be BV in G provided g is of class $L_1(G)$ and there is a set Z of measure zero in G such that the total variations $V_x(y)$ of $g(\cdot, y)$ in (a, b) and $V_y(x)$ of $g(x, \cdot)$ in (c, d) are measurable and L_1-integrable in $[c, d]$ and [*a, b*], respectively—where such total variations are computed completely disregarding the values of *g* on *Z*. The number

$$V_0 = V_0(g, G) = \int_a^b V_y(x)\, dx + \int_c^d V_x(y)\, dy$$

may well be taken as a definition of total variation of *g* in $G = (a, b; c, d)$. Analogous definitions hold for BV functions $g(t_1, \ldots, t_\nu)$ in an interval *G* of E_ν. We omit the more involved definition of a BV function in a general bounded open subset of E_ν.

If *g* is continuous in *G*, then no set *Z* need be considered and the concept reduces to Tonelli's concept of BV continuous functions. For discontinuous functions *g*, examples show how essential it is to disregard sets *Z* of measure zero. On the other hand, the concept obviously concerns equivalent classes in $L_1(G)$.

Cesari [22] proved that for a surface *S*: $z = g(x, y)$, $(x, y) \in G$, *g* of class $L_1(G)$, then $L(S)$ is finite if and only if *g* is BV.

To begin with, this result shows that the concept of BV functions is independent of the direction of the axes in E_ν. More than that, the concept of BV functions is actually invariant with respect to 1-1 continuous transformations in E_ν which are Lipschitzian in both directions.

In 1937 Cesari [23] proved that for $\nu = 2$, $G = (0, 2\pi; 0, 2\pi)$ and *g* BV in *G*, the double Fourier series of *g* converges to *g* (by rectangles, by lines, by columns) almost everywhere (a.e.) in *G*. Comparable, though weaker, results hold for BV functions of $\nu > 2$ independent variables and their multiple Fourier series [24].

In 1950 Cafiero and later in 1957 Fleming proved the relevant compactness theorem: any sequence $[g_k]$ of BV functions with equibounded total variations, say $V_0(g_k, G) \leqslant C$, and equibounded mean values in *G*, possesses a subsequence $[g_{k_s}]$ which is pointwise convergent a.e. in *G*, as well as strongly convergent in $L_1(G)$, toward a BV function *g*.

In 1967 Conway and Smoller [28] used these BV functions in connection with the weak solutions (shock waves) of conservation laws, a class of nonlinear hyperbolic partial differential equations in $E^+ \times E_\nu$. Indeed they proved that, if the Cauchy data on $(0) \times E_\nu$

are locally BV, then there is a unique solution on $E^+ \times E_\nu$ also locally BV and satisfying an entropy condition. Without any entropy condition there are in general infinitely many weak solutions. Later Dafermos [29] and Di Perna [30] characterized the properties of the BV weak solutions of conservation laws.

Meanwhile, in the fifties, distribution theory became known, and in 1957 Krickeberg [32] proved that the BV functions are exactly those $L_1(G)$ functions whose first order partial derivatives in the sense of distributions are finite measures in G.

Thus, a BV function $g(t)$, $t = (t_1, \ldots, t_\nu) \in G \subset E_\nu$, G a bounded open subset of E_ν, possesses first order partial derivatives in the sense of distributions, which are finite measures μ_j, $j = 1, \ldots, \nu$. On the other hand, if we think of the initial definition of BV functions, we see that g has also "usual" first order partial derivatives $D^j g$, $j = 1, \ldots, \nu$, a.e. in G, computed by usual incremental quotients disregarding the values taken by g on Z. These $D^j g$ are $L_1(G)$ integrable functions on G.

Much work followed on BV functions in terms of the new definition, that is, thought of as $L_1(G)$ functions whose first order partial derivatives are finite measures. We mention here Fleming [31], Volpert [34], Gagliardo, Anzellotti and Giaquinta, De Giorgi, Da Prato, Giusti, M. Miranda, Ferro, Caligaris, Oliva, Fusco, and Temam.

Now let us consider problems of the calculus of variations concerning the minimum of simple integrals

$$I(z) = \int_{t_1}^{t_2} f_0(t, z(t), z'(t)) \, dt$$

in a class Ω of BV functions $z(t) = (z_1, \ldots, z_n)$, $t_1 \le t \le t_2$, that is, each component z_i of z is BV. We assume here that the elements of the class satisfy the constraints

$$z(t_1) = z_1, \qquad z(t_2) = z_2, \qquad (t, z(t)) \in A, \qquad t \in [t_1, t_2],$$

$$z'(t) \in Q(t, z(t)), \qquad t \in [t_1, t_2] \quad \text{(a.e.)}.$$

Here t_1, t_2 are fixed, z_1, z_2 are fixed points in E_n, A is a given set in E_{n+1} whose projection on the t-axis contains $[t_1, t_2]$, and, for any $(t, z) \in A$, $Q(t, z)$ is a given set in E_n. The Lebesgue integral $I(z)$, however, does not yield stable and realistic values for I, and, therefore, a corresponding Serrin-type integral $\Im(z)$ [33] was proposed in [25], whose definition is in the same spirit of Lebesgue area (2.3). Namely, for every $z \in \Omega$ let $\gamma(z)$ denote the class of all sequences z_k of elements $z_k \in \Omega$ absolutely continuous (AC) in $[t_1, t_2]$ and such that $z_k \to z$ pointwise a.e. in $[t_1, t_2]$ as $k \to \infty$. If $\gamma(z)$ is empty, we take $\Im(z) = +\infty$; if $\gamma(z)$ is not empty we take

$$\Im(z) = \underset{\gamma(z)}{\mathrm{Inf}} \, \underset{k \to \infty}{\liminf} I(z_k).$$

Let M denote the set $M = [(t, z, \xi)|(t, z) \in A, \xi \in Q(t, z)] \subset E_{1+2n}$, and for any $(t, z) \in A$ let $\tilde{Q}(t, z)$ denote the "augmented" sets

$$\tilde{Q}(t, z) = [(\tau, \xi)|\tau \geqslant f_0(t, z, \xi), \xi \in Q(t, z)] \subset E_{1+n}.$$

The sets $\tilde{Q}(t, z)$ are said to satisfy property (Q) at $(t_0, z_0) \in A$ provided

$$\tilde{Q}(t_0, z_0) = \bigcap_\epsilon \mathrm{cl\,co}\Big[\bigcup \tilde{Q}(t, z), |t - t_0| + |z - z_0| \leqslant \epsilon\Big].$$

The following assumptions are relevant: (i) the sets A and M are closed; (ii) the sets $\tilde{Q}(t, z)$ satisfy property (Q) at every point $(t, z) \in A$; for every $z \in \Omega$ there is a constant C such that $V(z_k) \leqslant C$ for the elements z_k of $\gamma(z)$. Under these assumptions Cesari, Brandi, and Salvadori proved in [26] that $I(z) \leqslant \Im(z) \leqslant \liminf I(z_k)$.

The same authors also proved in [26] the following existence theorem: Under the same assumptions, if A is compact, if the class Ω is closed, and $\gamma(z)$ is nonempty for at least one $z \in \Omega$, then the integral $\Im(z)$ has an absolute minimum in Ω.

Analogous considerations hold for multiple integrals of the calculus of variations. Let $\nu \geqslant 1$, $n \geqslant 1$, and let $G \subset E_\nu$ be a bounded domain in the t-space E_ν, $t = (t_1, \ldots, t_\nu)$, possessing the cone property at every point of its boundary ∂G. Let $A \subset E_{\nu+n}$ be a compact subset of the tz-space $E_{\nu+n}$ whose projection on the t-space contains G. We deal here with a class Ω of vector valued functions $z(t) = (z_1, \ldots, z_n)$, $z_i \in BV$ in G, therefore possessing first order partial derivatives in the sense of distributions which are measures μ_{ij} and also generalized first order derivatives $D^j z^i$ a.e. in G which are functions of class $L_1(G)$, $j = 1, \ldots, \nu$, $i = 1, \ldots, n$. Let Dz denote the N-vector of these generalized derivatives, $N = n\nu$. For every $(t, z) \in A$ let $Q(t, z)$ be a given subset of E_N, let $M \subset E_{\nu+N+n}$ denote the set $M = [(t, z, \xi)|(t, z) \in A, \ \xi \in Q(t, z)]$, and let $f_0(t, z, \xi)$ be a given real valued function in M. We are interested in the minimum of multiple integrals with constraints

$$I(z) = \int_G f_0(t, z(t), Dz(t))\, dt, \qquad dt = dt_1 \ldots dt_\nu.$$

$$(t, z(t)) \in A, \qquad Dz(t) \in Q(t, z(t)), \qquad t \in G \ (\text{a.e.}).$$

We assume that all elements z of the class Ω satisfy these conditions, and for every $z \in \Omega$ we denote by $\gamma(z)$ the class of all sequences $[z_k]$ of functions $z_k \in (W^{1,1}(G))^n$ belonging to Ω, hence satisfying the same constraints, and such that $z_k \to z$ strongly in $(L_1(G))^n$. Again we introduce a Serrin type integral $\mathfrak{I}(z)$ by taking $\mathfrak{I}(z) = +\infty$ if $\gamma(z)$ is empty, and otherwise we take

$$\mathfrak{I}(z) = \operatorname*{Inf}_{\gamma(z)} \operatorname*{liminf}_{k \to \infty} I(z_k).$$

Again, augmented sets $\tilde{Q}(t, z)$ can be introduced for which property (Q) is requested at every point $(t, z) \in A$. Then, Cesari, Brandi, and Salvadori proved in [27] an existence theorem for multiple integrals similar to the one for simple integrals and under only slightly more restrictive conditions.

In a recent development the abstract approach to the Weierstrass integral E mentioned at the end of Section 9 (Cesari, Warner, Brandi, Salvadori) has been further extended (see Brandi and Salvadori [35]) so as to include BV varieties, both parametric and nonparametric. In particular, semicontinuity theorems in suitable topologies have been obtained, and representation theorems of the Weierstrass integral by Lebesgue-Stieltjes integrals with Radon-Nikodym derivatives as generalized Jacobians. At this level of generality the comparison between Serrin-type and Weierstrass-type integrals is being investigated.

On the general topic of this chapter the reader may consult the books [3ab] and [17] and the articles [4]–[7], as well as the other books and papers listed in the bibliography.

BIBLIOGRAPHY

CONTINUOUS PARAMETRIC AND NONPARAMETRIC SURFACES

1. Alexandroff, P., and H. Hopf, *Topologie*, Berlin, 1935.
2. Brandi, P. and A. Salvadori, (a) The nonparametric integral of the calculus of variations as a Weierstrass integral. I. Existence and representation. II. Some applications, *J. Math. Anal. Appl.*, 107 (1985), 67–95; 112(185), 290–313. (b) Sull'area generalizzata, *Atti Sem. Mat.-Fis. Univ. Modena*, 28 (1979), pp. 33–62. (c) On convergence in area in the generalized sense, *J. Math. Anal. Appl.*, 92 (1983), 119–138. (d) Sull'integrale debole alla Burkill-Cesari, *Atti Sem. Mat.-Fis. Univ. Modena*, 27 (1978), 14–38. (e) Sull'estensione dell'integrale debole alla Burkill-Cesari ad una misura, *Rend. Circ. Mat. Palermo*, 30 (1981), 207–234. (f) Martingale e integrale alla Burkill-Cesari, *Rend. Acad. Naz. Lincei*, 67 (1979), 197–203.
3. Cesari, L., (a) *Surface Area*. Princeton University Press, Princeton, N.J., 1956 (quoted in article as **SA**). (b) *Optimization–Theory and Applications*, Springer Verlag 1983.
4. ____, Variation, multiplicity, and semicontinuità, *American Mathematical Monthly*, 65 (1958), 317–332.
5. ____, Rectifiable curves and the Weierstrass integral, *American Mathematical Monthly*, 65 (1958), 485–500.
6. ____, Recent results in surface area theory, *American Mathematical Monthly*, 66 (1959), 173–193.
7. ____, Retraction, homotopy, integral. Invited address at the *International Congress of Mathematics*, Amsterdam, 1954, 3, 77–84.
8. ____, La nozione di integrale sopra una superficie in forma parametrica, *Annali Scuola Norm. Sup. Pisa*, (2) 13 (1944), 78–117.
9. ____, Condizioni sufficienti per la semicontinuità degli integrali sopra una superficie in forma parametrica, *Annali Scuola Normale Sup. Pisa*, (2) 14 (1945), 47–79.
10. ____, Condizioni necessarie per la semicontinuità degli integrali sopra una superficie in forma parametrica, *Annali Matematica pura appl.*, (4) 29 (1950), 199–224.

11. ____, An existence theorem of the calculus of variations for integrals on parametric surfaces, *Amer. J.. Math.*, 74 (1952), 265–295.

12. ____, (a) Quasi additive set functions and the concept of integral over a variety, *Transactions of the American Mathematical Society*, 102 (1962), 94–113. (b) Extension problem for quasi additive set functions and Radon-Nikodym derivatives, *Transactions of the American Mathematical Society*, 102 (1962), 114–46.

13. ____, Fine-cyclic elements of surfaces of type v, *Rivista Mat. Univ. Parma*, 7 (1956), 149–85.

14. Cesari, L. and L. H. Turner, Surface integral and Radon-Nikodym derivatives, *Rend. Cir. Mat. Palermo*, (2) 7 (1958), 143–154.

15. Hobson, E. W., *The Theory of Functions of a Real Variable*, vol. I. Cambridge University Press, Cambridge, 1927.

16. Neugebauer, C. J., *B*-sets and fine-cyclic elements, *Transactions of the American Mathematical Society*, 88 (1958), 121–36.

17. Rado, T., *Length and Area*, American Mathematical Society Colloquium Publication, Providence, vol. 30, 1948.

18. Warner, G. W., (a) The Burkill-Cesari integral, *Duke Math. J.*, 35 (1968), 61–78. (b) The generalized Weierstrass type integral, *Annali Scuola Norm. Sup. Pisa*, (3) 22 (1968), 163–192.

19. Williams, R. F., Lebesgue area zero, dimension and fine-cyclic elements, *Rivista Mat. Univ. Parma*, 10 (1959), 131–43.

20. Williams, R. F., Lebesgue area of maps from Hausdorff spaces, *Acta Mathematica*, 102 (1959), 33–46.

21. Whyburn, G. T., *Analytic Topology*, American Mathematical Society Colloquium Publication, Providence, vol. 28, 1942.

DISCONTINUOUS NONPARAMETRIC SURFACES

22. Cesari, L., Sulle funzioni a variazione limitata, *Annali Scuola Norm. Sup. Pisa*, (2) 5 (1936), 299–313.

23. ____, Sulle funzioni di due variabili a variazione limitata e sulla convergenza delle relative serie doppie di Fourier, *Rend. Sem. Mat. Univ. Roma* (4) 1 (1937), 277–294.

24. ____, Sulle funzioni di più variabili a variazione limitata e sulla convergenza delle relative serie multiple di Fourier, *Pont. Acad. Sci.*, Commentationes 3 (1939), 171–197.

25. Cesari, L., P. Brandi, and A. Salvadori, Discontinuous solutions in problems of optimization, a lecture read by Cesari at the Conference in honor of L. Tonelli, *Scuola Norm. Sup. Pisa*, 1986.

26. ____, Existence theorems concerning simple integrals of the calculus of variations for discontinuous solutions, *Archive Rat. Mech. Anal.*, 98 (1987), 307–328.

27. ____, Existence theorems for multiple integrals of the calculus of variations for discontinuous solutions, *Annali di Mat. pura appl.*, to appear.

28. Conway, E and J. A. Smoller, Global solutions of the Cauchy problem for quasi linear first order equations in several space variables, *Comm. Pure Appl. Math.*, 19 (1966), 95–105.

29. Dafermos, C. M., (a) Characteristics in hyperbolic conservation laws. A study of the structure and the asymptotic behavior of solutions, *Nonlinear Analysis and Mechanics*, Heriot Watt Symposium, Pitman 1 (1979), pp. 1–58. (b) Generalized characteristics and the structure of solutions of hyperbolic conservation laws, *Indiana Univ. Math. J.*, 26 (1977), 1097–1119.

30. Di Perna, R. J., (a) Singularities of solutions of nonlinear hyperbolic systems of conservation laws, *Arch. Rat. Mech. Anal.*, 60 (1975), 75–100. (b) The structure of solutions of hyperbolic conservation laws, *Nonlinear Analysis and Mechanics*, Heriot Watt Symposium, Pitman 4 (1977), pp. 1–16. (c) Convergence of approximate solutions to conservation laws, *Arch. Rat. Mech. Anal.*, 82 (1983), 27–70.

31. Fleming, W. H., Functions with generalized gradient and generalized surfaces, *Annali Matematica pura appl.*, 44 (1957), 93–103.

32. Krickeberg, K., Distributionen, Funktionen Beschränkter Variation, und Lebesguescher Inhalt nichtparametrischer Flächen, *Annali di Matematica pura appl.*, (4) 44 (1957), p. 92 and pp. 105–133.

33. Serrin, J., (a) On a fundamental theorem of the calculus of variations, *Acta Math.*, 102 (1959), 1–22. (b) A new definition of the integral for nonparametric problems in the calculus of variations, *Acta Math.*, 102 (1959), 23–32. (c) On certain definitions and properties of variational integrals, *Trans. Amer. Math. Soc.*, 101 (1961), 139–167.

34. Volpert, A. I., The space BV and quasi linear equations, *Math. USSR Sb.* 2 (1967), 225–267.

35. Brandi, P. and A. Salvadori, (a) A quasi-additivity type condition and the integral over a BV variety, to appear. (b) On the nonparametric integral over a BV surface, *Nonlinear Analysis*, to appear. (c) On the lower semicontinuity of certain integrals of the calculus of variations, *J. Math. Anal. Appl.*, to appear.

INTEGRAL GEOMETRY

L. A. Santalo

1. INTRODUCTION

We shall begin with three simple examples which will show the basic ideas on which integral geometry has been developed.

1.1. *Sets of points.* Let X be a set of points in the euclidean plane E_2. The measure (ordinary area) of X is defined by the integral

$$(1.1) \qquad m(X) = \int_X dx\, dy.$$

Let $\mathfrak{M}$ be the group of motions in E_2. With respect to an orthogonal Cartesian system of coordinates, the equations of a motion $u \in \mathfrak{M}$ are

$$(1.2) \qquad \begin{aligned} x' &= x \cos \varphi - y \sin \varphi + a \\ y' &= x \sin \varphi + y \cos \varphi + b. \end{aligned}$$

The fundamental property of the measure (1.1) is that of being

303

invariant under $\mathfrak{M}$. That is, if $X' = uX$ is the transform of X by u, we have

(1.3) $m(X') = \int_{X'} dx'\, dy' = \int_X dx\, dy = m(X)$

as follows immediately from (1.2). It is well known that this property characterizes the measure (1.1) up to a constant factor.

Because we are generally interested only in the differential form under the integral sign in (1.1), we shall write $dP = dx\, dy$, or, more precisely,

(1.4) $dP = dx \wedge dy$

to indicate that the differential form under a multiple integral sign is an exterior differential form [see, for example, Munroe (43)].

The exterior differential form (1.4) is called the density for points in E_2 with respect to $\mathfrak{M}$. We shall always take the densities in absolute value.

1.2. *Sets of lines.* Let X now be a set of lines in E_2—for example, the set of all lines G which intersect a given convex domain K. We ask for a measure of X invariant under $\mathfrak{M}$.

Let p be the distance from the origin O to G and θ the angle formed by the perpendicular to G through O and the x-axis. We maintain that this invariant measure is given by

(1.5) $m(X) = \int_X dp\, d\theta.$

For a proof, we observe that by the motion u [Relation (1.2)] the line coordinates p, θ transform according to

(1.6) $\theta' = \theta + \varphi, \qquad p' = p + a \cos\,(\theta + \varphi) + b \sin\,(\theta + \varphi)$

and putting $X' = uX$, we have

$$m(X') = \int_{X'} dp'\, d\theta' = \int_X dp\, d\theta = m(X)$$

which proves the invariance of $m(X)$. That this measure is unique, up to a constant factor, follows from the transitivity of the lines under $\mathfrak{M}$, since if $\int_X f(p, \theta)\, dp\, d\theta$ is invariant we must have $\int_{X'} f(p', \theta')\, dp'\, d\theta' = \int_X f(p, \theta)\, dp\, d\theta$, and, on the other hand, according to(1.6), $\int_{X'} f(p', \theta') \, dp' \, d\theta' = \int_X f(p', \theta') \, dp \, d\theta$. From the last two equalities, we obtain $\int_X f(p', \theta') \, dp \, d\theta = \int_X f(p, \theta) \, dp \, d\theta$. If this equality holds for any set X it must be true that $f(p', \theta') = f(p, \theta)$, and, since any line $G(p, \theta)$ can be transformed into any other $G(p', \theta')$ by a motion, we deduce $f(p, \theta) = $ constant.

The differential form

$$(1.7) \qquad dG = dp \wedge d\theta,$$

taken in absolute value, is called the density for lines in E_2 with respect to $\mathfrak{M}$.

Let us consider a simple application. To get the measure of the set of lines which cut a fixed segment S of length l, because of the invariance under $\mathfrak{M}$ we may take the origin of coordinates coincident with the middle point of S and the x-axis coincident with the direction of S; then we have

$$(1.8) \quad m(G; G \cap S \neq 0) = \int_{G \cap S \neq 0} dp \, d\theta = \int_0^{2\pi} \left| \frac{l}{2} \cos \theta \right| d\theta = 2l.$$

If instead of S we consider a polygonal line Γ composed of a finite number of segments S_i of lengths l_i, writing (1.8) for each S_i and summing we get

$$(1.9) \qquad \int n \, dG = 2L$$

where $n = n(G)$ is the number of points in which $G(p, \theta)$ cuts Γ and L is the length of Γ. The integral in (1.9) is extended over all lines of the plane, n being 0 if $G \cap \Gamma = 0$. By a limit process it is not difficult to prove that (1.9) holds for any rectifiable curve [Blaschke (3)].

Conversely, given a continuum of points Γ in the plane, if the integral on the left of (1.9) has a meaning, then it can be taken as a definition for the length of Γ, which is the so-called Favard length [Nöbeling (45)].

For a convex curve K we have $n = 2$ for all G which intersect K, except for the positions in which G is a supporting line of K, which are of zero measure. Consequently we have: The measure of the set of lines which intersect a convex curve is equal to its length.

1.3. *Kinematic density.* Let us now consider a set X of oriented congruent segments S of length l—for example, the set of those which intersect a fixed convex domain. The position of S in E_2 is determined by the coordinates of its origin $P(x, y)$ and the angle α formed by S and the x-axis. If we want to define a measure for X invariant under $\mathfrak{M}$, we must take

$$(1.10) \qquad m(X) = \int_X dx \, dy \, d\alpha.$$

To see this, we first observe that by a motion (1.2) the variables (x, y, α) transform according to (1.2) and $\alpha' = \alpha + \varphi$. Consequently the Jacobian of the transformation is 1, and we have

$$m(X') = \int_{X'} dx' \, dy' \, d\alpha' = \int_X dx \, dy \, d\alpha = m(X)$$

where $X' = uX$, which proves the invariance of $m(X)$. The uniqueness, up to a constant factor, follows from the transitivity of $\mathfrak{M}$ with respect to the congruent segments of the plane by the same argument previously given for the lines.

If instead of segments we want to measure sets of congruent figures K, since the position of such a figure is determined by the position of a point $P(x, y)$ rigidly bound to K and the angle α between a fixed direction PA in K and the x-axis, we can take the same integral (1.10). The differential form

$$(1.11) \qquad dK = dx \wedge dy \wedge d\alpha$$

is called the kinematic density for E_2 with respect to the group $\mathfrak{M}$. It is always taken in absolute value.

Another form for dK is obtained if instead of the coordinates (x, y, α) for the oriented segment S, we take the coordinates (p, θ) of the line G which contains S and the distance $t = HP$ from P to the foot H of the perpendicular drawn from the origin 0 to G. The transformation formulas are

$$(1.12) \qquad x = p \cos \theta + t \sin \theta, \quad y = p \sin \theta - t \cos \theta, \quad \alpha = \theta - \frac{\pi}{2}$$

and consequently, up to the sign, we have $dx \wedge dy \wedge d\alpha = dp \wedge d\theta \wedge dt$. We may then write

(1.13) $$dK = \overrightarrow{dG} \wedge dt$$

where we write $\overrightarrow{G}$ in order to indicate that G must be considered as oriented ($\overrightarrow{dG} = 2\,dG$).

From this expression for dK we easily deduce the measure of the set of segments of length l which intersect a given convex domain K of area F and perimeter L. In fact, calling λ the length of the chord determined by G on K, we have

$$m(S; S \cap K \neq 0) = 2 \int dp\, d\theta\, dt = 2 \int_{S \cap K \neq 0} (\lambda + l)\, dp\, d\theta$$
$$= 2\pi F + 2lL,$$

This formula can be generalized to surfaces [see (55)]; an application was given by Green (22).

If we ask for the measure of the set of segments S which are contained in K, the result is not simple; it depends largely on K. For instance, for a circle C of diameter $D \geq l$, we have

$$m(S; S \subset C) = \frac{\pi}{2}\left(\pi D^2 - 2D^2 \text{ arc sin}\frac{l}{D} - 2l\sqrt{D^2 - l^2}\right)$$

and for a rectangle R of sides a, b ($a \geq l$, $b \geq l$), we have

$$m(S; S \subset R) = 2(\pi ab - 2(a + b)l + l^2).$$

An unsolved problem is that of finding among all convex domains K with a given perimeter those which maximize the measure $m(S; S \subset K)$ of the segments of a given length which are contained in K. For $l = 0$ the problem is the classical isoperimetric problem and the solution is well-known to be the circle.

The preceding very simple examples show the three steps which constitute the so-called integral geometry in the original sense of Blaschke (3): (1) definition of a measure for sets of geometric objects with certain properties of invariance; (2) evaluation of this measure for some particular sets; and (3) application of the obtained result to get some statements of geometrical interest.

The same examples show the basic elements which are necessary to build the integral geometry from a general point of view: (1) a base space E in which the objects we consider are imbedded (in the preceding examples, E was the euclidean plane E_2); (2) a group of transformations $\mathfrak{G}$ operating on E (in the preceding examples $\mathfrak{G}$ was $\mathfrak{M}$; (3) geometric objects F contained in E which transform transitively by $\mathfrak{G}$ (in the preceding examples, the geometric objects were points, lines or congruent figures).

Given E, $\mathfrak{G}$, and F, the first problem of the integral geometry is to find a measure for sets of F invariant under $\mathfrak{G}$.

2. GENERAL INTEGRAL GEOMETRY

2.1. *Density and measure for groups of matrices.* Though the integral geometry deals with general Lie groups, from the geometrical point of view in which we are principally interested it suffices to consider Lie groups which admit a faithful representation, that is, which are isomorphic to a matrix group. We need some facts about groups of matrices, which we shall compile in this section. For a more general treatment, see Chevalley (12).

Let $\mathfrak{G}$ be a group of $n \times n$ matrices of dimension r, that is, each matrix $u \in \mathfrak{G}$ depends on r independent parameters a_1, a_2, $\cdots$, a_r; more precisely, each matrix $u \in \mathfrak{G}$ is determined by a point $a = (a_1, a_2, \cdots, a_r)$ of a differentiable manifold of dimension r, which we shall denote by the same letter $\mathfrak{G}$; $a_1, a_2, \cdots, a_r$ are then the coordinates of a in a suitable local coordinate system.

Let $e \in \mathfrak{G}$ be the unit matrix and u^{-1} the inverse of $u \in \mathfrak{G}$. If du denotes the differential of the matrix u, the equation

$$(2.1) \qquad\qquad u^{-1}(u + du) = e + \omega$$

defines a matrix $\omega = u^{-1}\,du$ of linear (pfaffian) differential forms which is called the matrix of Maurer-Cartan of $\mathfrak{G}$. The elements ω_{ij} of ω have the form $\omega_{ij} = \alpha_{ij1}\,da_1 + \cdots + \alpha_{ijr}\,da_r$, where the coefficients α_{ijk} are analytic functions of $a_1, a_2, \cdots, a_r$. From these n^2 pfaffian forms ω_{ij} there are r linearly independent (base of the vector space dual of the tangent space of $\mathfrak{G}$) which we shall denote by $\omega_1, \omega_2, \cdots, \omega_r$; they are called the forms of Maurer-Cartan of $\mathfrak{G}$ and are defined up to a linear combination with constant coefficients.

The fundamental property of the matrix ω is that of being left invariant under $\mathfrak{G}$. For if $u' = su$ (s is a fixed element of $\mathfrak{G}$), we have $du' = s\,du$, and therefore $\omega' = u'^{-1}\,du' = u^{-1}s^{-1}s\,du = u^{-1}\,du = \omega$.

As a consequence, the r forms of Maurer-Cartan are also left invariant under $\mathfrak{G}$, and this fact characterizes these forms up to a linear combination with constant coefficients. For a proof, we observe that since the forms of Maurer-Cartan $\omega_1, \cdots, \omega_r$ are independent, each pfaffian form Ω may be written $\Omega(a, da) = \sum_1^r A_i(a)\omega_i$. If Ω is left invariant under $\mathfrak{G}$, we have

$$\Omega' = \sum_1^r A_i(a')\omega_i' = \sum_1^r A_i(a)\omega_i$$

and since $\omega_i' = \omega_i$, we have

$$\sum_1^r (A_i(a') - A_i(a))\omega_i = 0.$$

Because of the independence of ω_i, it follows that $A_i(a') = A_i(a)$, which implies $A_i = $ constant. (Since we are interested only in the left invariance, we shall hereafter speak simply of invariance, understanding that it means left invariance.)

Notice that by exterior differentiation of $\omega = u^{-1}\,du$, taking into account that $du^{-1} = -u^{-1}\,du\,u^{-1}$, we get

$$(2.2) \qquad d\omega = -u^{-1}\,du\,u^{-1} \wedge du = -\omega \wedge \omega.$$

This matric equation includes the expression of the exterior differentials $d\omega_i$ of the forms of Maurer-Cartan as linear combinations with constant coefficients of the products $\omega_j \wedge \omega_k$; these expressions are called the equations of structure of Maurer-Cartan for the group $\mathfrak{G}$.

2.2. *Density and measure in homogeneous spaces.* Let $\mathfrak{H}$ be a subgroup of $\mathfrak{G}$ of dimension $r - h$. Suppose that $\mathfrak{H}$ itself is a Lie group isomorphic to a matrix group. We want to find the conditions for the existence of a density (that is, an element of volume) in the homogeneous space $\mathfrak{G}/\mathfrak{H}$ ($=$ set of left cosets $s\mathfrak{H}$, $s \in \mathfrak{G}$) invariant under $\mathfrak{G}$. For this purpose, we notice that the submanifold $\mathfrak{H}$ of the differentiable manifold $\mathfrak{G}$ and its left cosets $s\mathfrak{H}$ $(s \in \mathfrak{G})$ are the integral manifolds of a pfaffian system.

$$(2.3) \qquad \omega_1 = 0, \qquad \omega_2 = 0, \qquad \cdots, \qquad \omega_h = 0.$$

Because $\mathfrak{H}$ and its left cosets as a whole are invariant under $\mathfrak{G}$, the left side members of (2.3) will be linear combinations with constant coefficients of the forms of Maurer-Cartan of $\mathfrak{G}$, and, because these forms are defined up to a linear combination with constant coefficients, we may assume that they are the h first forms of Maurer-Cartan of $\mathfrak{G}$.

Because ω_i is invariant under $\mathfrak{G}$, the differential form

$$(2.4) \qquad \Omega_h = \omega_1 \wedge \omega_2 \wedge \cdots \wedge \omega_h$$

will be also invariant under $\mathfrak{G}$. However, Ω_h is not always a density for $\mathfrak{G}/\mathfrak{H}$ because its value can change when the points $a \in \mathfrak{G}$ displace on the manifolds $s\mathfrak{H}$. We shall now prove the following theorem.

THEOREM: *A necessary and sufficient condition for* Ω_h *to be a density for* $\mathfrak{G}/\mathfrak{H}$ *is that its exterior differential vanish, that is,*

$$(2.5) \qquad d\Omega_h = 0.$$

Proof: To prove this theorem, we observe that the submanifold $\mathfrak{H}$ and its left cosets fill up the manifold $\mathfrak{G}$ in such a way that for each point of $\mathfrak{G}$ passes one and only one submanifold. Thus, the system (2.3) is completely integrable and it is consequently equivalent to a system of the form

$$(2.6) \qquad d\xi_1 = 0, \qquad d\xi_2 = 0, \qquad \cdots, \qquad d\xi_h = 0,$$

where $\xi_i = \xi_i(a_1, a_2, \cdots, a_r)$ are functions of a_i such that the manifolds $s\mathfrak{H}$ are represented by $\xi_i = $ constant $(i = 1, 2, \cdots, h)$. We can make in $\mathfrak{G}$ the change of local coordinates $(a_1, a_2, \cdots, a_r)$ $\rightarrow (\xi_1, \xi_2, \cdots, \xi_h, x_{h+1}, \cdots, x_r)$. Since the systems (2.3) and (2.6) are equivalent, we have

$$(2.7) \qquad \Omega_h = A(\xi, x) \, d\xi_1 \wedge d\xi_2 \wedge \cdots \wedge d\xi_h,$$

where $A(\xi, x)$ denotes a function of $\xi_1, \cdots, \xi_h, x_{h+1}, \cdots, x_r$. When the point $a(\xi_1, \xi_2, \cdots, \xi_h, x_{h+1}, \cdots, x_r)$ varies on $s\mathfrak{H}$, the coordinates ξ_i are constant, and, therefore,

$$(2.8) \qquad \delta\Omega_h = \sum_{j=h+1}^{r} \frac{\partial A}{\partial x_j}\, dx_j \wedge d\xi_1 \wedge \cdots \wedge d\xi_h.$$

On the other side, by exterior differentiation of (2.7), we get

$$d\Omega_h = \sum_{j=1}^{h} \frac{\partial A}{\partial \xi_j}\, d\xi_j \wedge d\xi_1 \wedge \cdots \wedge d\xi_h$$

$$+ \sum_{j=h+1}^{r} \frac{\partial A}{\partial x_j}\, dx_j \wedge d\xi_1 \wedge \cdots \wedge d\xi_h = \delta\Omega_h,$$

because the first sum vanishes. Consequently, so that $\delta\Omega_h = 0$—that is, for Ω_h to be invariant by displacements on the manifolds $s\mathfrak{H}$, it is necessary and sufficient that $d\Omega_h = 0$. This proves the theorem.

If $\mathfrak{H}$ reduces to the identity, then $\mathfrak{G}/\mathfrak{H} = \mathfrak{G}$ and $\Omega_r = \omega_1 \wedge \omega_2 \wedge \cdots \wedge \omega_r$ gives the invariant density (= element of volume) of $\mathfrak{G}$, which in integral geometry takes the name of kinematic density of $\mathfrak{G}$. The integral of Ω_r gives an invariant measure for $\mathfrak{G}$ (Haar's measure) which is unique up to a constant factor.

2.3. *The examples of the introduction.* To exemplify these general results, we shall consider the examples appearing in the introduction.

The group of motions $\mathfrak{G} = \mathfrak{M}$ in E_2 can be represented by the group of 3-dimensional matrices,

$$(2.9) \qquad u = \begin{pmatrix} \cos\varphi & -\sin\varphi & a \\ \sin\varphi & \cos\varphi & b \\ 0 & 0 & 1 \end{pmatrix}$$

with the parameters $a_1 = a$, $a_2 = b$, $a_3 = \varphi$. We have

$$u^{-1} = \begin{pmatrix} \cos\varphi & \sin\varphi & -b\sin\varphi - a\cos\varphi \\ -\sin\varphi & \cos\varphi & -b\cos\varphi + a\sin\varphi \\ 0 & 0 & 1 \end{pmatrix}$$

$$du = \begin{pmatrix} -\sin\varphi\, d\varphi & -\cos\varphi\, d\varphi & da \\ \cos\varphi\, d\varphi & -\sin\varphi\, d\varphi & db \\ 0 & 0 & 0 \end{pmatrix}$$

and, therefore,

$$\omega = u^{-1}\,du = \begin{pmatrix} 0 & -d\varphi & \cos\varphi\,da + \sin\varphi\,db \\ d\varphi & 0 & -\sin\varphi\,da + \cos\varphi\,db \\ 0 & 0 & 0 \end{pmatrix}$$

The forms of Maurer-Cartan are

(2.10)

$$\omega_1 = \cos\varphi\,da + \sin\varphi\,db, \quad \omega_2 = -\sin\varphi\,da + \cos\varphi\,db, \quad \omega_3 = d\varphi,$$

and the equations of structure

$$d\omega = -\omega \wedge \omega = -\begin{pmatrix} 0 & 0 & -\omega_3 \wedge \omega_2 \\ 0 & 0 & \omega_3 \wedge \omega_1 \\ 0 & 0 & 0 \end{pmatrix}.$$

That is,

(2.11) $d\omega_1 = -\omega_2 \wedge \omega_3, \qquad d\omega_2 = -\omega_3 \wedge \omega_1, \qquad d\omega_3 = 0.$

The kinematic density of $\mathfrak{M}$ is

$$dK = \omega_1 \wedge \omega_2 \wedge \omega_3 = da \wedge db \wedge d\varphi,$$

which, up to the notation, coincides with (1.11).

Let $\mathfrak{H}_1$ be the subgroup of $\mathfrak{M}$ consisting of all motions which leave the line $G(p, \theta)$ invariant (equation of $G: x \cos\theta + y \sin\theta - p = 0$). There is a bijective mapping between the lines G of E_2 and the points of the space $\mathfrak{M}/\mathfrak{H}_1$. As density for lines, we take the density of $\mathfrak{M}/\mathfrak{H}_1$.

By the change of coordinates $(a, b, \varphi) \to (p, \theta, t)$ in $\mathfrak{M}$, given by the equations,

$$a = p \cos\theta + t \sin\theta, \qquad b = p \sin\theta - t \cos\theta, \qquad \varphi = \theta - \frac{\pi}{2}$$

$$p = a \cos\theta + b \sin\theta, \qquad t = a \sin\theta - b \cos\theta, \qquad \theta = \varphi + \frac{\pi}{2},$$

the points of $\mathfrak{M}/\mathfrak{H}_1$ are $p = $ constant, $\theta = $ constant. The system (2.6) is $dp = 0$, $d\theta = 0$, and the system (2.3) is

$$dp = \cos\theta\,da + \sin\theta\,db = -\sin\varphi\,da + \cos\varphi\,db = \omega_2 = 0,$$

$$d\theta = d\varphi = \omega_3 = 0.$$

Therefore, the density for lines takes the form

(2.12) $dG = \omega_2 \wedge \omega_3 = -\sin\varphi\,da \wedge d\varphi + \cos\varphi\,db \wedge d\varphi$

which is equivalent to

(2.13) $$dG = dp \wedge d\theta,$$

as stated in (1.7).

If $\mathfrak{H}_0$ is the subgroup of $\mathfrak{M}$ consisting of all motions which leave the point $P(a, b)$ invariant, there is a bijective mapping between the points (a, b) of E_2 and the points of the homogeneous space $\mathfrak{M}/\mathfrak{H}_0$. The system (2.6) is now $da = 0$, $db = 0$, and (2.3) gives $\omega_1 = 0$, $\omega_2 = 0$. The density (2.4) for points results in

(2.14) $$dP = \omega_1 \wedge \omega_2 = da \wedge db,$$

which coincides with (1.4). In both cases (2.13) and (2.14), the condition (2.5) is obviously satisfied.

To give an example in which the homogeneous space $\mathfrak{G}/\mathfrak{H}$ has not an invariant density, let us consider the 4-dimensional group $\mathfrak{G}$ of matrices of the form

$$u = \begin{pmatrix} a_1 & 0 & a_2 \\ 0 & a_3 & a_4 \\ 0 & 0 & 1 \end{pmatrix}, \quad a_1 a_3 \neq 0,$$

and the 2-dimensional subgroup $\mathfrak{H}$ of matrices of the form

$$u_1 = \begin{pmatrix} a_1 & 0 & 0 \\ 0 & a_3 & 0 \\ 0 & 0 & 1 \end{pmatrix}, \quad a_1 a_3 \neq 0.$$

To obtain the forms of Maurer-Cartan of $\mathfrak{G}$, we have

$$\omega = u^{-1} du = \begin{pmatrix} a_1^{-1} & 0 & -a_1^{-1}a_2 \\ 0 & a_3^{-1} & -a_3^{-1}a_4 \\ 0 & 0 & 1 \end{pmatrix} \begin{pmatrix} da_1 & 0 & da_2 \\ 0 & da_3 & da_4 \\ 0 & 0 & 0 \end{pmatrix}$$

$$= \begin{pmatrix} \omega_1 & 0 & \omega_2 \\ 0 & \omega_3 & \omega_4 \\ 0 & 0 & 0 \end{pmatrix},$$

where

$$\omega_1 = a_1^{-1} da_1, \quad \omega_2 = a_1^{-1} da_2, \quad \omega_3 = a_3^{-1} da_3, \quad \omega_4 = a_3^{-1} da_4.$$

The subgroup $\mathfrak{H}$ is characterized by $a_2 = 0$, $a_4 = 0$, and, therefore, the system (2.3) is now $\omega_2 = 0$, $\omega_4 = 0$. The differential form

$\Omega_2 = \omega_2 \wedge \omega_4$ is not a density, because $d\Omega_2 = -\omega_1 \wedge \omega_2 \wedge \omega_4 - \omega_3 \wedge \omega_2 \wedge \omega_4 \neq 0$.

3. INTEGRAL GEOMETRY IN THE THREE-DIMENSIONAL EUCLIDEAN SPACE

3.1. *The group of motions in E_3.* We shall consider in detail the integral geometry of the 3-dimensional euclidean space. The base space is E_3 and the group $\mathfrak{G}$ is the group of motions $\mathfrak{M}$ in it.

Let x represent the one-column matrix formed by the orthogonal coordinates x_1, x_2, x_3 of a point P. The matrix equation of a motion $x \rightarrow x'$ is

$$(3.1) \qquad\qquad x' = Ax + B,$$

where

$$(3.2) \qquad A = \begin{pmatrix} a_{11} & a_{12} & a_{13} \\ a_{21} & a_{22} & a_{23} \\ a_{31} & a_{32} & a_{33} \end{pmatrix}, \quad B = \begin{pmatrix} b_1 \\ b_2 \\ b_3 \end{pmatrix},$$

and A satisfies the conditions of orthogonality

$$(3.3) \qquad\qquad A^t = A^{-1} \quad (A^t = \text{transposed of } A).$$

The condition (3.3) reduces to 3 the number of independent parameters a_{ij} which, with b_1, b_2, and b_3, are the 6 parameters on which $\mathfrak{M}$ depends.

The group $\mathfrak{M}$ can be represented by the 4×4 matrices,

$$(3.4) \qquad u = \begin{pmatrix} A & \vdots & B \\ \cdots & \cdots & \cdots \\ 0 & \vdots & 1 \end{pmatrix}$$

with the ordinary rules,

$$u_2 u_1 = \begin{pmatrix} A_2 A_1 & \vdots & A_2 B_1 + B_2 \\ \cdots & \cdots & \cdots \\ 0 & \vdots & 1 \end{pmatrix}, \quad u^{-1} = \begin{pmatrix} A^{-1} & \vdots & -A^{-1}B \\ \cdots & \cdots & \cdots \\ 0 & \vdots & 1 \end{pmatrix}.$$

The matrix of Maurer-Cartan is

$$\omega = u^{-1} \, du = \begin{pmatrix} A^{-1} \, dA & \vdots & A^{-1} \, dB \\ \cdots\cdots\cdots\cdots \\ 0 & \vdots & 0 \end{pmatrix}.$$

If we introduce the two matrices

(3.5) $\qquad \omega_A = A^{-1} \, dA, \qquad \omega_B = A^{-1} \, dB$

of order 3×3 and 3×1, respectively, the equations of structure can be written

(3.6) $\qquad d\omega_A = -\omega_A \wedge \omega_A, \qquad d\omega_B = -\omega_A \wedge \omega_B.$

Since $\mathfrak{M}$ is a 6-parameter group, we must have 6 pfaffian forms of Maurer-Cartan. Effectively, from (3.3) and (3.5) we deduce $\omega_A = A^t \, dA = -dA^t \, A = -\omega_A^t$, and the 6 forms are the elements of the matrices,

$$\omega_A = \begin{pmatrix} 0 & \omega_{12} & \omega_{13} \\ -\omega_{12} & 0 & \omega_{23} \\ -\omega_{13} & -\omega_{23} & 0 \end{pmatrix}, \qquad \omega_B = \begin{pmatrix} \omega_1 \\ \omega_2 \\ \omega_3 \end{pmatrix},$$

which, explicitly, give

(3.7) $\qquad \omega_{ih} = -\omega_{hi} = \sum_{j=1}^{3} a_{ji} \, da_{jh}, \qquad \omega_i = \sum_{j=1}^{3} a_{ji} \, db_j.$

It is useful to give a more geometrical approach to the pfaffian forms ω_{ih} and ω_i. Let us consider in E_3 a fixed frame $(Q_0; e_1^0, e_2^0, e_3^0)$ composed of a point Q_0 and three orthogonal unit vectors e_i^0, and a moving frame $(Q; e_1, e_2, e_3)$ which results from the fixed frame by the motion u represented by (3.1). If we introduce the matrices

(3.8) $\qquad e^0 = (e_1^0, e_2^0, e_3^0), \qquad e = (e_1, e_2, e_3)$

we can write

(3.9) $\qquad Q = e^0 \, B, \qquad e = e^0 \, A,$

and, therefore,

(3.10) $\quad \begin{aligned} dQ &= e^0 \, dB = e \, A^{-1} \, dB = e\omega_B, \\ de &= e^0 \, dA = e \, A^{-1} \, dA = e\omega_A, \end{aligned}$

which may be written

$$(3.11) \qquad dQ = \sum_{j=1}^{3} \omega_j e_j, \quad de_i = \sum_{j=1}^{3} \omega_{ji} e_j.$$

These formulas are useful for the computation of densities, as we shall see in the next section. Because of the orthogonality of the unit vectors e_i, we have $e_i e_j = \delta_{ij}$, and from (3.11) we deduce

$$(3.12) \qquad \omega_j = e_j \, dQ, \quad \omega_{ji} = e_j \, de_i,$$

which are the vectorial form of the equations in (3.7).

3.2. *The area element of the unit sphere.* We need to remember two expressions for the element of area of the unit sphere. Let ν be the unit vector with the components

$$(3.13) \quad \nu_1 = \sin\theta\cos\varphi, \quad \nu_2 = \sin\theta\sin\varphi, \quad \nu_3 = \cos\theta$$

where θ, φ are the ordinary spherical coordinates corresponding to the endpoint of ν. The area element at this endpoint is known to be

$$(3.14) \qquad d\sigma = (\nu \, \nu_\theta \, \nu_\varphi) \, d\theta \wedge d\varphi = \sin\theta \, d\theta \wedge d\varphi$$

where $(\nu \, \nu_\theta \, \nu_\varphi)$ denotes the scalar triple product of the vectors ν, ν_θ, and ν_φ (subscripts denote partial derivation). Taking (3.13) into account, we have also

$$(3.15) \qquad d\sigma = \frac{d\nu_2 \wedge d\nu_3}{\nu_1} = \frac{d\nu_3 \wedge d\nu_1}{\nu_2} = \frac{d\nu_1 \wedge d\nu_2}{\nu_3},$$

and since $\nu_1^2 + \nu_2^2 + \nu_3^2 = 1$, we deduce

$$d\sigma = \nu_1 \, d\nu_2 \wedge d\nu_3 + \nu_2 \, d\nu_3 \wedge d\nu_1 + \nu_3 \, d\nu_1 \wedge d\nu_2.$$

On the other hand, if e_1, e_2, and e_3 are the 3 orthogonal unit vectors of a moving frame, we have

$$
\begin{aligned}
e_1 \, de_3 \wedge e_2 \, de_3 &= e_1(e_{3\theta} \, d\theta + e_{3\varphi} \, d\varphi) \wedge e_2(e_{3\theta} \, d\theta + e_{3\varphi} \, d\varphi) \\
&= (e_1 e_{3\theta} \cdot e_2 e_{3\varphi} - e_1 e_{3\varphi} \cdot e_2 e_{3\theta}) \, d\theta \wedge d\varphi \\
&= (e_1 \wedge e_2) \cdot (e_{3\theta} \wedge e_{3\varphi}) \, d\theta \wedge d\varphi \\
&= (e_3 e_{3\theta} e_{3\varphi}) \, d\theta \wedge d\varphi = d\sigma
\end{aligned}
$$

(3.16)

where $d\sigma$ denotes the area element of the unit sphere corresponding to the endpoint of e_3. From (3.12) and (3.16), we get

(3.17) $$d\sigma = \omega_{13} \wedge \omega_{23}.$$

We have now at our disposal all elements necessary to find the densities for points, lines and planes of E_3 invariant under $\mathfrak{M}$.

3.3. *Density for points.* Let $\mathfrak{H}_0$ be the set of motions which leave the point $Q(b_1, b_2, b_3)$ invariant; clearly it is a subgroup of $\mathfrak{M}$. According to (3.11), to keep Q fixed we must have

$$\omega_1 = 0, \qquad \omega_2 = 0, \qquad \omega_3 = 0,$$

which is the system (2.3), and, according to (2.4), the density for points will be $\omega_1 \wedge \omega_2 \wedge \omega_3 = db_1 \wedge db_2 \wedge db_3$ [applying (3.7) and taking into account the determinant $|a_{ij}| = 1$, because the matrix $A = (a_{ij})$ is orthogonal]. In general, for the point $P(x, y, z)$, we shall have

(3.18) $$dP = dx \wedge dy \wedge dz.$$

The condition (2.5) is obviously satisfied.

3.4. *Density for planes.* Let $\mathfrak{H}_2$ be the set of motions which leave the plane $E(e_1, e_2)$ invariant; clearly it is a subgroup of $\mathfrak{M}$.

By the motions of $\mathfrak{H}_2$ the unit vector e_3 remains fixed and the point Q can only move on the plane e_1, e_2; therefore, according to (3.11), the pfaffian system which characterizes the planes is

$$\omega_3 = 0, \qquad \omega_{13} = 0, \qquad \omega_{23} = 0,$$

and the density for planes results:

(3.19) $$dE = \omega_3 \wedge \omega_{13} \wedge \omega_{23}.$$

If θ, φ are the spherical coordinates of the endpoint of e_3, (3.14) and (3.17) give

(3.20) $$\omega_{13} \wedge \omega_{23} = d\sigma = \sin\theta \, d\theta \wedge d\varphi.$$

If p is the distance from the origin Q_0 of the fixed frame to the plane E, and $a_{13} = \sin\theta \cos\varphi$, $a_{23} = \sin\theta \sin\varphi$, $a_{33} = \cos\theta$ are the components of e_3 (normal to E), we have $p = a_{13}b_1 + a_{23}b_2 + a_{33}b_3$, and, according to (3.7),

(3.21) $$\omega_3 = \sum_{j=1}^{3} a_{j3} \, db_j = dp + R \, d\theta + S \, d\varphi.$$

Here, R, S are functions of θ, φ, b_i, the explicit form of which has no interest for us. From (3.19) and (3.20) we get

(3.22) $\qquad dE = \sin \theta \, dp \wedge d\theta \wedge d\varphi = dp \wedge d\sigma.$

The condition (2.5) is obviously satisfied, and hence we have: If a plane E is determined by its normal e_3 and its distance p to a fixed origin, the density is given by (3.22), where $d\sigma$ denotes the area element of the unit sphere corresponding to the endpoint of the unit vector e_3.

As an exercise, prove that if the plane is given by the equation $ux + vy + wz + 1 = 0$, its density takes the form

$$dE = \frac{du \wedge dv \wedge dw}{(u^2 + v^2 + w^2)^2}.$$

Example

Let S be a fixed segment of length l. To compute the measure of the set of planes E which intersect S, we take S on the e_3^0-axis and the middle point of S as the origin of coordinates. Then we have

(3.23) $\quad m(E; E \cap S \neq 0) = \displaystyle\int_{E \cap S \neq 0} dE$

$$= \frac{l}{2} \int_0^{2\pi} d\varphi \int_0^\pi |\cos \theta| \sin \theta \, d\theta = \pi l.$$

If Γ is a polygonal line of length L, writing (3.23) for all sides of Γ and adding, we obtain

(3.24) $\qquad \displaystyle\int n \, dE = \pi L,$

where n denotes the number of intersection points of E with Γ. By a limit process it is not difficult to prove that (3.24) holds for any rectifiable curve. The integral in (3.24) is extended over all planes of E_3, n being 0 for the planes which do not intersect Γ.

3.5. *Density for straight lines.* Let $\mathfrak{H}_1$ be the set of motions leaving the line G which contains the unit vector e_3 invariant; clearly $\mathfrak{H}_1$ is a subgroup of $\mathfrak{M}$.

By a motion of $\mathfrak{H}_1$, the point Q can only move in the direction of e_3, and, therefore, (3.11) gives $\omega_1 = 0$, $\omega_2 = 0$. Moreover, because e_3 is fixed, from (3.11) we deduce $\omega_{13} = 0$, $\omega_{23} = 0$. The pfaffian system (2.3) for the lines of E_3 becomes

$$(3.25) \qquad \omega_1 = 0, \qquad \omega_2 = 0, \qquad \omega_{13} = 0, \qquad \omega_{23} = 0,$$

and the density for lines is

$$(3.26) \qquad dG = \omega_1 \wedge \omega_2 \wedge \omega_{13} \wedge \omega_{23}.$$

According to (3.12), $\omega_1 \wedge \omega_2$ equals the area element of the plane (e_1, e_2) at the point Q, and we have seen that $\omega_{13} \wedge \omega_{23}$ is the area element of the unit sphere corresponding to the endpoint of e_3, that is, to the direction of G. If G is determined by its direction e_3 and its intersection point (x, y) with a fixed plane, denoting by ψ the angle between e_3 and the normal to the fixed plane, we have $\omega_1 \wedge \omega_2 = |\cos \psi| \, dx \wedge dy$, and we can write (3.26) in the form

$$(3.27) \qquad dG = |\cos \psi| \, dx \wedge dy \wedge d\sigma.$$

From (3.26) and (3.6) it is easy to show that the condition (2.5) is satisfied.

As an exercise, prove that if G is given by the equations $x = az + p$, $y = bz + q$, then its density is

$$dG = \frac{da \wedge db \wedge dp \wedge dq}{(1 + a^2 + b^2)^2}.$$

Example

Let Σ be a fixed surface of class C^1 (= with a continuous tangent plane). If P denotes a point of the intersection $G \cap \Sigma$ and df denotes the area element of Σ at P, the density for lines can be written $dG = |\cos \psi| \, df \wedge d\sigma$, where ψ denotes the angle between G and the normal to Σ at P. Fixed P, the integral of $|\cos \psi| \, d\sigma$ extended over all the lines which pass through P, gives the projection of one-half the unit sphere upon a diametral plane—that is, π. The integration of df over the whole Σ gives the area F of Σ. Therefore, taking into account that each line has been counted as many times n as it has intersection points with Σ, we get

$$(3.28) \qquad \int n \, dG = \pi F,$$

where the integral is extended over all lines of E_3, n being 0 for the lines which do not intersect Σ.

3.6. *Kinematic density.* The kinematic density is

$$(3.29) \qquad dK = \omega_1 \wedge \omega_2 \wedge \omega_3 \wedge \omega_{12} \wedge \omega_{13} \wedge \omega_{23}.$$

To give a geometrical interpretation to $\omega_{12} = e_1\, de_2$, we observe that if we take on the plane e_1, e_2 two fixed orthogonal unit vectors e_1^*, e_2^* and call α the angle between e_1 and e_1^*, we can write $e_1 = \cos\alpha\, e_1^* + \sin\alpha\, e_2^*$, $e_2 = -\sin\alpha\, e_1^* + \cos\alpha\, e_2^*$; therefore, $e_1\, de_2 = -d\alpha$. That is, ω_{12} means an elementary rotation about the e_3-axis. Consequently, according to (3.17) and (3.29), if a motion is determined by the position of the moving frame $(Q; e_1, e_2, e_3)$, the kinematic density has the form

$$(3.30) \qquad dK = dP \wedge d\sigma \wedge d\alpha,$$

where dP is the volume element of E_3 at the origin Q of the moving frame, $d\sigma$ is the area element of the unit sphere corresponding to the endpoint of e_3, and $d\alpha$ is the element of rotation about e_3. We remember that we always consider the densities in absolute value; thus, there is no question of sign.

Let us do an application of (3.30). Let Γ be a fixed curve with continuous tangent at every point and finite length L and let Σ be a moving surface of class C^1 and finite area F. Let Q be a point of $\Gamma \cap \Sigma$ and let e_3 be the normal to Σ at Q. If θ denotes the angle between e_3 and the tangent to Γ at Q (which we may take as the e_3^0-axis of the fixed frame) and df denotes the area element of Σ at Q, we have $dP = |\cos\theta|\, df \wedge ds$ ($s =$ arc length of Γ). Putting this value in (3.30) and integrating over all the positions of Σ in which it has common point with Γ, because each position of Σ will be counted as many times n as intersection points have Σ and Γ, we get

$$(3.31) \qquad \int n\, dK = 4\pi^2 FL.$$

Notice that the same formula holds if we suppose Σ fixed and Γ moving with density dK.

If Σ is the unit sphere, we can take the origin of the moving frame at the center of Σ; then we have $\int n \, dK = 8\pi^2 \int n \, dP$, and (3.31) gives

$$(3.32) \qquad\qquad \int n \, dP = 2\pi L,$$

which is valid for any rectifiable curve (51).

3.7. *A differential formula.* In Section 5 we will need an important auxiliary formula which derives from (3.30). Let Σ_0 be a fixed surface of class C^1. At each point Q of Σ_0 we consider an orthogonal frame $(Q; e_1^0, e_2^0, e_3^0)$ with origin at Q and with e_3^0 normal to Σ_0. If the displacement vector on Σ_0 at Q is $\omega_1 e_1^0 + \omega_2 e_2^0$, the area element is $df = \omega_1 \wedge \omega_2$. To the unit vector e^0 tangent to Σ_0 at Q which forms with e_1^0 the angle τ_0, is attached the differential form $dL_0 = \omega_1 \wedge \omega_2 \wedge d\tau_0$ called the density for line elements $(Q; e^0)$ on Σ_0, and the pfaffian form $ds = \cos \tau_0 \, \omega_1 + \sin \tau_0 \, \omega_2$ called the element of length corresponding to the direction e^0.

Now let Σ_1 be a moving surface of class C^1, and assume that the intersection $\Sigma_0 \cap \Sigma_1$ is a rectifiable curve Γ. Let Q be a point of Γ and $(Q; e_1, e_2, e_3)$ be an orthogonal frame with e_3 perpendicular to Σ_1. Let ds be the length element of Γ at Q and ds_0, ds_1 those normal to Γ on Σ_0 and Σ_1, respectively. Let θ be the angle between the normals e_3^0, e_3. If df_0, df_1 are the elements of area of Σ_0, Σ_1 at Q and dP denotes the element of volume of E_3 at Q, we have $dP = \sin \theta \, df_0 \wedge ds_1$ and $df_1 = ds \wedge ds_1$. The element of area of the unit sphere at the endpoint of e_3 may be written $d\sigma = \sin \theta \, d\theta \wedge d\tau_0$. Putting now $d\tau_1 = d\alpha$ to unify the notation of (3.30), from this equation and the preceding relation, we deduce immediately (up to the sign)

$$(3.33) \qquad ds \wedge dK = \sin^2 \theta \, df_0 \wedge d\tau_0 \wedge df_1 \wedge d\tau_1 \wedge d\theta$$

$$= \sin^2 \theta \, dL_0 \wedge dL_1 \wedge d\theta,$$

which is the differential formula we want.

An immediate consequence is obtained by integrating both sides over all positions of the moving surface Σ_1. We get

$$(3.34) \qquad\qquad \int L \, dK = 2\pi^3 F_0 F_1,$$

where L denotes the length of the curve $\Sigma_0 \cap \Sigma_1$, and F_0, F_1 are the areas of Σ_0, Σ_1, respectively.

If Σ_1 is the unit sphere and we take the origin of the moving frame at the center of Σ_1, we have

$$\int L \, dK = 8\pi^2 \int L \, dP_1$$

and (3.34) gives

$$(3.35) \qquad\qquad \int L \, dP = \pi^2 F_0.$$

3.8. *A definition of area.* Let now Σ_1, Σ_2 be two moving unit spheres and Σ_0 a fixed surface. Let N be the number of points of the intersection $\Sigma_0 \cap \Sigma_1 \cap \Sigma_2$. If dP_i denotes the volume element at the center of Σ_i $(i = 1, 2)$, we get from (3.32) and (3.35)

$$\int N \, dP_1 \, dP_2 = 2\pi \int L \, dP_1 = 2\pi^3 F_0.$$

Conversely, this result conduces to define the area of a continuum of points by the formula

$$F_0 = \frac{1}{2\pi^3} \int N \, dP_1 \, dP_2,$$

provided the integral of the right-hand side exists [see (52)]. Applications of the integral geometry to the definition of area for k-dimensional surfaces have been made by Federer (17–19) and Hadwiger (23) and (25). See also Nöbeling (45) and (46).

3.9. *Planes through a fixed point.* Let us now consider the set of planes E_0 which pass through a fixed point O. The density for sets of E_0 invariant under the group $\mathfrak{M}_0$ of the rotations about 0, is clearly $dE_0 = d\sigma$, where $d\sigma$ denotes the area element of the unit sphere corresponding to the direction perpendicular to E_0. In fact, this differential form is invariant under $\mathfrak{M}_0$, and, because of the transitivity of the planes E_0 with respect to $\mathfrak{M}_0$, it is unique up to a constant factor. The planes E_0 are considered non-oriented; therefore, the measure of all the planes through O will be

$$(3.36) \qquad\qquad \int dE_0 = \int_{\frac{1}{2}Z} d\sigma = 2\pi,$$

where $\frac{1}{2}Z$ denotes the half of the unit sphere.

Let S be a fixed arc of great circle on the unit sphere of center O of length α. The measure of the set of planes E_0 which intersect S (= measure of the set of great circles which intersect S) will be the area of the lune bounded by the great circles the poles of which are the endpoints of S—that is, $m(E_0; S \cap E_0 \neq 0) = 2\alpha$. If instead of S we have a spherical polygonal line Γ the sides of which have the lengths α_i, we have, writing the last formula for each side and adding,

$$(3.37) \qquad \int n \, dE_0 = 2L,$$

where L denotes the total length of Γ. The integration is extended over all (non-oriented) planes through O—that is, according to $dE_0 = d\sigma$, over half the unit sphere. By a limit process we can prove that (3.37) holds for any rectifiable spherical curve of the unit sphere.

Following Fenchel (20), we want to apply (3.37). Let K be a closed space curve of class C^2 without multiple points and let Γ be the spherical indicatrix of it (= the curve $T = T(s)$, where T is the tangent unit vector to K). The arc length element of Γ is $ds_t = |\varkappa| \, ds$, where $\varkappa$ denotes the curvature and s the length of K. Consequently, (3.37) yields

$$(3.38) \qquad \int n \, dE_0 = 2 \int_K |\varkappa| \, ds.$$

Every closed space curve K has at least 2 tangents which are parallel to an arbitrary plane. This means that every plane E_0 intersects Γ in at least 2 points. Hence, $n \geqq 2$, and (3.36) and (3.38) give

$$(3.39) \qquad \int_K |\varkappa| \, ds \geqq 2\pi,$$

a classical inequality of Fenchel.

If K is knotted, it is easy to see that it has at least 4 tangents parallel to an arbitrary plane. Hence, $n \geqq 4$, and (3.36) and (3.38) give the following inequality of Fáry (for knotted curves) (16),

$$(3.40) \qquad \int_K |\varkappa| \, ds \geqq 4\pi.$$

These results have been generalized to closed varieties in E_n by Chern and Lashof (11).

4. APPLICATIONS TO CONVEX BODIES

The integral geometry is closely related to the theory of convex bodies. We compile in this section some simple facts on this theory from many sources—for example, Bonnesen and Fenchel (4), Busemann (5), Hadwiger (24), and Vincensini (72).

Let k be a plane convex set of area f placed in E_3. Let f_σ be the area of the orthogonal projection of k on a plane perpendicular to the direction σ, and let θ be the angle between σ and the normal to the plane which contains k; we have $f_\sigma = |\cos \theta| f$. If $d\sigma$ denotes the area element of the unit sphere Z corresponding to the direction σ, we have

$$(4.1) \qquad \int_Z f_\sigma \, d\sigma = f \int_0^{2\pi} d\varphi \int_0^\pi |\cos \theta| \sin \theta \, d\theta = 2\pi f,$$

and, therefore,

$$(4.2) \qquad\qquad f = \frac{1}{2\pi} \int_Z f_\sigma \, d\sigma.$$

Now let K be a convex body of E_3; we shall denote by ∂K the convex surface bounding K. Let F be the area of ∂K and F_σ the area of the orthogonal projection of K on a plane perpendicular to the direction σ. Applying (4.2) to each element of area of ∂K and integrating over all ∂K, we get

$$(4.3) \qquad\qquad F = \frac{1}{\pi} \int_Z F_\sigma \, d\sigma,$$

known as Cauchy's formula for the area of a convex body.

Let O be an interior point of K and $p = p(\sigma) = p(\theta, \varphi)$ be the supporting function of K with respect to O (= distance from O to the supporting plane perpendicular to the direction σ of spherical coordinates θ, φ). The convex body K_h parallel to K at distance h has the supporting function $p_h = p(\sigma) + h$, and if R_1, R_2 are the principal radii of curvature of ∂K, those of ∂K_h at corresponding points are $R_1 + h$ and $R_2 + h$. Between the area element df of

∂K and the area element $d\sigma$ of its spherical image, there is the relation $df/d\sigma = R_1 R_2$, and consequently, we have

$$(4.4) \qquad\qquad F = \int_Z R_1 R_2 \, d\sigma.$$

Applying this formula to ∂K_h, we get

$$(4.5) \quad F_h = \int_Z (R_1 + h)(R_2 + h) \, d\sigma = F + 2Mh + 4\pi h^2,$$

where

$$(4.6) \qquad M = \frac{1}{2} \int_Z (R_1 + R_2) \, d\sigma = \frac{1}{2} \int_{\partial K} \left(\frac{1}{R_1} + \frac{1}{R_2} \right) df$$

is the integral of mean curvature of ∂K. If V denotes the volume of K and V_h that of K_h, from (4.5) we deduce

$$(4.7) \qquad V_h = V + \int_0^h F_h \, dh = V + Fh + Mh^2 + \tfrac{4}{3}\pi h^3,$$

which is the so-called Steiner's formula for parallel convex bodies in E_3.

For plane convex sets, the formula analogous to (4.7) is

$$(4.8) \qquad\qquad f_h = f + uh + \pi h^2,$$

where $u = $ length of ∂k. Applying (4.8) to the orthogonal projection of K on a plane perpendicular to the direction σ, we have

$$F_{\sigma,h} = F_\sigma + u_\sigma h + \pi h^2,$$

and by Cauchy's formula,

$$(4.9) \quad F_h = \frac{1}{\pi} \int_Z F_{\sigma,h} \, d\sigma = \frac{1}{\pi} \int_Z F_\sigma \, d\sigma + \frac{h}{\pi} \int_Z u_\sigma \, d\sigma + 4\pi h^2.$$

Comparing (4.9) with (4.5), we get (since both formulas hold for any h)

$$(4.10) \qquad\qquad M = \frac{1}{2\pi} \int_Z u_\sigma \, d\sigma,$$

which is a very useful expression for the integral of mean curvature of the boundary of a convex body.

On the other side, considering the volume V of K as a sum of pyramids with the common vertex O, we have

(4.11) $V = \frac{1}{3} \int_{\partial K} p \, df = \frac{1}{3} \int_Z p R_1 R_2 \, d\sigma.$

Applying this formula to K_h, we get

$$V_h = \frac{1}{3} \int_Z (p + h)(R_1 + h)(R_2 + h) \, d\sigma$$

$$= V + \frac{h}{3} \int_Z (R_1 R_2 + p(R_1 + R_2)) \, d\sigma$$

$$+ \frac{h^2}{3} \int_Z (p + R_1 + R_2) \, d\sigma + \frac{4}{3} \pi h^3.$$

Comparison with (4.7) yields

(4.12) $F = \frac{1}{2} \int_Z p(R_1 + R_2) \, d\sigma, \qquad M = \int_Z p \, d\sigma.$

The last formula allows definition of M for any convex body without the conditions of regularity necessary to define the principal radii of curvature of ∂K. A practical way to compute M for convex surfaces ∂K not sufficiently smooth is to compute the integral of mean curvature M_h of the parallel surface ∂K_h (which is smooth) and then to pass to the limit for $h \to 0$. This method yields the following results easily. (1) For a convex polyhedron the edges of which have lengths a_i and the corresponding dihedral angles of which are α_i, we have

$$M = \frac{1}{2} \Sigma \, (\pi - \alpha_i) a_i.$$

(2) For a right cylinder of height h and radius r,

$$M = \pi h + \pi^2 r.$$

(3) For a plane convex domain, considered as a flattened convex body of E_3, we have

$$M = \frac{\pi}{2} u,$$

where u is the length of the boundary of the domain.

Notice that, according to (3.22), the second formula (4.12) gives the measure of the set of planes E which cut K—that is, we have the formula

(4.13)
$$\int_{E \cap K \neq 0} dE = M.$$

On the other side, applying (3.28) to convex surfaces ($n = 2$), we get

(4.14)
$$\int_{G \cap K \neq 0} dG = \frac{\pi}{2} F.$$

We may therefore state:

The volume V of a convex body K is the measure of the points contained in it; the area F of ∂K is (up to the constant factor $\pi/2$) the measure of the lines which intersect K; the integral of mean curvature M is the measure of the planes which intersect K.

These integral geometric interpretations of V, F, and M have been generalized to convex bodies of the n-dimensional euclidean space [(60) and Hadwiger (23) and (25)].

5. THE KINEMATIC FUNDAMENTAL FORMULA IN E_3

5.1. *The Euler characteristic of a domain.* Let Σ be a closed surface in E_3 which is of class C^2 and bounds a domain D of volume V. If df is the area element of Σ and $d\sigma$ the area element of the corresponding spherical image, we know the formulas

(5.1)
$$\frac{d\sigma}{df} = \frac{1}{R_1 R_2}, \quad I(\Sigma) = \int_\Sigma \frac{1}{R_1 R_2} df = 4\pi\chi,$$

where R_1, R_2 are the principal radii of curvature, $I(\Sigma)$ denotes the area of the spherical image of Σ, and $\chi = \chi(D)$ is the Euler characteristic of D. Because Σ is closed, its spherical image covers the unit sphere an integer number of times, and therefore $\chi = I(\Sigma)/4\pi$ is an integer. For example, for domains topologically equivalent to the solid sphere, $\chi = 1$, and for domains which are topologically equivalent to a torus, $\chi = 0$ [see, for example, Struik (69, p. 159)].

If Σ is not of class C^2 but consists of a finite number of faces (= pieces of class C^2) which intersect along edges (= closed curves of class C^2), the Euler characteristic is obtained adding to the area of the spherical image of the faces (5.1) the area of the spherical image corresponding to the edges, which we shall now compute. Let Γ be an edge of Σ and let T, N, B, denote its unit vectors tangent, principal normal, and binormal; let s be the arc length of Γ. If e_3, e_3' are the outward normal unit vectors to the faces of Σ at the points of Γ and we call θ_1, θ_1' the angles which they form with $-N$, the spherical image corresponding to Γ is the portion of unit sphere defined by the equation,

$$Y(s, \theta) = -\cos \theta\, N + \sin \theta\, B \quad (\theta_1 \leqq \theta \leqq \theta_1', 0 \leqq s \leqq L),$$

where L is the length of Γ.

Using Frenet's formulas, we have $Y_\theta^2 = 1$, $Y_s Y_\theta = -\tau$, $Y_s^2 = \varkappa^2 \cos^2 \theta + \tau^2$, $(Y_\theta^2 Y_s^2 - (Y_s Y_\theta)^2)^{1/2} = \varkappa \cos \theta$, where $\varkappa$ and τ are the curvature and the torsion of Γ. The area $I(\Gamma)$ of the spherical image corresponding to Γ will be

$$(5.2) \qquad I(\Gamma) = \int_\Gamma \varkappa \cos \theta\, d\theta\, ds = \int_\Gamma (\sin \theta_1' - \sin \theta_1)\, \varkappa ds.$$

Under the assumption that Σ has no vertices (= points in which more than two different faces intersect), the Euler characteristic of Σ is given by the second formula (5.1); we take into account that at the left side, the integral analogous to (5.2) for all the edges of Σ should be added.

5.2. *The kinematic formula.* Let D_0, D_1 be two domains of E_3 bounded respectively by the surfaces Σ_0, Σ_1, which we assume to be of class C^2. Let V_i, χ_i be the volume and the Euler characteristic of D_i and let F_i, M_i be the area and the integral of mean curvature of Σ_i ($i = 0, 1$). Suppose D_0 is fixed and D_1 is moving, and let dK be the kinematic density for D_1. If $\Phi(D_0 \cap D_1)$ denotes a function of the intersection $D_0 \cap D_1$, one of the main purposes of the integral geometry is the evaluation of integrals of the type

$$(5.3) \qquad\qquad J = \int \Phi(D_0 \cap D_1)\, dK$$

over all positions of D_1. For example, if $\Phi = V_{01} =$ volume of $D_0 \cap D_1$, we can easily prove that $\int V_{01}\, dK = 8\pi^2 V_0 V_1$, and if

$\Phi = F_{01}$ is the area of the boundary of $D_0 \cap D_1$, the formula $\int F_{01} \, dk = 8\pi^2(V_0 F_1 + V_1 F_0)$ holds (50). The most important case corresponds to $\Phi = \chi(D_0 \cap D_1)$ is the Euler characteristic of $D_0 \cap D_1$. Surprisingly enough, the integral $\int \chi(D_0 \cap D_1) \, dK$ over all positions of D_1 can be expressed by only V_i, χ_i, F_i, M_i $(i = 0, 1)$. The result is the following:

$$(5.4) \quad \int \chi(D_0 \cap D_1) \, dK = 8\pi^2(V_0\chi_1 + V_1\chi_0) + 2\pi(F_0 M_1 + F_1 M_0).$$

This result is the so-called kinematic fundamental formula, which we shall now prove.

We need to compute $\chi(D_0 \cap D_1)$. The boundary of $D_0 \cap D_1$ consists in a part Σ_{01} of Σ_0 which is interior to D_1 and a part Σ_{10} of Σ_1 which is interior to D_0. Both Σ_{01} and Σ_{10} are of class C^2 and are joined by an edge $\Gamma = \Sigma_0 \cap \Sigma_1$, composed of one or more closed curves, of the boundary of $D_0 \cap D_1$. According to (5.1) we will have

$$(5.5) \qquad 4\pi\chi(D_0 \cap D_1) = I(\Sigma_{01}) + I(\Sigma_{10}) + I(\Gamma),$$

and we can write

$$(5.6) \quad 4\pi \int \chi(D_0 \cap D_1) \, dK = \int I(\Sigma_{01}) \, dK$$
$$+ \int I(\Sigma_{10}) \, dK + \int I(\Gamma) \, dK,$$

where the integrals are extended over all positions of D_1.

The first two integrals on the right-hand side of (5.6) are easily evaluated. Taking the first integral, let P be a point of $\Sigma_0 \cap D_1$ and let $d\sigma_P$ denote the area element of the unit sphere at the spherical image of P. By first fixing D_1 and then letting P vary over $\Sigma_0 \cap D_1$, we get

$$\int_{P \in \Sigma_0 \cap D_1} d\sigma_P \, dK = \int I(\Sigma_{01}) \, dK,$$

and by first fixing P and then rotating D_1 about this point and letting it vary over D_1 and Σ_0,

$$\int_{P \in \Sigma_0 \cap D_1} d\sigma_P \, dK = \int_{P \in \Sigma_0} d\sigma_P \int_{P \in D_1} dK = 8\pi^2 V_1 \int_{P \in \Sigma_0} d\sigma_P$$

$$= 8\pi^2 V_1 I(\Sigma_0) = 32\pi^3 V_1 \chi_0.$$

Thus, we have

(5.7) $$\int I(\Sigma_{01}) \, dK = 32\pi^3 V_1 \chi_0.$$

Similarly, by the evident invariance of the kinematic density under the inversion of the motion, we have

(5.8) $$\int I(\Sigma_{10}) \, dK = 32\pi^3 V_0 \chi_1.$$

It remains to evaluate the third integral in (5.6). Let Q be a point of Γ. By Meusnier's theorem, if ρ is the radius of curvature of Γ and R, r are the radii of normal curvature of Σ_0 and Σ_1 in the direction of the tangent to Γ at Q, we have

(5.9) $$\rho = R \cos \theta_1 = r \cos \theta_1'$$

where θ_1, θ_1' are the angles between the outward normals e_3, e_3' to Σ_0, Σ_1 at Q and the vector $-N$ opposite to the principal normal N of Γ at Q. Taking into account the identity

(5.10) $$\frac{\sin \theta_1' - \sin \theta_1}{\cos \theta_1' + \cos \theta_1} = \tan \frac{1}{2} (\theta_1' - \theta_1),$$

and putting $\theta_1' - \theta_1 = \theta$, we deduce from (5.9) and (5.10)

(5.11) $$\sin \theta_1' - \sin \theta_1 = \rho \left(\frac{1}{R} + \frac{1}{r} \right) \tan \frac{1}{2} \theta.$$

If τ_0, τ_1 denote the angles between the tangent to Γ at Q and the first principal direction of Σ_0, Σ_1 at Q, by Euler's theorem we have

(5.12) $$\frac{1}{R} = \frac{\cos^2 \tau_0}{R_1} + \frac{\sin^2 \tau_0}{R_2}, \qquad \frac{1}{r} = \frac{\cos^2 \tau_1}{r_1} + \frac{\sin^2 \tau_1}{r_2},$$

where R_1, R_2 are the principal radii of curvature of Σ_0, and r_1, r_2 are those of Σ_1 at Q. By (5.2), (5.11), and (3.33) we have

(5.13) $$\int I(\Gamma) \, dK = \int \left(\frac{\cos^2 \tau_0}{R_1} + \frac{\sin^2 \tau_0}{R_2} + \frac{\cos^2 \tau_1}{r_1} + \frac{\sin^2 \tau_1}{r_2} \right)$$

$$\tan \tfrac{1}{2}\theta \, \sin^2 \theta \, df_0 \, d\tau_0 \, df_1 \, d\tau_1 \, d\theta,$$

where the limits of integration for the angles are

$$0 \leqq \tau_0 \leqq 2\pi, \qquad 0 \leqq \tau_1 \leqq 2\pi, \qquad 0 \leqq \theta \leqq \pi.$$

Computing the integral in (5.13), we get

(5.14) $$\int I(\Gamma) \, dK = 8\pi^2(F_0 M_1 + F_1 M_0).$$

Adding (5.7), (5.8), and (5.14), and considering (5.6), we get the desired result (5.4).

The formula (5.4) is the work of Blaschke (3). It has been generalized to E_n by Chern (8). For the generalization to spaces of constant curvature (noneuclidean geometry) see Wu (76) and (54), (57), and (58). For another kind of proof valid for more general domains than those considered here, see Hadwiger (23).

Notice that if D_0, D_1 are convex bodies, we have $\chi(D_0) = \chi(D_1) = \chi(D_0 \cap D_1) = 1$ if $D_0 \cap D_1 \neq 0$, and $\chi(D_0 \cap D_1) = 0$, if $D_0 \cap D_1 = 0$. The formula (5.4) yields

(5.15) $$\int_{D_0 \cap D_1 \neq 0} dK = 8\pi^2(V_0 + V_1) + 2\pi(F_0 M_1 + F_1 M_0),$$

which gives the measure of the set of congruent convex bodies D_1 having a common point with a fixed convex body D_0.

If D_1 is a sphere of radius r, we can take the origin of the moving frame at the center of D_1; then we have $\int dK = 8\pi^2 \int dP$, and (5.15) gives

$$\int dP = V_0 + F_0 r + M_0 r^2 + \tfrac{4}{3}\pi r^3,$$

the Steiner's formula (4.7).

6. INTEGRAL GEOMETRY IN COMPLEX SPACES

6.1. *The unitary group.* The integral geometry of complex spaces has not been developed very much, and it deserves further study. We shall give a simple typical example.

Let P_n be the n-dimensional complex projective space with the homogeneous coordinates $z_i(i = 0, 1, \cdots, n)$, so that $z = (z_0, z_1, z_2, \cdots, z_n)$ and $\lambda z = (\lambda z_0, \lambda z_1, \cdots, \lambda z_n)$, where λ is a nonzero complex number; define the same point. Let $\bar{z}_i$ denote the complex conjugate of z_i. We assume the homogeneous coordinates z_i are normalized so that

$$(6.1) \qquad\qquad (z\bar{z}) = \sum_{0}^{n} z_i\bar{z}_i = 1,$$

which determine z_i up to a factor of the form $\exp{(i\alpha)}$.

We consider the group $\mathfrak{U}$ (unitary group) of linear transformations

$$(6.2) \qquad\qquad z' = Az$$

which leaves the form (6.1) invariant. The matrices A satisfy

$$(6.3) \qquad A\overline{A}^t = E, \qquad A^{-1} = \overline{A}^t, \qquad \overline{A}^tA = E,$$

where E is the unit matrix. These relations show that $\mathfrak{U}$ depends upon $(n+1)^2$ real parameters. If we interpret the elements a_{hk} ($h = 0, 1, \cdots, n$) of the matrix A as the homogeneous coordinates of a point $a_k \in P_n$, the conditions (6.3) give

$$(6.4) \qquad\qquad (a_j\bar{a}_k) = \delta_{jk},$$

which show that the points a_k are normalized; they form the vertices of an autoconjugate n-simplex with respect to the quadric $(z\bar{z}) = 0$. Because a_k and $a_k \exp{(i\alpha_k)}$ are the same geometric point, to determine an element $u \in \mathfrak{U}$ we must give the $n+1$ geometric points a_k [with the conditions of (6.4)], as well as the $n+1$ real parameters α_k.

The invariant matrix of Maurer-Cartan is

$$(6.5) \qquad\qquad \omega = A^{-1}\,dA = \overline{A}^t\,dA,$$

which satisfies, in consequence of (6.3),

$$(6.6) \qquad\qquad \omega + \bar{\omega}^t = 0.$$

The invariant pfaffian forms are

$$(6.7) \qquad\qquad \omega_{jk} = \sum_{h=0}^{n} \bar{a}_{hj}\,da_{hk} = (\bar{a}_j\,da_k),$$

and (6.6) gives

$$(6.8) \qquad\qquad \omega_{jk} + \bar{\omega}_{kj} = 0.$$

The kinematic density of $\mathfrak{U}$, up to a constant factor, is

(6.9) $du = [\Pi \, \omega_{jk}\bar{\omega}_{jk} \, \Pi \, \omega_{hh}], \quad j < k, \quad 0 \leqq j, k, h \leqq n,$

where the product is exterior.

We have all necessary elements for the study of the integral geometry of the unitary group. We shall restrict ourselves to the case $n = 2$ (complex projective plane).

6.2. *Meromorphic curves.* A complex analytic mapping $E_1 \rightarrow P_2$ of the complex euclidean line E_1 into the complex projective plane P_2 defines a meromorphic curve in the sense of J. Weyl, H. Weyl (75), and L. Ahlfors (1); it is defined by three analytic functions $z_i = z_i(t), \; (i = 0, 1, 2)$. Every such curve Γ has an invariant integral with respect to $\mathfrak{U}$, which we shall call the order of Γ. When the homogeneous coordinates z_i are normalized such that the condition (6.1) is satisfied, the order of Γ is defined by the following integral (up to the sign which depends upon the orientation assumed for Γ),

(6.10) $$J = \frac{1}{2\pi i} \int_\Gamma \Omega$$

where $i = \sqrt{-1}$ and

(6.11) $\Omega = [dz \, d\bar{z}] = dz_0 \wedge d\bar{z}_0 + dz_1 \wedge d\bar{z}_1 + dz_2 \wedge d\bar{z}_2.$

If Γ is an algebraic curve, we shall see that J coincides with its ordinary order or grad.

If the coordinates z_i are not normalized, we set $Z_i = z_i/(z\bar{z})^{1/2}$, and an easy calculation gives

(6.12) $$\Omega = [dZ \, d\bar{Z}] = \frac{|z \wedge z'|^2}{z^4} \; dt \wedge d\bar{t},$$

where $z \wedge z'$ denotes the vector with the components $z_1 z'_2 - z_2 z'_1$, $z_2 z'_0 - z_0 z'_2$, and $z_0 z'_1 - z_1 z'_0$.

For some purposes, it is convenient to write Ω in another form. Let c be a point on the tangent to Γ at the point z such that

(6.13) $(c\bar{c}) = 1, \qquad (\bar{c}z) = 0.$

We will have (since c is on the tangent to Γ at z),

(6.14) $dz = \alpha z + \beta c, \qquad d\bar{z} = \bar{\alpha}\bar{z} + \bar{\beta}\bar{c}.$

where α, β are the pfaffian forms

(6.15) $$\alpha = (\bar{z}\, dz), \qquad \beta = (\bar{c}\, dz).$$

From (6.13) and (6.14), we deduce

(6.16) $$\Omega = [dz\, d\bar{z}] = \alpha \wedge \bar{\alpha} + \beta \wedge \bar{\beta},$$

and because $\alpha = -\bar{\alpha}$, we have $\alpha \wedge \bar{\alpha} = 0$. Therefore,

(6.17) $$\Omega = \beta \wedge \bar{\beta} = (\bar{c}\, dz) \wedge (c\, d\bar{z}),$$

a formula which will be useful in the following discussion.

As an application, we shall use (6.17) to obtain the order of a complex straight line. Since J is invariant by unitary transformations and any line can be transformed into the axis $z_1 = 0$, it suffices to compute the order in this case. We take, in order to satisfy (6.1) and (6.13),

$$z = (\rho e^{i\varphi}(1 + \rho^2)^{-1/2}, 0, (1 + \rho^2)^{-1/2}),$$
$$c = (-e^{i\varphi}(1 + \rho^2)^{-1/2}, 0, \rho(1 + \rho^2)^{-1/2}),$$

and we get

$$(\bar{c}\, dz) = -\frac{d\rho + i\rho\, d\varphi}{1 + \rho^2}, \qquad (c\, d\bar{z}) = -\frac{d\rho - i\rho\, d\varphi}{1 + \rho^2}$$

and

(6.18) $$\Omega = (\bar{c}\, dz) \wedge (c\, d\bar{z}) = \frac{2i\rho}{(1 + \rho^2)^2}\, d\varphi \wedge d\rho.$$

The order of the segment $a \leqq \rho \leqq b$, $0 \leqq \varphi \leqq 2\pi$ will be

$$J = \frac{1}{2\pi i} \int_a^b \int_0^{2\pi} \frac{2i\rho}{(1 + \rho^2)^2}\, d\varphi\, d\rho = \frac{b^2 - a^2}{(1 + a^2)(1 + b^2)}.$$

For $a = 0$, $b = \infty$, we obtain $J = 1$, which is the order of a line.

6.3. *A generalization of the theorem of Bezout.* Let Γ_1, Γ_2 be two meromorphic curves of P_2 of orders J_1, J_2, respectively. Let $u\Gamma_2$ be the transform of Γ_2 by $u \in \mathfrak{U}$. In the theory of meromorphic curves it is important to determine the difference between the product $J_1 J_2$ and the number $N(\Gamma_1 \cap u\Gamma_2)$ of points of intersection of Γ_1 and $u\Gamma_2$, each counted with its proper multiplicity [Ahlfors (1), Chern (9) and (10), and H. Weyl (75)].

Our goal is more simple. We wish to obtain the mean value of $N(\Gamma_1 \cap u\Gamma_2)$ for all $u \in \mathfrak{U}$. First, we will compute the integral

(6.19) $$I = \int_{\mathfrak{U}} N(\Gamma_1 \cap u\Gamma_2) \, du$$

where the element of volume du is given by (6.9). In our case, $n = 2$, making use of (6.8), and considering only the absolute value, we have

(6.20) $$du = (\bar{a}_0 \, da_1) \wedge (\bar{a}_1 \, da_0) \wedge (\bar{a}_0 \, da_2) \wedge (\bar{a}_2 \, da_0) \wedge (\bar{a}_1 \, da_2)$$
$$\wedge (\bar{a}_2 \, da_1) \wedge (\bar{a}_0 \, da_0) \wedge (\bar{a}_1 \, da_1) \wedge (\bar{a}_2 \, da_2).$$

Inasmuch as we are only interested in the transformations u such that $\Gamma_1 \cap u\Gamma_2 \neq 0$, we may choose the points a_0, a_1, and a_2, which determine u, so that: a_0 = point of $\Gamma_1 \cap u\Gamma_2$; a_1 = point on the tangent to $u\Gamma_2$ at a_0; a_2 is then determined by the relations (6.4), which we now write

(6.21) $$(a_0\bar{a}_0) = (a_1\bar{a}_1) = (a_2\bar{a}_2) = 1, \quad (a_0\bar{a}_1) = (a_0\bar{a}_2) = (a_1\bar{a}_2) = 0.$$

Let s be the point in which the line determined by a_1, a_2 intersects the tangent to Γ_1 at a_0. We shall have

(6.22) $$(s\bar{s}) = 1, \quad (s\bar{a}_0) = 0, \quad (\bar{s}a_0) = 0.$$

According to (6.17), the differential form which gives the order of $u\Gamma_2$ is

(6.23) $$\Omega_2 = (\bar{a}_1 \, da_0) \wedge (a_1 \, d\bar{a}_0) = (\bar{a}_0 \, da_1) \wedge (\bar{a}_1 \, da_0).$$

Since we always take a_0 on Γ_1, we have $da_0 = \alpha a_0 + \beta s$, where $\alpha = (\bar{a}_0 \, da_0)$, $\beta = (\bar{s} \, da_0)$. Consequently, we have

$$(\bar{a}_2 \, da_0) = \beta(\bar{a}_2 s), \quad (a_2 \, d\bar{a}_0) = \bar{\beta}(a_2\bar{s}),$$

and, by exterior multiplication,

(6.24) $$(\bar{a}_2 \, da_0) \wedge (a_2 \, d\bar{a}_0) = (\bar{a}_0 \, da_2) \wedge (\bar{a}_2 \, da_0)$$
$$= (\beta \wedge \bar{\beta})(\bar{a}_2 s)(a_2\bar{s}) = (\bar{a}_2 s)(a_2\bar{s})\Omega_1,$$

where Ω_1 is the differential form which gives the order of Γ_1.

From (6.20), (6.23), and (6.24), we have

(6.25) $$du = \Omega_2 \wedge \Omega_1(\bar{a}_2 s)(a_2\bar{s}) \wedge (\bar{a}_1 \, da_2) \wedge (\bar{a}_2 \, da_1) \wedge (\bar{a}_0 \, da_0)$$
$$\wedge (\bar{a}_1 \, da_1) \wedge (\bar{a}_2 \, da_2).$$

We first keep fixed the geometric points a_0, a_1, and a_2. With the normalization (6.21), their homogeneous coordinates a_{hj} ($h = 0, 1, 2$) are determined up to an exponential factor $\exp(i\alpha_j)$; the parameters $\alpha_j (j = 0, 1, 2)$ are variables in (6.25). Putting $a_j = a_j^*\exp(i\alpha_j)$, we have $da_j = a_j i\, d\alpha_j$, $(\bar{a}_j\, da_j) = i\, d\alpha_j$, and, consequently, $\int (\bar{a}_j\, da_j) = 2\pi i$ ($j = 0, 1, 2$).

From the right side of (6.25) it remains to evaluate (a_0 being fixed) $\int (\bar{a}_2 s)(a_2 \bar{s})(\bar{a}_1\, da_2) \wedge (\bar{a}_2\, da_1)$, where a_1, a_2 describe the line $(\bar{a}_0 z) = 0$ which contains the point s. We can assume, because of the invariance of the integrand by unitary transformations, that this line is the axis $z_1 = 0$. According to (6.18), we then have

$$(6.26) \qquad \int (\bar{a}_1\, da_2) \wedge (\bar{a}_2\, da_1) = \int \frac{2i\rho}{(1+\rho^2)^2}\, d\rho\, d\varphi,$$

where we have put $a_2 = (\rho e^{i\varphi}(1+\rho^2)^{-1/2}, 0, (1+\rho^2)^{-1/2})$, $a_1 = (-e^{i\varphi}(1+\rho^2)^{-1/2}, 0, \rho(1+\rho^2)^{-1/2})$. Taking $s = (0, 0, 1)$, we obtain

$$(6.27) \qquad\qquad (\bar{a}_2 s)(a_2 \bar{s}) = \frac{1}{1+\rho^2},$$

and, therefore,

$$(6.28) \quad \int (\bar{a}_2 s)(a_2 \bar{s})(\bar{a}_1\, da_2) \wedge (\bar{a}_2\, da_1)$$

$$= \int_0^\infty \int_0^{2\pi} \frac{2i\rho}{(1+\rho^2)^3}\, d\rho\, d\varphi = \pi i.$$

From (6.25) and (6.28), we obtain the integral of du extended over all u such that $\Gamma_1 \cap u\Gamma_2 \neq 0$, each u counted $N(\Gamma_1 \cap u\Gamma_2)$ times. We get (up to the sign which is unessential),

$$(6.29) \qquad\qquad \int_u N(\Gamma_1 \cap u\Gamma_2)\, du = 32\pi^6 J_1 J_2,$$

where J_1 and J_2 are the orders of Γ_1 and Γ_2, respectively.

To obtain the mean value of $N(\Gamma_1 \cap u\Gamma_2)$, we need the total measure of $\mathfrak{U}$. Taking for Γ_1 and Γ_2 two straight lines, we know that $J_1 = J_2 = 1$ and $N = 1$; therefore (6.29) gives $\int_{\mathfrak{u}} du = 32\pi^6$. Consequently, the mean value of N is

(6.30) $$\overline{N} = J_1 J_2.$$

For algebraic curves, N is constant and (6.30) gives the classical theorem of Bezout; therefore our result may be considered a generalization of this theorem to meromorphic curves. For the extension to analytic manifolds of P_n see (56).

7. INTEGRAL GEOMETRY IN RIEMANNIAN SPACES

7.1. *Geodesics which intersect a fixed surface.* The methods of the integral geometry can be also applied to Riemannian spaces, mainly to spaces of constant curvature or other spaces which admit a group of transformations into themselves. The case of surfaces is simple and well known (55). Here, we want to consider the case of 3-dimensional spaces.

Let R_3 be a 3-dimensional Riemannian space defined by $ds^2 = g_{ij}\, dx_i\, dx_j$, where the summation convention is adopted; i, j are summed from 1 to 3. Let us introduce the notations,

(7.1) $$F = (g_{ij}x_i'x_j')^{1/2}, \qquad p_i = \frac{\partial F}{\partial x_i'},$$

where $x_i' = dx_i/dt$. As we know, a geodesic of R_3 is determined by a point x_i and a direction x_i', which is equivalent to give x_i, p_i ($i = 1, 2, 3$). The density for sets of geodesics is defined by the following exterior differential form, taken always in absolute value:

(7.2) $$dG = dp_2 \wedge dx_2 \wedge dp_3 \wedge dx_3 + dp_3 \wedge dx_3 \wedge dp_1 \wedge dx_1$$
$$+ dp_1 \wedge dx_1 \wedge dp_2 \wedge dx_2.$$

The measure of a set of geodesics is the integral of dG extended over the set. The density (7.2) is the second power of the differential invariant $\sum_1^3 dp_i \wedge dx_i$, which constitutes the invariant integral of Poincaré of the dynamics (6, pp. 19 and 78), and it therefore possesses the following two properties of invariance: (1) it is invariant with respect to a change of coordinates in the space; (2) it is invariant under displacements of the elements (x_i, p_i) on the respective geodesic.

To give a geometrical interpretation of dG, let us consider a fixed surface Σ and a set of geodesics which intersect Σ. Let G be such a geodesic and P its intersection point with Σ. In a neighborhood of P we may assume that the equation of Σ is $x_3 = 0$ and that the coordinate system is orthogonal, that is, $ds^2 = g_{11}\,dx_1^2 + g_{22}\,dx_2^2 + g_{33}\,dx_3^2$, and thus $p_i = g_{ii}(dx_i/ds)$. If ν_i represents the cosine of the angle between G and the x_i-coordinate curve at P, we have

$$(7.3) \quad \nu_i = \sqrt{g_{ii}}\,\frac{dx_i}{ds}, \quad p_i = \sqrt{g_{ii}}\,\nu_i, \quad dp_i = \sqrt{g_{ii}}\,d\nu_i + \frac{\partial\sqrt{g_{ii}}}{\partial x_h}\,\nu_i\,dx_h.$$

To determine G according to the second property of invariance of dG, we may choose its intersection point P with Σ. At this point we have $x_3 = 0$, $dx_3 = 0$, and, consequently, (7.2) takes the form

$$(7.4) \qquad\qquad dG = dp_1 \wedge dx_1 \wedge dp_2 \wedge dx_2,$$

or, according to (7.3),

$$(7.5) \qquad\qquad dG = \sqrt{g_{11}g_{22}}\,d\nu_1 \wedge dx_1 \wedge d\nu_2 \wedge dx_2.$$

On the other hand, to each set of direction cosines ν_1, ν_2, and ν_3 corresponds a point of the unit euclidean sphere and the area element in it has the value (3.15)

$$(7.6) \qquad\qquad d\sigma = \frac{d\nu_1 \wedge d\nu_2}{\nu_3}.$$

Hence, we have, in absolute value,

$$(7.7) \qquad\qquad dG = |\cos\varphi|\,d\sigma \wedge df,$$

where φ is the angle between the tangent to G and the normal to Σ at P, and $df = \sqrt{g_{11}g_{22}}\,dx_1 \wedge dx_2$ is the element of area Σ at P.

Integrating over all geodesics which intersect Σ, on the left side each geodesic is counted a number of times equal to the number n of intersection points of G and Σ; on the right, the integral of $|\cos\varphi|\,d\sigma$ gives one-half the projection of the unit sphere upon a diametral plane ($= \pi$). Consequently, we get the integral formula

$$(7.8) \qquad\qquad \int n\,dG = \pi F,$$

where F is the area of Σ. This formula generalizes (3.28) to Riemannian spaces.

7.2. *Sets of geodesic segments.* Let t be the arc length on the geodesic G. From (7.7) we deduce

$$(7.9) \qquad dG \wedge dt = |\cos \varphi| \, d\sigma \wedge df \wedge dt.$$

The product $|\cos \varphi| \, dt$ equals the projection of the arc element dt upon the normal to Σ at P; consequently, $|\cos \varphi| \, df \wedge dt$ equals the element of volume dP of the space at P, and (7.9) can be written in the form,

$$(7.10) \qquad dG \wedge dt = dP \wedge d\sigma.$$

An oriented segment S of geodesic is determined either by G, t (G = geodesic which contains S; t = abscissa on G of the origin of S) or by P (= origin of S) and the point of the unit euclidean sphere which gives the direction of S. The two equivalent forms (7.10) may therefore be taken as density for sets of segments of geodesic lines.

For example, let us consider the set of oriented segments S with the origin inside a fixed domain D. The integral of the left of (7.10) gives $2 \int \lambda \, dG$, where λ denotes the length of the arc of G which lies inside D (the factor 2 appears as a consequence that dG means the density for non-oriented geodesic lines). The integral of the right is equal to $4\pi V$, where V is the volume of D. Consequently, we have the following integral formula

$$(7.11) \qquad \int \lambda \, dG = 2\pi V,$$

where the integral is extended over all geodesics which intersect D.

7.3. *Some integral formulas for convex bodies in spaces of constant curvature.* Let R_3 now be a 3-dimensional space of constant curvature k. With respect to a system of geodesic polar coordinates, it is known that the element of length can be written in the form

$$(7.12) \qquad ds^2 = d\rho^2 + \frac{\sin^2 \sqrt{k}\rho}{k} \, d\tau,$$

where ρ denotes the geodesic distance from a fixed point (origin of coordinates) and $d\tau$ represents the length element of the 2-dimensional unit euclidean sphere. The volume element has the form

$$(7.13) \qquad dP = \frac{\sin^2 \sqrt{k}\rho}{k} \, d\rho \wedge d\sigma,$$

where $d\sigma$ denotes the element of area on the unit sphere.

Let P_1, P_2 be two points in R_3 such that there is only one geodesic G which unites them. Let ρ_1, ρ_2 be the abscissas on G of P_1 and P_2. With respect to a system of geodesic polar coordinates with the origin at P_1, the element of volume dP_2 has the form

$$(7.14) \qquad dP_2 = \frac{\sin^2 \sqrt{k} \, |\rho_2 - \rho_1|}{k} \, d\rho_2 \wedge d\sigma.$$

By exterior multiplication by dP_1, we have, in consequence of (7.10),

$$(7.15) \qquad dP_1 \wedge dP_2 = \frac{\sin^2 \sqrt{k} \, |\rho_2 - \rho_1|}{k} \, d\rho_1 \wedge d\rho_2 \wedge dG.$$

This formula is the work of Haimovici (27).

Let D be a convex domain of volume V (that is, it contains, with each pair of its points, the arc of geodesic, assumed unique, determined by them) and consider all the pairs P_1, P_2 inside D. The integral of the left side of (7.15) is equal to V^2. If λ denotes the length of the arc of G which lies inside D, then by calculating the integral of the right side we have

$$\int_0^\lambda \int_0^\lambda \sin^2 \sqrt{k} \, |\rho_2 - \rho_1| \, d\rho_1 \, d\rho_2 = \frac{1}{2} \left(\lambda^2 - \frac{1}{k} \sin^2 \sqrt{k}\lambda \right).$$

Hence, we have the integral formula

$$(7.16) \qquad \frac{1}{k} \int \left(\lambda^2 - \frac{1}{k} \sin^2 \sqrt{k}\lambda \right) dG = 2V^2,$$

where the integral is extended over all geodesics which intersect D.

For the elliptic space ($k = 1$), this formula reduces to

$$(7.17) \qquad \int (\lambda^2 - \sin^2 \lambda) \, dG = 2V^2,$$

and for the hyperbolic space ($k = -1$),

(7.18) $$\int (\sinh^2\lambda - \lambda^2)\, dG = 2V^2.$$

For the euclidean space ($k = 0$), passing to the limit for $k \to 0$ in (7.16) we get

(7.19) $$\int \lambda^4\, dG = 6V^2,$$

which is a formula of Herglotz [Blaschke (3)].

Formulas of this kind referring to convex figures in the plane or to convex bodies in the euclidean space were first obtained by Crofton (7), considered the creator of the integral geometry. A great deal of them were given successively by several authors: Lebesgue (34), Blaschke (3), Massoti Biggiogero (38–42). Paper (38) contains an extensive bibliography.

The generalization to spaces of constant curvature is less known. However for certain types of formulas, the treatment in elliptic space is more satisfactory than that in euclidean space, owing to the possibility of dualization. Let us consider the following examples.

In the elliptic 3-dimensional space, all geodesics are closed and have the finite length π. The planes have finite area 2π. Since any geodesic intersects a fixed plane in one and only one point, the formula (7.8) gives the measure of the set of all geodesics of the space:

(7.20) $$\int dG = 2\pi^2.$$

Let D be a convex body of area F and volume V and let us consider the set of geodesic segments of length π which intersect D. The integral on the left of (7.10) extended over this set making use of (7.8) for $n = 2$, has the value

(7.21) $$\int dG\, dt = \pi \int dG = \frac{\pi^2}{2} F,$$

and the integral on the right is

(7.22) $$\int dP \wedge d\sigma = 2\pi V + \int\limits_{P \notin D} \Phi\, dP,$$

where Φ denotes the solid angle under which D is seen from P

(*P* exterior to *D*). From (7.21) and (7.22), we deduce the integral formula

$$(7.23) \qquad \int_{P \notin D} \Phi \, dP = \tfrac{1}{2}\pi^2 F - 2\pi V.$$

Let us now see which formula corresponds to (7.11) by duality. Let M, F be the integral of mean curvature and the area of the boundary of D. For the dual convex body D^* it is known that we have

$$(7.24) \quad F^* = 4\pi - F, \qquad M^* = M, \qquad V^* = \pi^2 - M - V.$$

By duality to each straight line (geodesic) G corresponds another straight line G^* and, hence, if we use (7.24), formula (7.11) gives

$$\int_{G^* \cap D^* = 0} (\pi - \varphi^*) \, dG^* = 2\pi(\pi^2 - M^* - V^*),$$

where φ^* denotes the angle between the two supporting planes of D^* through G^* and the integral is extended over all geodesics G^* exterior to D^*. Taking into account (7.20) and (7.8), and replacing G^* by G, we get the integral formula

$$(7.25) \qquad \int_{G \cap D = 0} \varphi \, dG = 2\pi(M + V) - \tfrac{1}{2}\pi^2 F,$$

which has no analogue in the euclidean geometry.

Similarly, as dual of the formula (7.17), we have

$$(7.26) \qquad \int_{G \cap D = 0} (\varphi^2 - \sin^2 \varphi) \, dG = 2(M + V)^2 - \tfrac{1}{2}\pi^3 F,$$

where, as in (7.25), φ denotes the angle between the two supporting planes of D through G and the integral is extended over all geodesics which do not intersect D. For the integral geometry in spaces of constant curvature, see Petkantschin (48), and (53), (54), and (59).

8. SUPPLEMENTARY REMARKS AND BIBLIOGRAPHICAL NOTES

8.1. *General integral geometry.* The integral geometry has its origin in the theory of geometrical probabilities [Crofton (13),

Deltheil (14), and Herglotz (29), and it was widely developed by Blaschke and his school in a series of papers quoted in Reference (3). The inclusion of the methods and results of the integral geometry within the framework of the theory of homogeneous spaces (as we have done in Section 2) is the work of Weil (73) and (74), and Chern (7). After their work, the measure theory in groups and homogeneous spaces became of fundamental interest in integral geometry. Every new result in that direction can be applied and probably exploited with success to get integral geometric statements; at least, it is sure that the integral geometry constitutes the most abundant source of examples [Nachbin (44) and Helgason (28, Chap. X)].

The inverse problem of finding a general formulation of certain particular formulas of integral geometry (Crofton's formulas) is also an interesting one [Hermann (30) Legrady (36)]. A very simple example follows. We have seen that the kinematic density for the group of motions $\mathfrak{M}$ of the plane is $dK = dP \wedge d\alpha$ (1.11). From the point of view of the homogeneous spaces, dP is the density of the space $\mathfrak{M}/\mathfrak{M}_1$, where $\mathfrak{M}_1$ denotes the group of rotations about a fixed point and $d\alpha$ is the density of $\mathfrak{M}_1$. If we write, symbolically, $dK = d\mathfrak{M}$, $dP = d(\mathfrak{M}/\mathfrak{M}_1)$, $d\alpha = d\mathfrak{M}_1$, the formula (1.11) gives $d\mathfrak{M} = d(\mathfrak{M}/\mathfrak{M}_1) \wedge d\mathfrak{M}_1$, which induces us to ask if it will hold for a general group $\mathfrak{G}$ and its subgroup $\mathfrak{g}$. In this particular example, it is well known that the formula $d\mathfrak{G} = d(\mathfrak{G}/\mathfrak{g}) \wedge d\mathfrak{g}$, in fact, holds for any locally compact topological group $\mathfrak{G}$ and any closed subgroup $\mathfrak{g}$ of $\mathfrak{G}$ [Weil (73, pp. 42–45) and Ambrose (2)].

8.2. Sets of manifolds. Some problems of integral geometry may also be presented under the following form. Let V denote a differentiable manifold and F a family of submanifolds in it. First we ask for the existence of a transformation group $\mathfrak{G}$ of V onto itself which transforms the elements of F onto elements of F. Then, if such a group exists, we ask for a measure of sets of varieties of F invariant under $\mathfrak{G}$. We shall give two simple examples.

Examples

1. Let V be the euclidean plane E_2 and F the family of all circles of it. The group $\mathfrak{G}$ is known to be the group of similitudes

(8.1) $\quad x' = \rho(x \cos \varphi - y \sin \varphi) + a,$

$$y' = \rho(x \sin \varphi + y \cos \varphi) + b,$$

which depends on the 4 parameters a, b, ρ, and φ. This group can be represented by the group of matrices,

$$u = \begin{pmatrix} \rho \cos \varphi & -\rho \sin \varphi & a \\ \rho \sin \varphi & \rho \cos \varphi & b \\ 0 & 0 & 1 \end{pmatrix},$$

and by the method of Section (2.2), we find immediately that the forms of Maurer-Cartan are

$$\omega_1 = \frac{d\rho}{\rho}, \qquad \omega_2 = d\varphi, \qquad \omega_3 = \frac{\cos \varphi}{\rho} da + \frac{\sin \varphi}{\rho} db,$$

$$\omega_4 = \frac{\sin \varphi}{\rho} da + \frac{\cos \varphi}{\rho} db.$$

The similitudes which leave invariant a given circle are characterized by a, b, ρ = constants, and, consequently, the system (2.3) is $\omega_1 = 0$, $\omega_3 = 0$, $\omega_4 = 0$. The density for sets of circles (of center a, b and radius ρ) invariant under the group of similitudes results:

$$dC = \frac{da \wedge db \wedge d\rho}{\rho^3}.$$

2. Let V be the real projective plane and F the family of nondegenerate conics in it. Then the group G is the projective group and the density for conics is (61),

$$dC = \frac{da_{00} \wedge da_{01} \wedge da_{02} \wedge da_{11} \wedge da_{12}}{3\Delta^2}$$

where $\Delta = \det(a_{ij})$ and the equation of the conic is assumed to be

$$a_{00}x_0^2 + 2a_{01}xy + a_{11}y^2 + 2a_{02}x + 2a_{12}y + 1 = 0.$$

Other examples of this kind have been given by Stoka (63–68). For sets of degenerate conics, see Luccioni (37).

8.3. *Integral geometry of special groups.* The metric (euclidean and noneuclidean) integral geometry is the best known; however, other cases have also been investigated. The integral geometry of the unimodular affine group of the euclidean space onto itself leads to certain affine invariants for convex bodies (62). The integral geometry of the projective group has been considered by Varga (70) and is pursued in (55); that of the symplectic group has been studied by Legrady (35).

In the last years, Gelfand and his school have largely generalized the ideas of the integral geometry and used them in problems of group representation (21).

REFERENCES

The following list contains almost exclusively the papers mentioned in the text. References (3), (23), (38) contain a more complete bibliography.

1. Ahlfors, L. V., "The theory of meromorphic curves," *Acta Soc. Sc. Fenn.*, A-III (1941), No. 4, 1–31.

2. Ambrose, W., "Direct sum theorem for Haar measures," *Transactions of the American Mathematical Society*, 61 (1947), 122–27.

3. Blaschke, W., *Vorlesungen über Integralgeometrie*, 3rd ed. Berlin: 1955.

4. Bonnesen, T., and W. Fenchel, *Theorie der konvexen Körper*. Berlin: Ergebnisse der Math., 1934.

5. Busemann, H., *Convex surfaces*. New York: Interscience, 1958.

6. Cartan, É., *Leçons sur les invariants intégraux*. Paris: Hermann, 1922.

7. Chern, S. S., "On integral geometry in Klein spaces," *Ann. of Math.*, 43 (1942), 178–89.

8. ———, "On the kinematic formula in the euclidean space of n dimensions," *American Journal of Mathematics*, 74 (1952), 227–36.

9. ———, "Differential geometry and integral geometry," *Proceedings of the International Congress of Mathematics*, Edinburgh (1958), 441–49.

10. ———, "The integrated form of the first main theorem for complex analytic mappings in several complex variables," *Ann. Math.*, 71 (1960), 536–51.

11. Chern, S. S., and R. K. Lashof, "On the total curvature of immersed manifolds," I, *American Journal of Mathematics*, 79 (1957), 306–18; II, *Michigan Mathematical Journal*, 5 (1958), 5–12.

12. Chevalley, C., *Theory of Lie Groups*. Princeton, N.J.: Princeton University Press, 1946.

13. Crofton, M. W., "On the theory of local probability," *Phil. Trans. R. Soc. London*, 158 (1868), 181–99.

14. Deltheil, R., *Probabilités géométriques*. Paris: Albin Michel, 1926.

15. Fáry, I., "Functionals related to mixed volumes," *Illinois Journal of Mathematics*, 5 (1961), 425–30.

16. ———, "Sur la courbure totale d'une courbe gauche faisant un noeud," *Bull. Soc. Math. France*, 77 (1949), 128–38.

17. Federer, H., "Coincidence functions and their integrals," *Transactions of the American Mathematical Society*, 59 (1946), 441–66.

18. ———, "The (φ, k)-rectifiable subsets of n space," *Transactions of the American Mathematical Society*, 62 (1947), 114–92.

19. ———, "Some integral geometric theorems," *Transactions of the American Mathematical Society*, 77 (1954), 238–61.

20. Fenchel, W., "On the differential geometry of closed space curves," *Bulletin of the American Mathematical Society*, 57 (1951), 44–54.

21. Gelfand, I. M., "Integral geometry and its relations to the theory of representations," *Uspehi Mat. Nauk*, 15 (1960), 155–64.

22. Green, L. W., "Proof of Blaschke's sphere conjecture," *Bulletin of the American Mathematical Society*, 67 (1961), 156–58.

23. Hadwiger, H., *Vorlesungen über Inhalt, Oberflache und Isoperimetrie*. Berlin: Springer, 1957.

24. ———, *Altes und neues über konvexe Körper*. Basel and Stuttgart: Birkhauser Verlag, 1955.

25. ———, "Normale Körper im Euklidischen Raum und ihre topologischen und metrischen Eigenschaften," *Math. Zeits.*, 71 (1959), 124–40.

26. Haimovici, M., "Géométrie intégrale sur les surfaces courbes," *C. R. Acad. Sc. Paris*, 203 (1936), 230–32.

27. ———, "Généralisation d'une formule de Crofton dans un espace de Riemann à n dimensions," *C. R. Acad. Sc. Roumanie*, 1 (1936), 291–96.

28. Helgason, S., *Differential Geometry and Symmetric Spaces*. New York: Academic Press Inc., 1962.

29. Herglotz, G., *Lectures on Geometric Probabilities* (mimeographed notes). Göttingen: 1933.

30. Hermann, R., "Remarks on the foundation of integral geometry," *Rend. Circ. Mat. Palermo, Ser. II,* 9 (1960), 91–96.

31. Kurita, M., "On the volume in homogeneous spaces," *Nagoya Math. J.,* 15 (1959), 201–17.

32. ——, "An extension of Poincaré formula in integral geometry," *Nagoya Math. J.,* 2 (1951), 55–61.

33. ——, *Integral Geometry.* Nagoya, Japan: Nagoya University, 1956.

34. Lebesgue, H., "Exposition d'un mémoire de M. W. Crofton," *Nouvelles Ann. de Math. S. IV,* 12 (1912), 481–502.

35. Legrady, K., "Symplektische Integralgeometrie," *Annali di Mat. Serie IV,* 41 (1956), 139–59.

36. ——, "Sobre la determinación de funcionales en geometría integral," *Rev. Union Mat. Argentina,* 19 (1960), 175–78.

37. Luccioni, R. E., "Sobre la existencia de medida para hipercuadricas singulares en espacios proyectivos," *Rev. Mat. y Fis. Univ. Tucuman,* 14 (1962), 269–76.

38. Masotti Biggiogero, G., "La geometria integrale," *Rend. Sem. Mat. e Fis. Milano* 25 (1953–54), 3–70.

39. ——, "Nuove formule di geometria integrale relative agli ovaloidi," *Rend. Ist. Lombardo* A 96 (1962), 666–85.

40. ——, "Nuove formule di geometria integrale relative agli ovali," *Annali di Mat.,* 4 58 (1962), 85–108.

41. ——, "Su alcune formule di geometria integrale," *Rend. di Mat.* (5) 14 (1955), 280–88.

42. ——, "Sulla geometria integrale: nuove formule relative agli ovaloidi," *Scritti Mat. in onore di Filippo Sibirani,* Bologna: Cesari Zuffi (1957), 173–79.

43. Munroe, M. E., *Modern Multidimensional Calculus.* Reading, Mass.: Addison-Wesley Publishing Co., Inc., 1963.

44. Nachbin, L., *Integral de Haar, Textos de Matematica.* Recife, Brazil: 1960.

45. Nöbeling, G., "Über die Flachenmasse im Euklidischen Raum," *Math. Ann.,* 118 (1943), 687–701.

46. ——, "Über den Flacheninhalt dehnungsbeschränkten Flächen," *Math. Z.,* 48 (1943), 747–71.

47. Owens, O. G., "The integral geometric definition of arc length for two-dimensional Finsler spaces," *Transactions of the American Mathematical Society*, 73 (1952), 198–210.

48. Petkantschin, B., "Zusammenhänge zwischen den Dichten der linearen Unterräume im *n*-dimensionales Raum," *Abh. Math. Sem. Univ. Hamburg* 11 (1936), 249–310.

49. Rohde, H., "Unitäre Integralgeometrie," *Abh. Math. Sem. Univ. Hamburg*, 13 (1940), 295–318.

50. Santalo, L. A., "Über das kinematische Mass im Raum," *Actualités Sci. Ind.* No. 357. Paris: Hermann, 1936.

51. ———, "A theorem and an inequality referring to rectifiable curves," *American Journal of Mathematics*, 63 (1941), 635–44.

52. ———, "Unas formulas integrales y una definicion de area *g*-dimensional de un conjunto de puntos," *Rev. de Mat. y Fis. Univ. Tucuman*, 7 (1950), 271–82.

53. ———, "Measure of sets of geodesics in a Riemannian space and applications to integral formulas in elliptic and hyperbolic spaces," *Summa Brasil Math.*, 3 (1952), 1–11.

54. ———, "Geometría integral en espacios de curvatura constante," *Com. E. Atom. Buenos Aires*, No. 1 (1952).

55. ———, "Introduction to integral geometry," *Actualités Sci. Ind.*, No. 1198. Paris: Hermann, 1953.

56. ———, "Integral geometry in Hermitian spaces," *American Journal of Mathematics*, 74 (1952), 423–34.

57. ———, "On the kinematic formula in spaces of constant curvature," *Proceedings of the International Congress of Mathematics, Amsterdam*, 2 (1954), 251–52.

58. ———, "Sobre la formula fundamental cinematica de la geometría integral en espacios de curvatura constante," *Math. Notae*, 18 (1962), 79–94.

59. ———, "Cuestiones de geometría diferencial y integral en espacios de curvatura constante," *Rend. Sem. Mat. Torino*, 14 (1954–55), 277–95.

60. ———, "Sur la mesure des espaces linéaires que coupent un corps convexe et problèmes qui s'y rattachent," *Colloque sur les questions de réalité en géométrie*, Liège (1956), 177–90.

61. ———, "Two applications of the integral geometry in affine and projective spaces," *Publ. Math. Debrecen*, 7 (1960), 226–37.

62. ———, "On the measure of sets of parallel subspaces," *Canadian Journal of Mathematics*, 14 (1962), 313–19.

63. Stoka, M., "Masure unei multimi de varietati dintr-un spatiu R_n," *Bull. Math. Soc. Sci. Ac. R. P. R.*, 7 (1955), 903–37.

64. ———, "Asupra grupurilor G_r masurabili dintri-un spatiu R_n," *Com. Acad. R. P. R.*, 7 (1957), 581–85.

65. ———, "Geometria integrale in uno spazio euclideo E_n," *Boll. Un. Mat. Italiana*, 13 (1958), 470–85.

66. ———, "Famiglie di varieta misurabili in uno spazio E_n," *Rend. Circ. Mat. Palermo, II*, 8 (1959), 1–14.

67. ———, "Géométrie intégrale dans un espace E_n," *Acad. R. P. R. IV*, (1959), 123–56.

68. ———, "Integralgeometrie in einen Riemannschem Raum V_n," *Acad. R. P. R. V* (1960), 107–20.

69. Struik, D. J., "Lectures on classical differential geometry," Cambridge, Mass.: Addison-Wesley, 1950.

70. Varga, O., "Über Masse von Paaren linearer Mannigfaltigkeiten im projektiven Raum, P_n," *Rev. Mat. Hispano-Americana*, (1935), 241–278.

71. Vidal Abascal, E., "Sobre algunos problemas en relacion con la medida en espacios foliados," *Primer Coloquio de Geometría Diferencial, Santiago de Compostela*, (1963), 63–82.

72. Vincensini, P., "Corps convexes, séries linéaires, domaines vectoriels," *Mémorial des Sciences Mathématiques*, No. 94. Paris: 1938.

73. Weil, A., "L'intégration dans les groupes topologiques et ses applications," *Actualités Sci. Ind.* No. 869. Paris: Hermann, 1940.

74. ———, Review of the paper 7 of Chern, *Mathematical Reviews*, 3 (1942), 253.

75. Weyl, H., "Meromorphic functions and analytic curves," Princeton University Press (1943).

76. Wu, T. J., "Über elliptische Geometrie," *Math. Z.*, 43 (1938), 212–27.

Since the publication of the first edition of this book, the following books on the subject have appeared. In most of them, a rather extensive bibliography on particular topics of the recent integral geometry may be found.

1. Ambartzumian, R. V., *Combinatorial Integral Geometry with Applications to Mathematical Stereology*, J. Wiley, Chichester, 1982.

2. Blaschke, W., *Gesammelte Werke*, Bd. 2, Thales Verlag, Essen, 1985.

3. Harding, E. F., and Kendall, D. G., *Stochastic Geometry*, J. Wiley, 1974.

4. Matheron, G., *Random Sets and Integral Geometry*, Wiley, New York, 1975.

5. Santaló, L. A., *Integral Geometry and Geometric Probability*, Encyclopedia of Mathematics and its Applications, Addison-Wesley, Reading, Mass. 1976.

6. Schneider, R., *Integralgeometrie*, Mathem. Institut der Universität, Freiburg, 1979.

7. Solomon, H., *Geometric Probability*, Society for Industrial and Applied Mathematics, Philadelphia, Pa., 1978.

8. Stoka, M. I., *Géométrie Intégrale*, Memorial des Sciences Mathematiques, Fasc. 165, Gauthier-Villars, Paris, 1968.

9. Stoka, M. I., *Geometrie Integrala*, Ed. Acad. R.S. Romania, 1967.

10. Stoyan, D., and Mecke, J., *Stochastische Geometrie*, Akademie Verlag, Berlin, 1983.

11. Vranceanu, G., and Filipescu, D., *Elemente de Geometrie Integrala*, Ed. Acad. R.S. Romania, Bucaresti, 1982.

12. Stoyan, D., Kendall, W. S., and Mecke, J., *Stochastic Geometry and its Applications*, Akademie Verlag, Berlin, 1987.

13. Bryant, R. L., Guillemin, V., Helgason, S., and Wells, R. O., editors, *Integral Geometry*, American Mathematical Society, Contemporary Mathematics, vol 63, 1987.

INDEX

$\pi_i(M, p)$, 203
$\tau(\infty)$, 232
τ_{pq}, 227
τ_{px}, 233
τ_v, 226
Φ_p, 224, 227, 252

absolute differentiation, 6
action of isometries on $\tilde{M}(\infty)$, 235
Aleksandrow-Toponogov angle
  comparison theorems, 197
analysis in the large, 259
angle sum law, 228
area element, 321
area of a continuum, 332
area element of the unit sphere, 316
atlas, 39

Ballmann, W., 255
Berger, M., 253
Bernstein's theorem, 78, 83, 136
Bieberbach, 251
Bishop, R., 223
Bonnet, 60
Bonnet's theorem, 152
Brouwer fixed point theorem, 55
Burns, K., 255
Busemann functions, 192

Calabi-Hopf Maximum Principle, 195
Cartan, E., 229
Cauchy's formula, 324, 325
chart, 39
Chern classes, 7
Chern forms, 7
closed form, 45

compact totally convex exhaustion,
  193
complete, 225
complex line bundle, 18
complex projective space, 215, 331
  distance spheres of, 220
  isometries of, 217
  natural embedding of, 216
  submanifold metric, 216
  totally geodesic submanifolds, 217
cone topology, 234
conjugate, 148, 149
conjugate point, 140, 148
connection, 1
convex, 230
convex sets, 223
convex bodies in elliptic space, 340
convex bodies in hyperbolic space, 340
convex functions, 223
convexity, 229
covariant differentiation, 6
Crofton's theorem, 116
curvature matrix, 6
curvature tensor, 219, 226, 243
cut locus, 144
cut locus estimates, 207
cut point, 140, 143, 144

de Rham cohomology, 45
de Rham cohomology group, 10
de Rham theorem, 45
deckgroup, 235
deformation of a space curve, 118
density for circles, 344
density for conics, 344
density for geodesic segments, 339

density for geodesics, 337
density for lines in E_2, 305, 312
density for lines in E_3, 318
density for planes, 317
density for points in E_2, 304
density for points in E_3, 317
density in homogeneous spaces, 309
differentiable manifold, 39
differential form, 27, 40, 304
discrete, 236
distance functions, 173
  critical points of, 205
dual convex body, 342
duality condition, 249, 254

eigenvalue-flag pairs, 247
elliptic, 235
equations of structure, 309, 315
Euler characteristic, 16, 327
Euler's theorem, 342
Euler-Poincare characteristic, 63
exact form, 45
exponential map, 142, 173, 225
exterior derivative, 32
exterior differential form, 304
exterior product, 29, 30

farthest point, 206
Fary-Milnor theorem, 118
Favard length, 305
Fenchel's theorem, 113
fixed frame, 315
fixed point, 55
fixed points of isometries, 188
flags in $\mathbb{R}^n$, 246
flats, 246
forms of Maurer-Cartan, 308, 309, 310, 312
formula of Haimovici, 340
formula of Herglotz, 341
formulas of Frenet, 32
four-vertex theorem, 106
frame, 58
Fujimoto's theorem, 90, 96
fundamental group, 189, 235

G/K, 240
G^*, 224, 252
gauge potential, 6
Gauss-Bonnet formula, 122
Gauss-Bonnet theorem, 16, 60
Gauss's theorem, 33
Gaussian curvature, 63
general integral geometry, 308
generalized Rauch estimates, 184
geodesic flow, 251, 254
geodesic symmetries, 224
geodesic symmetry, 252
geodesics, 337
Gromov, 255
groups of matrices, 308
groups of motions, 303, 314
groups of similitudes, 344

Haar's measure, 311
Hadamard's theorem, 128
Hessian estimates, 193
holonomy, 243, 245
holonomy group, 224
holonomy group at p, 227
homogeneous spaces, 309
Hopf bundle, 19
Hopf circle, 215, 220
hyperbolic space, 231
hyperboloid model, 231

in opposition, 247
inequality of Fary, 323
inequality of Fenchel, 323
infinitesimal variation, 147
integral geometry, 303
integral geometry in complex spaces, 331
integral of mean curvature, 325
integral of mean curvature of a plane convex domain, 326
integral of mean curvature of a polyhedron, 326
integral of mean curvature of a right cylinder, 326
integration of manifolds, 41

isoperimetric inequality, 78
isoperimetric inequality for plane
    curves, 108

Jacobi field, 149, 174
Jacobi's theorem, 127
joining points at infinity, 234

k-flat, 243
Karpelevic, 223
kinematic density in E_n, 306
kinematic density in E_3, 320
kinematic density fundamental
    formula, 326

lattice, 236, 250, 255
law of cosines, 228

$\tilde{M}$, 227
$\tilde{M}(\infty)$, 224
$\tilde{M}_n$, 224, 232, 241, 244
$\tilde{M}_n(\infty)$, 247
manifold, 38, 39
matrix of Maurer-Cartan, 308, 315,
    332
maximal compact subgroup, 242
Maxwell's equations, 23
mean curvature, 75
mean curvature vector, 77
meromorphic curves, 333
Meusnier theorem, 330
minimal surface, 73, 77
minimal surface equation, 73
minimizing geodesic, 142
Minkowski's uniqueness theorem, 133
modified distance function, 178
Morse's inequality, 266
Mostow rigidity theorem, 255
moving frame, 58, 315

Nirenberg's conjecture, 83, 86
nonpositively curved manifolds, 187
nonwandering modulo Γ, 250
normal bundle, 4

$\frac{1}{4}$-pinching, 212
O'Neill, B., 223
Ostrogradsky's theorem, 33

parallel convex bodies, 324, 325
Pfaffian, 29
Pfaffian forms, 308, 309, 315
Poincare lemma, 34, 35
Poincare lemma, converse of, 36
points at infinity, 246
Pontrjagin class, 13
Pontrjagin form, 13
Prasad, 256
Preissman, 237
principle curvatures, 175, 177
    bounds, 183

$r(\Gamma)$, 256
Raghunathan, 256
rank, 246, 254
rank of a symmetric space, 243
rank of the fundamental group, 256
rank (Γ), 256
Rauch's comparison theorem, 153
"Rauch hinge", 197
regular and singular points at infinity,
    243
Riccati comparison, 182
Riccati equation, 176
    inequality, 179
Ricci diameter bound, 190
Ricci splitting theorem, 195
rotation index, 101

$S\tilde{M}$, 227
second variation of area, 77
second fundamental forms, 14
section, 2
sectional curvature, 226
semisimple, 240
sets of geodesic segments, 339
sets of manifolds, 343
shape operator, 174
"short" set of generators, 203
$SL(n, \mathbb{R})/SO(n, \mathbb{R})$, 244

s^p, 224, 239, 252
Spatzier, R., 255
sphere theorems, 212
stability, 78
Steiner's formula, 325, 331
Stokes's theorem, 31
Study-Fubini metric, 21
support functions, 192
symmetric spaces of noncompact type, 223, 239
symmetry diffeomorphisms, 239
Synge's lemma, 153, 209

tangent bundle, 3
theorem of turning tangents, 99
Topogonov, 198, 228
total curvature, 63
total curvature of a space curve, 112
total twist, 18
transition functions, 2

transvections, 241
truncated cones, 234

uniqueness theorem of Cohn-Vossen, 129
upper half-space model, 231

$V(p, x)$, 233
variation, 147
vector bundle, 1
vector field, gradient like, 205
volume bounds, 185
volume growth rate, 181

Weierstrass representation, 81
Wirtinger's inequality, 80
Wirtinger's theorem, 22

Yang-Mills' equations, 23, 25